THE PLANTATION MACHINE

THE EARLY MODERN AMERICAS

Peter C. Mancall, Series Editor

Volumes in the series explore neglected aspects of
early modern history in the western hemisphere.
Interdisciplinary in character, and with a special
emphasis on the Atlantic World from 1450 to 1850,
the series is published in partnership with the
USC-Huntington Early Modern Studies Institute.

THE PLANTATION MACHINE

ATLANTIC CAPITALISM
IN FRENCH SAINT-DOMINGUE
AND BRITISH JAMAICA

TREVOR BURNARD
AND
JOHN GARRIGUS

PENN

UNIVERSITY OF PENNSYLVANIA PRESS

PHILADELPHIA

Copyright © 2016 University of Pennsylvania Press

All rights reserved. Except for brief quotations used
for purposes of review or scholarly citation, none of this
book may be reproduced in any form by any means without
written permission from the publisher.

Published by
University of Pennsylvania Press
Philadelphia, Pennsylvania 19104-4112
www.upenn.edu/pennpress

Printed in the United States of America on acid-free paper
1 3 5 7 9 10 8 6 4 2

Cataloging-in-Publication Data is available from the
Library of Congress.
ISBN 978-0-8122-4829-6

CONTENTS

1. A Comparative History of Jamaica and Saint-Domingue   1

2. The Plantation World   25

3. Urban Life   50

4. The Seven Years' War in the West Indies   82

5. Dangerous Internal Enemies   101

6. Racial Reconfigurations Before the American Revolution   137

7. The Golden Age of the Plantocracy   164

8. The American Revolution in the Greater Antilles   192

9. Recovery and Consolidation in the 1780s   219

10. The Ancien Régime in the Greater Antilles   244

Notes   269

Index   335

Acknowledgments   349

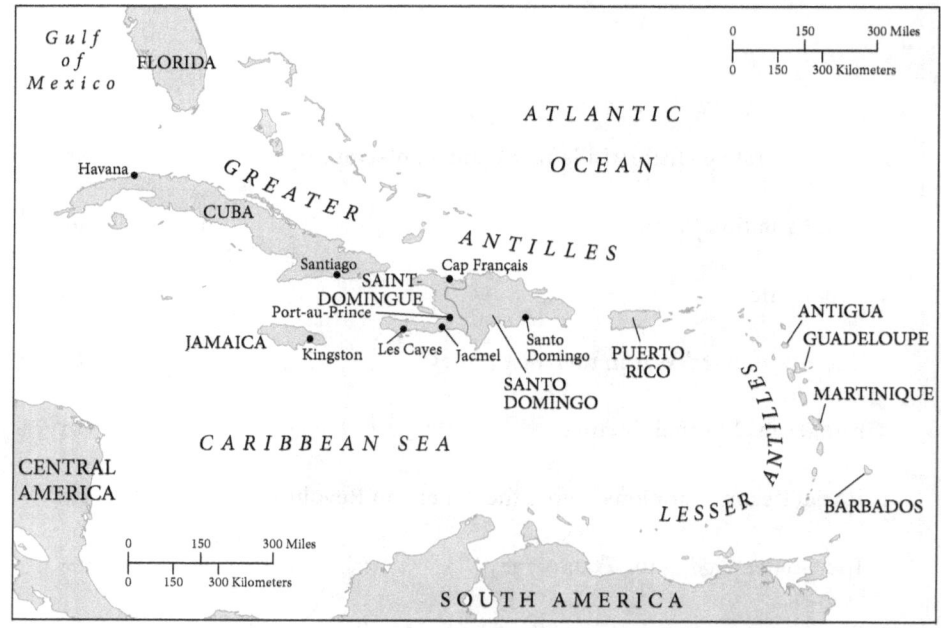

Map 1. The Greater Antilles at the Time of the Seven Years' War

CHAPTER 1

A Comparative History of Jamaica and Saint-Domingue

This book is about social, political, and economic transformations in eighteenth-century Jamaica and Saint-Domingue, two extremely profitable but socially monstrous slave societies. These islands, we argue, were more than minor colonies in distant parts of the world, far removed from European consciousness and without global importance. Rather, they were Europe's most successful plantation societies, and we examine them during the years when they were at their absolute peak, between 1740 and 1788. This study therefore chronicles the two most important (and not coincidentally the two most brutal) slave societies within the plantation complex that shaped the Atlantic World from its fourteenth-century origins to the end of slavery in Cuba and Brazil in the 1880s.

That plantation complex depended on a vibrant Atlantic slave trade, supplying large numbers of Africans to the tropical regions of the Americas, where they were coerced into producing luxury commodities originally developed in Asia for a European market. In the years between 1740 and 1788, this plantation complex—which we prefer to call the plantation machine, adopting the mechanistic metaphor common in the age—had reached an apogee of sorts in the Greater Antilles. In other words, Jamaica and Saint-Domingue came close to perfecting a form of economic organization that operated on a global scale, using specialized laborers from one continent who did not merely farm but in fact manufactured products in a second continental zone for consumers in a third. This revolutionary form of social and economic organization bears careful study, not only for how it functioned but for the social and political mechanisms that sustained it. This work provides that close-grained, empirically based investigation of the plantation machine within the social and political contexts of Saint-Domingue and Jamaica between 1748 and 1788.

We take as our starting point that in the seventeenth and eighteenth centuries the plantation machine was the main driving force shaping most aspects of European colonization in the Atlantic World. Plantation colonies were peculiar and distinctive in that they were not based on community production opening up opportunities for trade. Rather their plantations were specialized producers that relied on exchange systems developed elsewhere, in Europe and especially in Africa. The plantation machine was quintessentially colonial, barely using local resources while depending heavily on goods from elsewhere—capital from Europe and labor from Africa. The colonies that perfected it, as did Jamaica and Saint-Domingue by the 1780s, exemplify a particular kind of colonialism that proved influential not just in the Americas but in European empires generally. Thus, this book is not just a local study of two interesting but unusual colonies on the margins of the Atlantic World. That might appear to be the case when we think of Jamaica and Saint-Domingue's later history in the nineteenth and twentieth centuries, when they were shriveled—if somewhat more egalitarian—versions of their former selves. Jamaica and Saint-Domingue were much more important in the eighteenth century than their later global insignificance might suggest.[1]

The Caribbean was the world region that exemplified more than any other the complex forces of early modern imperialism, forces that helped make colonialism a truly global phenomenon. The Caribbean was the first area in the New World to experience European imperialism and was battered by those forces for longer than anywhere else. Starting around 1500, Europeans largely eliminated the Amerindian population; they converted tropical hardwood forests into grazing and farmland; and they imposed a nightmarish system of slave labor to work this new land. These events, especially the nature of the sugar plantation system, left the West Indies as a diminished and regressive world.[2] But this book draws attention to another aspect of Caribbean history that at first glance seems contradictory: despite their brutal nature, indeed because of that brutality, Jamaica and Saint-Domingue were central to a developing eighteenth-century Atlantic capitalism. As Abbé Raynal commented in 1770, "The labours of the colonists settled in these long-scorned islands are the sole basis of the African trade, extend the fisheries and cultivation of North America, provide advantageous outlets for the manufacture of Asia, double perhaps triple the activity of the whole of Europe. They can be regarded as the principal cause of the rapid movement which stirs the universe."[3]

At the heart of this book is the argument that the Caribbean sugar plantation of the eighteenth century was an industrial, or at least proto-industrial, operation.[4] Jamaica and Saint-Domingue were important in the Atlantic World because in the 1700s sugar planters there were able to maximize the scale—and the social and cultural consequences—of the "integrated" plantation model developed first in the Lesser Antilles in the 1600s. The eighteenth-century integrated sugar plantation, described below, led planters and their employees to develop new attitudes toward wealth, society, and production. Sugar plantations were among the first Western institutions to implement industrial-style production, deploying hundreds of workers factory-like in an array of complex interdependent tasks with careful attention to time. Their managers did not recognize or analyze their use of capital or time in ways that would become common in later manufacturing enterprises. Planters' extraordinary focus on wealth meant that material success, allied to a heightened racial sensibility that elevated "whiteness" to a position of great importance, supplemented class standing as a primary basis of social standing. White colonial residents of Jamaica and Saint-Domingue developed individualistic behaviors that might be described as egalitarian if these societies had not been so heavily dependent on enslaved laborers. Facing the threat of slave rebellion, foreign attack, and this new individualism, colonial elites invented new ways to create or reinforce some sense of community. They also developed new racial legislation in the 1760s that established "whiteness" as a defining characteristic of full citizenship. Wealth was the defining feature of the two colonies. What was harder to develop was a sense of belonging—awareness by colonists that they belonged to places with a distinct social and corporate identity.

In order to understand how Jamaica and Saint-Domingue were central to the development of eighteenth-century Atlantic capitalism, we need to understand the workings of the central institution in each colony: the plantation. The "integrated" sugar plantation dominated the social and economic worlds of both Saint-Domingue and Jamaica. Sugar was not the only business in the two colonies: each possessed a diversified economy with a vibrant commercial culture, producing a variety of tropical crops unlike the small-island sugar colonies of the Lesser Antilles such as the British colonies of Barbados and Antigua and the French colonies of Martinique and Guadeloupe. Saint-Domingue especially produced a range of other plantation commodities, most notably coffee. But sugar was central to colonial culture. It employed more workers, and generated more profits, than any other crop on either island. A sugar plantation required more land, labor, and technology than any other type of enterprise in the Caribbean. Moreover, white people in both colonies regarded sugar planters as the social elite. In 1746 Jamaica had 455 sugar estates, and in 1753 Saint-Domingue had 556.[5]

An "integrated" sugar plantation is one that not only grows sugarcanes but also transforms them into sugar crystals. Until roughly the middle of the seventeenth century, sugar production in Europe's Atlantic colonies was often split between cane growers and sugar manufacturers. This was especially true in northern Brazil, where mills in the 1600s produced three to six times as much sugar as their equivalents in the Atlantic colonies of Madeira and Sao Tomé or on the Spanish Caribbean islands of Cuba and Santo Domingo.[6] Specialization allowed Brazilian sugar makers to solve technical problems, for example, adopting a vertical crushing mill in place of the inefficient horizontal millstones used by Mediterranean sugar makers.[7]

Brazilian methods, including the split between cane farmers and refiners, shaped how English and French colonists began making sugar in the Caribbean in the 1640s and 1650s.[8] Up to the 1670s, Barbados and Martinique sugar estates were unlikely to have their own mills.[9] It took some forty years before cane growers on these two islands began to build their own sugar works. A hurricane in 1676 hastened the process in Barbados because many small and medium landowners sold their farms rather than rebuild.[10] But the most important aspect of the transition to an integrated plantation model was the rise of the gang system of slave labor. In the gang system managers grouped field workers according to their physical stamina, rather than assign tasks to individual slaves. This policy allowed them to enforce the rigid cultivation schedules that mills and boiling houses required. They charged the first gang with the hardest tasks like planting or harvesting cane, for it was important that workers move through the fields at approximately the same pace, and chop harvested canes into similarly sized pieces to be fed through the mill. Planters grouped adolescents and workers past their physical prime into a second gang, used to do less difficult work, like weeding. The French word for these slave gangs—*ateliers* or "workshops"—illustrates the connection contemporaries saw between this new way of organizing field work and the manufacturing process. Gang labor was first mentioned in connection with Barbados in the 1650s, about the time that the first integrated plantation arose. But gang labor required dozens of workers. Planters who wanted to build gangs needed capital that would roughly equal what they spent to acquire land and build a refining operation. Sources are not clear about the rate at which planters adopted gang labor, but by the eighteenth century the system was ubiquitous on Caribbean sugar estates.[11]

When Lesser Antilles planters integrated planting and manufacturing, and restructured their work regimes accordingly, they discovered that this approach was very profitable. An estate that successfully timed its harvest schedule to match the speed of its mill and boiling houses saved a lot of canes from rotting in its fields and mill yard. The owner of an integrated plantation could thus time the planting of his fields to keep his manufacturing operations occupied the better part of the year. As sugar makers came to participate in harvest decisions, they understood the kinds of cane juice that were sluicing into their cauldrons, whether they were from a field that gave high or low yields. A growing mastery of the details of raw materials, manufacturing constraints, and labor management produced more salable sugar. This "new plantation" model also brought significant efficiencies of scale. Larger slave forces and more land allowed planters in Jamaica and Saint-Domingue to train more slave artisans and specialists, establish livestock pens and provision grounds, and generally cut their operating costs, at least in theory. Antigua's Samuel Martin, in the 1765 edition of his *Essay upon Plantership*, used the metaphor of a machine to convey the importance of regularity in plantation management. He noted that "a plantation ought to be considered as a well-constructed machine, compounded of various wheels, turning different ways, and yet all contributing to the great end proposed; but if any one part runs too fast or too slow in proportion to the rest, the main purpose is defeated."[12] It was in these Greater Antilles colonies that the full economic, social, and cultural impact of this mode of sugar planting was expressed.

For the integrated model to work, its agricultural operations required an extraordinary degree of labor, capital investment, and attention to the passage of time. Setting aside problems that all early modern farmers faced—bad weather, plant disease, invading weeds, declining fertility, and pest incursions—planting and harvesting sugarcane was significantly more difficult than for any other early modern food crop, in terms of labor, capital equipment, and timing. Early modern sugar workers planted a field by burying thousands of sugarcane cuttings two to three feet deep into the soil. Planting or "holing" a cane field required extraordinary exertion by a team of workers. In the sugar-producing regions of the Americas these men and women were almost always enslaved and of African descent. Depending on the plant variety and the environmental conditions, these canes required roughly twelve to eighteen months to grow to maturity.

Harvesting the cane required backbreaking labor as each worker cut thousands of tall stalks several inches above the soil level and chopped each of them into three similarly sized pieces. Planters in Brazil in 1689 required their slaves to cut forty-two hundred canes per day.[13] Heavy with juice, those canes had to be immediately hauled to a mill because they would rot in twenty-four to forty-eight hours if they were not crushed. In fact, planters often found canes rotting in the field during harvest, if they did not have enough labor or expertise to start this process at the proper moment and to complete it promptly. In addition to having a large human labor force, cane farmers needed to be keenly aware of time. They also needed teams of oxen, mules, and carts. After harvest, the root system established by the initial cutting produced more cane offshoots, known in English as "rattoons" and in French as "*rejettons*." Most eighteenth-century plantations took advantage of this regrowth to reap as many as five harvests from a single planting, a method known in English as "rattooning." Successive generations of rattooned cane, however, though they mature faster, contain progressively less sucrose.

The same three factors that shaped cane growing—time sensitivity plus access to human labor and machinery—also determined whether canes could be transformed into saleable sugar. Time is especially critical in the sugar manufacturing process and not only because hundreds of thousands of canes must be crushed one or two days after harvesting. After a mill pressed the juice from a cane, spoilage was still a danger. Early modern sugar makers boiled the sugar juice, up to the point when they could see sucrose crystals emerging from the watery solution. This evaporation stage, however, was necessarily slow, because sugar syrup caramelizes if it receives too much heat at once. To control this process, sugar makers moved the hot sugar solution through five progressively smaller cauldrons as it thickened. Once their expert touch could discern sugar crystals, sugar refiners poured the syrup into hundreds of conical clay forms. These containers sat in a dry place for four to six weeks as the remaining water drained out, in the form of a dark, sweet molasses. To make a more refined "clayed sugar" or "*sucre terré*," a sugar master covered the open tops of these pots with wet clay. The clay gradually released its water into the crystals below, carrying away impurities. The resulting sugar was lighter and sold for approximately twice the price of brown sugar. The claying process also produced a larger amount of molasses, which was an additional source of profit for a planter who distilled it into rum himself or sold it to a distillery. Making sugar profitably in such conditions required that an eighteenth-century planter be an accomplished farmer and a manager of a complex manufacturing process. He needed to have access to sufficient credit to buy and replace agricultural tools. He constantly worried about making sure sugar-making supplies were available at the times when the process needed them. Most of all, he had to carefully manage a large and troublesome labor force, which, moreover, was always in a state of flux, as sugar workers were frequently unhealthy or dying, consumed by an estate's hard-driving operating schedule. In both societies, slave numbers were maintained only by frequent purchases of new slaves, usually direct from West or West Central Africa.

To an extent, these sugar plantations were "factories in the field." Scholars have pointed out a number of characteristics that disqualify eighteenth-century sugar estates from being labeled Europe's first "industrial" producers.[14] Yet by the second half of the eighteenth century, these nonindustrial characteristics were fading. One key argument against a plantation being akin to an industrial enterprise is that eighteenth-century planters with their slave forces viewed labor as a fixed, not a variable cost, as factory owners did in a later period. For this reason, planters have been seen as resistant to a key aspect of the Industrial Revolution, the adoption of labor-saving technology, for example new plow designs. Planters did resist using plows to plant cane, though as chapter ten of this volume makes clear, Jamaican and especially Saint-Domingue sugar planters became increasingly interested after 1750 in technologies that could make their estates more profitable. Their lack of interest in plows stemmed from the fact that they always needed large workforces to cut the cane, so they saw no compelling reason to save labor in the planting process. Even in the early twenty-first century, workers with machetes, not machines, harvest much of the world's sugar. Nevertheless, eighteenth-century planters used the gang system to make an enslaved cane cutter into a type of machine.[15]

It is true that sugar planters mostly did not deal with a free labor market, though many of them did lease "jobbing gangs" from local entrepreneurs for special tasks, like planting.[16] The critical point, however, is that sugar planters saw their laborers not only as workers but as highly flexible capital investments. Slaves in these colonies were viewed both as property—capital that could be used in a variety of ways—and also as people—workers who could be forced to produce income for owners. Planters could sell them, use them as collateral, or hire them out to planters who needed additional labor. Slaves thus represented highly liquid forms of capital, as well as being people who produced income for their owners. Despite this "fixed-cost" labor situation, sugar planters' attitude that their enslaved workers were flexible and expendable capital investments was consistent with an emerging industrial mentality. Planters' skillful and increasingly flexible deployment of "human resources" (to give a modern term to this emerging way of seeing people in capital investment terms) demonstrated their place on the cutting edge of financial and industrial capitalism.

Why have we chosen to write about two colonies within separate imperial regimes rather than two islands under the same monarch, like Barbados and Jamaica, or Saint-Domingue and Martinique? The simple answer is that we believe the unusual place of the Caribbean within the early modern Atlantic World is best revealed when its two largest slave societies are examined side by side. Such juxtaposition highlights differences between the British and French imperial systems and cultures. But this book illustrates that more unites Jamaica and Saint-Domingue than separates them. This is less a comparative history, therefore, than a twin portrait of societies moving along parallel pathways.

The question of comparison is always a vexed one, as the twentieth-century history of New World slavery illustrates. In 1947, Frank Tannenbaum, an American scholar of Latin American history, wanted to understand why "the adventure of the Negro in the New World had been structured differently in the United States than in other parts of the hemisphere." His conclusion—that the colonizing European culture determined the levels and nature of slavery and racism—inspired many comparative works, some of which were on Saint-Domingue and Jamaica.[17] Such comparative studies have not resolved the question of whether slaves' experience varied significantly according to the nationality of the masters. One reason is that it is difficult to find societies with similar plantation systems, at equivalent levels of demographic and economic development. The nature of a slave economy has much more influence on slaves' lives than any imperial law or institution. A second reason is that such comparisons require precise definitions of slave "treatment."[18] To create such a definition and then research it in two different colonial systems is extraordinarily difficult. "Treatment" might consist of material considerations like access to food, clothing, and medical care, to be revealed by plantation records; it might be defined in legal terms, like protection from torture or overwork, and whether those legal protections were actually enforced; or it might involve the way slaves were treated in the wider culture, like their ability to leave slavery, maintain kin networks, or join religious communities.

Scholars may never be able to answer these types of comparative questions in a rigorous way. But direct comparison is not what interests us. Tannenbaum and those he inspired tried to show how imperial cultures shaped slave societies. An essential truth of the West Indies in this period, however, is that "geography outweighed nationality."[19] Greater Antilles colonies, like Jamaica and Saint-Domingue, shared more physical characteristics with each other than with Lesser Antilles colonies within the same empire. Barbados, for example, was the island most often linked to Jamaica in the eighteenth century; both colonies had a sugar plantation economy, a society built around slavery, and a prominent place in the British Empire. Nevertheless, there were stark differences between the two places. Barbados was physically much smaller, had a larger and more settled white population, and, most important, was the sole British sugar colony to experience natural increase in its slave population before slavery was abolished in 1834. It is as appropriate to compare Barbados with Virginia or South Carolina as it is to look at what it shared with Jamaica.[20] A similar argument can be made in regard to comparing Martinique to Saint-Domingue. This Lesser Antilles colony was less than one-twentieth the size of Saint-Domingue and after 1750 no longer received significant numbers of European or African newcomers. The two colonies were thus very different from each other.

Jamaica and Saint-Domingue in the late eighteenth century resembled each other in being societies in which economic success for the few was matched by grotesque poverty for the majority. Both islands were charnel houses in which neither the white nor black population was demographically self-sustaining. Consequently, on both islands the population structure was peculiarly skewed toward recent migrants from either Africa or Europe. Adults between twenty and forty years old were a disproportionate presence while young and old people were comparatively absent. White men greatly outnumbered white women, but the great majority of the population was of African descent. Culturally, these colonies were extensions of Africa in the New World, despite whites' political and social control. The men and women who flourished in these places were profoundly shaped by the experience of death. European visitors were either horrified or exhilarated by colonists' lack of restraint, moderation, and deference. Saint-Domingue may have been French, and Jamaica may have been British, but they were remarkably similar in cultural terms, as observers to both societies noted.

In 1733 the Jesuit Jean-Baptiste Le Pers described Saint-Domingue's creoles (native-born whites) as differing markedly from the metropolitan French in personality and even in physique. He noted that "They are commonly well-built and easy going, though somewhat flighty and inconstant. They are frank, energetic, proud, haughty, presumptuous, [and] intrepid; in religious matters they are criticized for having very little aptitude and much indulgence but we have seen that a good upbringing easily corrects most of their faults."[21] At the end of the century, Moreau de Saint-Méry described Saint-Domingue's island-born whites as having "a host of admirable qualities; frank, good natured, generous, perhaps ostentatiously, confident, brave, steadfast friends and good fathers, they are exempt from the crimes that degrade humanity."[22] Both writers identified hospitality as the principal virtue of Saint-Domingue whites.

These descriptions were strikingly similar to the portrait of the Jamaican planter made by Edward Long, the foremost mid-eighteenth-century historian of Jamaica. Long's Jamaican planters were "of quick apprehension, brave, good-natured, affable, generous, temperate, and sober . . . lovers of freedom . . . tender fathers, humane and indulgent masters, firm and sincere friends." They were notorious for their abundant hospitality, to friends and strangers alike, and celebrated for their devotion to "gaiety and diversions." It is an attractive portrait. Even planter faults—Long declared them indulgent, heavy drinkers, poor economists, notoriously fickle and desultory in their pursuits, and addicted to both venery and also conspicuous consumption—point them out as agreeable companions. The faults and virtues that Long discerned in white women in Jamaica—their fidelity, good looks, chastity, modesty, and fondness of dancing, music, and other female genteel pursuits being virtues, and their overindulgence of children, ignorance and lack of education, and over familiarity with enslaved female domestics being principal defects—were similarly echoed by propagandists in Saint-Domingue.[23]

Long and Moreau were, of course, determined to put a positive spin on the character of white settlers on the two islands. But less favorable modern accounts also suggest that the similarities between Frenchmen and Britons in the tropics were more significant than the differences. The brutality of the Jamaican slave overseer Thomas Thistlewood, whose life can be chronicled in some detail, is matched by the savagery toward enslaved people of Nicholas Lejeune in Saint-Domingue, whose defiant defense of his inexcusable behavior toward slave women, examined in Chapter 10, laid bare the absolute power of white masters on Caribbean sugar plantations. The psychologies of such men were shaped by the nature of the societies in which they lived. Underneath the surface shimmer of civil institutions, routinized hospitality, and the languid sensuality of colonial life, Jamaica and Saint-Domingue were societies at war. A free population that was outnumbered ten to one by enslaved people had to establish its authority through arbitrary and brutal measures. The ethos of these societies was a peculiar combination of libertinism and authoritarianism, in which the authority of masters over slaves was frequently greater than the power of the imperial state and its legal institutions. Oppressors and victims lived in uneasy tension, which on occasion exploded into open and violent resistance, as in Tacky's 1760 revolt in Jamaica, and in the hysteria surrounding the poisoning campaign allegedly undertaken by the escapee Macandal in 1757 in Saint-Domingue. Just as hurricanes, the most feared feature of the West Indian environment, could destroy everything in their path, so too the white colonists of Jamaica and Saint-Domingue lived on the edge of a social hurricane. Slave revolts could be as deadly in their implications for white society as hurricanes were for the physical landscape.

Comparative history has not been a feature of scholarship on the Caribbean since the 1980s, but a couple of important lessons emerged from those earlier works. One is the danger of writing a comparative study without a deep immersion in the archives, which in practical terms makes coauthorship necessary. The material conditions of plantation life, in particular the mortality rates of enslaved workers, show a similarity among sugar plantation colonies that overrides many of the apparent legal and cultural differences.[24] Political, religious, legal, and other cultural differences shaped the lives of enslaved and free people, but we cannot reliably judge those influences until we understand the material conditions of transportation, nutrition, work, and illness. For this reason a comparison of Saint-Domingue and Jamaica helps us understand how geography constrained and liberated development on both islands and meant that there would always be significant differences between them. These differences will emerge several times in this joint portrait.

    Saint-Domingue and Jamaica occupied similar positions within their respective empires because of the way each was different from older Lesser Antilles colonies. English planters developed the integrated plantation model in the 1660s in Barbados. It spread gradually to French Martinique, where planters adapted it as the principal mode of social and economic organization. But the amount of land available in the Greater Antilles provided the optimal scale for eighteenth-century sugar technology. Jamaica and

Saint-Domingue were each twenty-five times the size of Barbados and Martinique, respectively, the Lesser Antilles colonies to which they were most frequently compared. These smaller islands are located where the Atlantic trade winds enter the Caribbean Sea, so for sailing ships they were closer to Europe and Africa than Jamaica or Saint-Domingue. The location of the Greater Antilles, however, gave them easier communication with the North American mainland. Within their respective empires, the two Greater Antilles colonies developed considerably later than their smaller counterparts. By the early 1670s, Barbados was as settled as southern England. Governor Sir Jonathan Atkins declared that "there was not a foot of land in Barbados that is not employed even to the very seaside." Visitors thought that it looked like one continuous green garden.[25] By contrast, Jamaica in this period was barely inhabited. French Martinique, founded in 1635, had definitively shifted to sugar by 1669, when two-thirds of all arable land was owned by sugar planters.[26] Saint-Domingue's sugar "take-off" occurred only between 1715 and 1748. At the turn of the seventeenth century, when Barbados and Martinique were looking like fully settled islands, Saint-Domingue was still mostly a wilderness.

Despite these similarities in geography and in settlement patterns, there were important differences in the physical geography and economic chronology of Jamaica and Saint-Domingue. Modern Haiti, whose borders somewhat exceed those of Saint-Domingue, is roughly two and a half times larger than Jamaica, with a total surface area of 10,714 square miles (27,750 square kilometers) compared to 4,213 square miles (10,911 square kilometers). Both land areas are part of a common limestone formation that makes up the Greater Antilles from Cuba into Puerto Rico. Both therefore have high and rugged interiors that were difficult for eighteenth-century cartographers to depict. In a 1763 map of Jamaica drawn by Thomas Craskell and James Simpson, the island's chief engineer and chief surveyor created a detailed image of the colony that would not be surpassed until 1804. They depicted the island's recently settled western highlands as far more forbidding terrain than the Blue Mountains in the long-settled eastern parishes.[27] In fact the Blue Mountains are Jamaica's only true mountain range, with elevations reaching 7,402 feet (2,256 meters). Much of the rest of the interior is a rugged and elevated limestone plateau, with peaks as high as 2,770 feet (844 meters).[28]

Cartographers were even less prepared to portray Saint-Domingue's terrain. The Pic de la Selle, at 8,793 feet above sea level (2,680 meters), is 15 percent taller than Jamaica's highest point. This mountain is only the highest of a complex collection of mountain systems, composed of many different ranges that divide the country into at least four distinctive regions.[29] By the 1760s, French mapmakers were technically able to measure these heights and to depict them in two dimensions. Most published maps of Saint-Domingue, however, either relied on older symbolic depictions of mountains, or offered topographic detail only in specific regions, as in the case of a 1776 map created to accompany a memorandum by Charles d'Estaing, a former governor.[30] Only in the twentieth century would Haiti's topography be fully mapped.

Because of their mountainous interiors, in both colonies maritime transportation from one region to another was often easier than overland travel. The 1763 Craskell/Simpson map of Jamaica shows far more attention to plantations and to anchorage points than to the road network. Saint-Domingue's maps, like one drawn in 1722 by Guillaume Delisle, show roads more clearly, but because these images contained only vague topographical detail they exaggerate the ease of overland travel.[31] Only about 14 percent of Jamaica is level land, some of it in alluvial plains, the largest of which lie along the southern coast. The remainder is found in valleys enclosed by the central limestone plateaus. As late as the 1930s, geographers considered only 10 percent of the island to be arable.[32] In modern Haiti, similarly, experts conclude that only one-fifth of land is appropriate for farming.[33]

Jamaica's geography, despite the challenges it posed to colonists, allowed a greater commercial and administrative centralization than was possible in Saint-Domingue. By the early eighteenth century the parishes of Kingston and St. Catherine, on the island's eastern leeward side, had emerged as the colony's political and mercantile center. The political capital at Spanish Town, the commercial center in Kingston, and smuggling and pirate town of Port Royal were all located within a twenty-mile radius. Saint-Domingue's size, unusual coastline, and mountainous interior prevented such coherence. The most important commercial port, Cap Français, on the Atlantic coast, was so vulnerable to attack that French administrators created a series of capitals on the western coast. In the 1750s they finally settled on Port-au-Prince. Like Jamaica's Spanish Town, this served as the official residence of the governor-general and the meeting place of the Superior Council of Port-au-Prince. Cap Français was so far away that it had its own Superior Council, and its own provincial governor. The colony's southern coast also had its own governor, though never a Council. In terms of communication and administration, therefore, Saint-Domingue was "three colonies in one."[34]

Saint-Domingue's dispersed and divided terrain made it more economically diverse than the English island. By the 1720s sugar was firmly established as the dominant product of Jamaican agriculture. As the 1763 Craskell/Simpson map shows, colonists there also planted cacao, ginger, cotton, and pimentos and raised livestock. No other crop, however, rivaled sugar for export earnings, capital investment, or enslaved labor. In Saint-Domingue, where sugar planting was slower to emerge, many planters produced indigo or cotton. Coffee, introduced from Martinique in the 1730s, came to rival sugar by the 1760s as a Dominguan export. It could be grown in the interior mountains, which were useless for sugar, and it required a far smaller investment in labor, animals, and machinery.

Physical and geopolitical conditions in Jamaica and Saint-Domingue meant that both territories sheltered what Franklin Knight dubbed "transfrontier" populations even before the English and French took power.[35] Spanish colonists bought enslaved Africans early in the 1500s, and by the mid-seventeenth century, communities of escaped black men and women—maroons—were already living in the mountain interiors of both colonies. In Jamaica, newly imported Africans joined the former slaves of the Spanish so that by the 1730s the island had about one thousand maroons.[36] They fought the British throughout the 1730s, winning treaties in 1738 and 1739 that recognized their right to exist in the interior and to govern themselves. Spanish Santo Domingo had more maroons than white settlers in 1542, and maroon villages remained in place in much of the island in the mid-1600s as the French established themselves in the western third.[37] From the 1720s through to the 1740s, French colonists in Saint-Domingue fought maroon bands in the mountains. In the 1750s some observers estimated that the colony had three thousand maroons.[38] Saint-Domingue, perhaps because of its larger size and the distance between its mountain regions, never had a maroon war like that of Jamaica. By 1785, Saint-Domingue had only one large maroon population, the Le Maniel maroons, who were living on the southernmost part of the French-Spanish colonial border when they signed a treaty with the French.

Although the English and French established their claims differently on the two colonies, Jamaica and Saint-Domingue started with a significant number of inhabitants who might be described as white maroons—escaped sailors and indentured servants seeking freedom. An English fleet attacked and conquered Spanish Jamaica in 1655, after failing to take Santo Domingo. The island quickly became a major center for smuggling and piracy. France claimed Saint-Domingue earlier but less forcefully. In the 1640s, French authorities in the Lesser Antilles began sending governors to the western third of Hispaniola, which the Spanish had forcibly evacuated in 1605. Officials of the royally chartered companies that administered this frontier territory in the seventeenth century gradually claimed control over the population of castaways, escapees, and deserters who lived here hunting feral cattle, farming tobacco, and attacking passing ships. Company governors brought in new colonists from the metropole as well as from the Lesser Antilles. The French navy, with fewer sailors and ships than the English, allied with these pirates in the 1680s and 1690s for raids on the Spanish mainland and even on Jamaica in 1694, carrying away many slaves. English forces out of Jamaica attacked the French colony the following year.

Even with the growth of sugar, Jamaica remained a major commercial center for illegal trade with Spanish colonies, especially the transshipment of enslaved Africans. In 1700 the prospect of placing a Bourbon monarch on the Spanish throne led French authorities to imagine that Saint-Domingue could take over Jamaica's role as commercial gateway to Spanish American market. But over a twenty-year period a French monopoly company charged with this task failed to dislodge the English from this trade. As French colonists noted, their kingdom's slave trade could not supply enough workers even to them, let alone to the Spanish. Throughout the eighteenth century, England's naval superiority to France and the far greater volume of its African trade, at least until the 1780s, when slave imports to Saint-Domingue far surpassed that of Jamaica, would be two essential points of difference between Saint-Domingue and Jamaica.[39]

Another essential difference between the two colonies was the way in which their empires' respective mercantile policies were configured. England's Navigation Acts, passed in 1651, were designed to reclaim shipping and commerce among English colonies and the metropole from Dutch shippers. Both England and France initially formed joint-stock companies, which received royal monopolies on trade with certain geographical regions, like, for example, England's Royal African Company (1660–1752), or France's Company of the West Indies (1664–74) or the French Company of the Indies (1719–69), all loosely modeled on the Dutch West Indies Company. Both kingdoms asserted a national monopoly on trade with their colonies, including the all-important commerce in captive Africans bound for the sugar fields. In the 1720s and 1730s, both kingdoms opened up the African and West Indian trades, eventually allowing any subject to participate. The national monopoly, however, remained in place.

In general, metropolitan officials believed their closed colonial trade systems were essential to their kingdom's international strength. Nevertheless, French colonists in Saint-Domingue (far more than their counterparts in Jamaica) came to see this mercantilist system as a major obstacle to economic development. Jamaica benefited from the strength of English commercial shipping. Its proximity to the British North American mainland allowed its colonists to legally trade food, timber, animals, fish, and other supplies from the north for tropical commodities. In 1733 England prohibited its West Indian colonists from refining sugar on their plantations, in order to protect the profits of metropolitan sugar refiners, which meant that planters could export only muscovado. This brown sugar was worth far less per pound than semirefined white and golden varieties. English monopoly law, however, prevented metropolitan consumers from buying cheaper foreign sweeteners. In this way mercantilism guaranteed a higher price for West Indian sugars in Britain than could be found in Continental European markets.[40]

Saint-Domingue's planters had to sell to France, which reexported about half of all colonial products to other continental ports, especially ports in the North Sea. The French colonial monopoly did not protect sugar prices, but by the middle of the eighteenth century it allowed colonists to refine their product from muscovado or brown sugar to a more valuable clayed or golden sugar. French planters complained they were poorly served by French commerce. The difficulty of travel between French Canada and the Antilles meant that French colonists mostly had to import provisions from France, at a much higher cost than if they could have come legally from ports like Philadelphia or New York. Smuggling was widely practiced in Saint-Domingue, where colonists sold sugar but also cotton and indigo to English captains from Jamaica and North America. The weak supply of African slaves from French slavers was also a special grievance for Saint-Domingue planters.

Just as France and Britain operated under different political conditions during the eighteenth century, so too their colonies were governed in quite separate ways. Jamaica had a much greater degree of self-governing autonomy than did Saint-Domingue, even though London often fought this tendency.[41] By comparison with Saint-Domingue, gubernatorial authority was comparatively weak. Governors did not control the money supply, or at least did not do so very effectively, even after 1726, when the Assembly agreed to a perpetual tax supporting the government. Jamaica's governors had to continuously negotiate with an Assembly elected by the island's planter class. This body saw itself as the functional equivalent of Parliament, able to make laws as its members saw fit.

What this meant, in short, was that Jamaican planters and merchants were prepared to exercise local authority based on what they believed were their immutable rights as freeborn Englishmen. In the mid-1760s, for example, in an extensive controversy between the Assembly and the governor, Nicholas Bourke defended Jamaicans as "men zealous for the constitution and liberties of their country" against the governor's supposed support for "the absurd and slavish Doctrines of DIVINE and HEREDITARY RIGHT and PASSIVE OBEDIENCE and NON-RESISTANCE."[42]

Saint-Domingue, in contrast, had no such legislative body where patriotic colonists could voice their discontent. All laws came from the colony's governor-general and often from the Naval Ministry at Versailles. To the extent that the colonial elite had a voice in the government, this voice came from connections to judges sitting on one of two Superior Councils in Port-au-Prince and Cap Français. These bodies began in the seventeenth century as councils of leading planters, but by the eighteenth century they had become formal courts of law, modeled on France's thirteen regional *parlements*. Judges were appointed by the Crown and had to have legal training that was available only in France. These Councils were courts of appeal, but, like metropolitan parlements, they had the ceremonial right to register all laws, giving them legal standing. In France's absolutist tradition, they did not have the right to deny registration, or even to voice complaints about these new laws. But in reality, the Councils saw themselves as representing or protecting the colony against the excesses and errors of royal administrators.

One important way in which Jamaicans exercised their "zealous" regard for their inherited rights as Englishmen was in insisting on almost absolute control—a tyrant's charter, in fact—over how they treated their enslaved population. There were effectively no constraints in Jamaica limiting slave owners' behavior toward enslaved people. Saint-Domingue, in contrast, had a comprehensive slave law, the Code Noir of 1685, which theoretically governed relations between slaves and masters in the French colonies. In practice, however, the Code Noir was rarely used to punish masters. Throughout the eighteenth century Saint-Domingue's planters claimed the right to control and punish slaves as they saw fit. In many ways, the attitudes and claims of Saint-Domingue's planters resembled those of their Jamaican counterparts, as this book will demonstrate.

The varying geographies of the two colonies; the nature of settlement, commerce, and migration; and, most important, the different configurations of imperial governance made differences between the two islands significant.

Yet the similarities were also profound. One major similarity was how each society developed new racial classificatory systems following France's defeat in the Seven Years' War and Tacky's Revolt in Jamaica in 1760. These events were sufficiently traumatic for white colonists to abandon their previous social definition of race in favor of a racism that prefigured the explicitly biological racism of nineteenth-century imperialism. "Whiteness" became more important: "passing" from black to white became more difficult as the state required more documentation from people of mixed ancestry and as new laws restricted their wealth and demeaned their social status. Both Jamaica and Saint-Domingue transformed themselves from places that had a degree of fluidity in their social hierarchies to societies with an almost caste-like racial rigidity. This subjugation of people of color lingered in history for a long time, influencing colonial attitudes to race in places as diverse as nineteenth-century Australasia and South Africa and twentieth-century Southeast Asia.[43]

We do not claim that Jamaican and Saint-Domingue colonists invented racism. The construction of race is a complex matter, especially in a period where old and new ideas of social identity mixed uneasily. As we explain in Chapters 6 and 7, colonists' efforts to define who was "white" and who was not arose from short-term political needs as much as from beliefs that blacks were racially inferior. Nevertheless, the emergence of a heightened legal and social awareness of "whiteness" in Jamaica from the 1760s and in Saint-Domingue in the 1770s was an important step in the wider history of race and Atlantic capitalism.[44] Before the Seven Years' War, a literate, nominally Christian man or woman who had freedom, wealth, and connections to a prominent colonial family was considered a member of the colonial elite. By the 1760s and 1770s, however, both societies came to embrace the idea that "whiteness" was more important than mere freedom, or even wealth, in determining social and political position.

In this twin portrait, we are conscious that while the historical trajectories of the two colonies merged together at times, in the end the historical experience of the two places led to different futures. Jamaica flourished during the twenty five years that followed Britain's victory in the Seven Years' War. But Saint-Domingue experienced an even greater transformation, catapulting it above Jamaica in wealth and imperial importance. The British may have won the war in the Greater Antilles but the French most definitely won the peace. As the fourth chapter of this book details, the irony of the Seven Years' War was that the long term result of the conflict, in the Caribbean as much as in North America, was that the winner (Britain) lost while the loser

(France) triumphed. By allowing France to rebuild in the Caribbean after the Seven Years' War, Britain threw away the advantages it had gained for itself in its victories of the late 1750s and early 1760s. But another irony lurked in French recovery after 1763. Saint-Domingue became too successful for its own good. Whether the growth of Saint-Domingue would have continued if the French Revolution had not occurred is the subject of debate.[45] What is undeniable, however, is that the consequences of metropolitan and internal challenges to white planter dominance in the 1790s were more dramatic than in Jamaica. Saint-Domingue became Haiti by the early nineteenth century. It was by then a radically different place than before. Jamaica remained Jamaica, even if diminished in wealth and importance.[46] What we hope readers will gain from the twinned histories of these two eighteenth-century colonies is that Jamaica and Saint-Domingue were at the forefront of social, economic, and political development in the eighteenth-century Atlantic World. How they developed, and how they prospered, in the period between the Seven Years' War and the start of the French Revolution was not incidental but central to the nature of French and British imperialism in the New World and in imperial settings generally.

This introduction has outlined the major themes we explore in this work. But it might be worth briefly outlining the plan of what follows. Chapters 2, 3, and 4 set out the background necessary to understand the peculiar societies of Jamaica and Saint-Domingue, places that seem as strange to modern readers as they were to European contemporaries. These chapters describe the most salient features of each society, focusing on the plantation and the town. They provide some glimpses into what colonists experienced and what Africans suffered in Jamaica and Saint-Domingue during the period of the Seven Years' War.

The narrative section of the book focuses around the three great global wars that transformed the Caribbean in the second half of the eighteenth century. Chapter 4 provides a Caribbean-focused narrative of the Seven Years' War, which is key to our argument in the book's most important section, Chapters 5, 6, and 7. These chapters describe the new legal definitions of "whiteness" formulated in Jamaica and Saint-Domingue. In the book's third and final section, Chapters 8, 9, and 10, we describe the results of this transformation in racial thinking and document how the American Revolutionary War shaped life for white colonists and by extension their enslaved property. We conclude just before the start of the final and most momentous global war in the second half of the eighteenth century. This war, lasting from 1789 through to 1815, was occasioned by the French Revolutions and led to the most significant turbulence in the French Antilles since the Columbian Encounter of the early sixteenth century. But when we end, in 1788, that turbulence was ahead of colonists in Jamaica and Saint-Domingue and was largely unimaginable to anyone living in either place.

Every writer thinks his or her topic of enquiry is especially important. We are no different. We believe that if you are to understand the making of the modern world in the crucial years encompassing the various political and economic revolutions (American, French, Haitian, and Industrial) that transformed Western and then global society in the second half of the eighteenth century, then what happened in Saint-Domingue and Jamaica between 1748 and 1788 cannot be ignored. They were examples of success in the modern world, in that the plantation machine was a fundamental step forward in the organization of labor for the enrichment of the fortunate owners of large-scale enterprises. The plantation was a great success because it was a precursor to the industrial factory in its management of labor, its harvesting of resources, and its scale of capital investment and output.[47] As the preceding quote from Abbé Raynal illustrates, contemporaries were well aware of how the economic potential of the plantation system, including its methods of management and economic organization, made places like Jamaica and Saint-Domingue extremely valuable imperial possessions. They were also geopolitically significant, and imperial officials devoted large resources to defending them. In the American Revolution, for example, keeping Jamaica safe from French attack was so important that Britain compromised the defense of its American mainland possessions by withdrawing its navy from the Battle of Yorktown in 1781 in order to send it southward. Saint-Domingue was even more important geopolitically to France, as shown by its expenditure of vast amounts of money and huge reserves of well-trained European soldiers during the calamitous (for the French, anyway) Haitian Revolution between 1791 and 1804 in an ultimately fruitless campaign to restore slavery and the plantation system in its former Greater Antillean "jewel."[48] Haitians remembered how they had humiliated the French. One of Jean-Jacques Dessalines's officers attended patriotic plays celebrating Haitian independence wearing a large hat on which was written, in large red letters, "Haiti, the tomb of the French." The French remembered their humiliation also, inflicting on Haiti in 1825 an enormous and crippling indemnity of 150 million francs for the privilege of engaging in international trade.[49]

The significance of these remarkably successful and terrifyingly brutal slave societies is increasingly recognized in an Atlantic-inflected historiography in which Jamaica and Saint-Domingue are seen not just as important within Caribbean and American history but as vital parts of eighteenth-century British and French imperial history. As authors with a long-standing interest in the history of the Greater Antilles, we have watched how over the last twenty years the history of this part of the world and the story of the plantation machine has moved from the margins of historical interest to become a more central component of the story of how the modern world came into being.[50] We believe that our cautious approach to the importance of Jamaica and Saint-Domingue to French and British social, political, and economic development in the second half of the eighteenth century allows for a defensible appreciation of the links between Caribbean slavery and European industrialism. Wealth from Jamaica proved beneficial to Britain, enhancing manufacturing and urbanization, and providing an impetus to Britain's powerful mercantile class. So too wealth from Saint-Domingue helped move French eighteenth-century merchant capitalism toward industrialism, though the American and French Revolutions delayed and shaped this process.[51] Exports from Britain to the Americas made a powerful contribution to British economic growth, although the most significant area to contribute to British wealth was not West Indian slave societies as much as the northern mainland colonies in which growing European populations stimulated demand for British manufactures.[52] In the absence of slave colonies, the northern colonies may still have imported similar quantities of British manufactures, but the existence of places like Jamaica allowed northern merchants to acquire the finance necessary to buy British goods through money gained in trade with West Indian colonies.

The wealth derived from Jamaica was significant in other ways. A few Jamaican politicians, notably William Beckford II, who was an intimate of William Pitt the elder, were able to influence British politics in respect to the governance of Atlantic colonies.[53] But Jamaica was most important to Britain as a source of wealth and prestige for a substantial portion of Britain's ruling elite and as an incubator of culturally important institutions that helped define Britain as a nation. Jamaican money and Jamaican planters who were resident in Britain, especially in southeastern England, altered consumption patterns and changed cultural practices, particularly in regard to house building, connoisseurship, and philanthropy.[54] In addition, the enslaved and free people of color that Jamaicans brought with them into Britain altered irrevocably, as we discuss in later chapters of this book, the character of British race relations, just as they also did in France. Indeed, Britons' self-conception of themselves as modern people, living in diverse societies, hinged, as Kathleen Wilson has argued, on a developing historical consciousness shaped by contact and exchange in the metropolis between Britons and people of African descent brought in by Jamaican planters.[55]

Less work has been done on the contribution of planters in the French Caribbean to France in the late eighteenth century, with Atlantic history relatively marginalized within French historiography. Moreover, the collapse of Saint-Domingue during the Haitian Revolution has tended to hide the dynamic influence of wealthy planters in French commerce and social life prior to the French Revolution.[56] But some recent work has shown that, as in Britain, the wealth of the Antilles in the second half of the eighteenth century penetrated into the highest reaches of French society, notably in the port towns of Bordeaux and Nantes and into the financial and political capital of Paris. French metropolitan investment in the Antilles was even more extensive than in Britain, with large merchant houses such as the Chaurands in Nantes investing 3 million livres tournois into West Indian planters.[57] They were backed up by leading Parisian banks, which saw investment in Saint-Domingue as potentially lucrative. Caribbean planters made a sizeable contribution to France, with imperial products amounting to as much as 15 percent of overall economic growth in France during the expansionary years between 1716 and 1787.[58] Much of this money ended up in Paris, which in the eighteenth century was transformed from an administrative and manufacturing city into a financial powerhouse. Money flooded into the capital from everywhere in France and also from Saint-Domingue. For example, in the 1770s, Jean-Joseph de Laborde, the wealthiest man in France, invested heavily in Saint-Domingue, after retiring from court finance and rebuilding the La Grange-Batelière district of Paris into a stylish *quartier*. Laborde spent 1.2 million livres to acquire a collection of contiguous plantations, and 750,000 livres as his share of a complex irrigation system in the region. He populated his estates with thousands of captives, many of whom traveled from Africa on his slave ships. By 1789 his estates had fourteen hundred enslaved workers.[59]

As the French metropolis became increasingly indebted after the American Revolution, its economic reliance on Saint-Domingue deepened and intensified. Allan Potofsky has detailed how Jean-Baptiste Hotten (1741–1802) used his profits from his large sugar estate with 219 enslaved laborers to become a Parisian real estate mogul. The son of a Bordelais sugar merchant, Hotten was unusual in carrying little debt at a time when private and public debt was spiraling out of control. This strong financial position allowed him to splurge on Paris real estate, constructing an elaborate complex in northwest Paris, consisting of a *hôtel particulier*, rental homes, and a large private garden. Interestingly, he continued his purchases well into the 1790s, showing that Caribbean planters were not left as destitute by the Haitian Revolution as they pretended.[60]

Hotten's prosperity in France, at least until he died violently during an ill-advised return to the colony, shows that we should not base our judgments about late eighteenth-century Saint-Domingue on Haiti's later history. We know what contemporaries did not, namely that in a few short years after the late 1780s Saint-Domingue society would explode into the revolutionary conflagration that led to the creation of the first independent nation in the Western Hemisphere led by people of African ancestry. The situation in Jamaica in the late 1780s and 1790s was not so dramatic. Indeed, the destruction of Saint-Domingue proved a boon to Jamaica, which profited from the economic vacuum created by the Haitian Revolution. In the 1790s it enjoyed a prolonged if unsustainable boom in sugar production and slave importation that few would have predicted a decade previously. But Jamaica faced long-term problems after the 1780s, mostly initiated from actions taken in the metropolis, all of which were designed, in the minds of proslavery Jamaicans, to undermine an immensely productive plantation system. They thought it "madness."[61] Unlike Saint-Domingue, Jamaica's slide from great wealth and geopolitical importance happened gradually, occasioned by the growing power of a force that Jamaican planters barely recognized before 1788—abolitionists campaigning for the end of the slave trade and, by the 1820s, for the abolition of slavery itself.

We ask readers to try and forget what they know is going to happen to these two colonies. After 1788 each confronted an unusual combination of outside events and revolutionary movements that transformed not just the Greater Antilles but also the Atlantic World. To describe these transformations is beyond the scope of this book. But we will show that in 1788 neither Jamaica nor Saint-Domingue were disasters waiting to happen, as an earlier generation of historians suggested. Rather, they were remarkable economic successes and considerable contributors to imperial wealth in the dramatic years between the start of the Seven Years' War and the calling of the French Estates General in 1788. In Jamaica, the travails of the American

Revolutionary years and the multiple hurricanes that afflicted the island from 1780 to 1786 had ended by 1788. The slave trade had started to flow again; plantation profits were once again buoyant; and productivity gains on plantations were making slavery ever more efficient, but just as deadly to workers as before. Britain remained strongly committed to Jamaica as the centerpiece of its British Atlantic strategy. By any standard except one Jamaica was doing very well.[62] The exceptional standard was Saint-Domingue. By 1788, Saint-Domingue's culture of relentless slave exploitation and highly capitalized agro-industry had been bolstered by a doubling and tripling of the French trade in African slaves, making it the most valuable colonial territory on earth.

The French government, even more than the British government in Jamaica, recognized Saint-Domingue's importance and devoted ever increasing resources to its protection and development. It gave beneficial terms to slave traders and spent massive sums on rebuilding its navy. The results were remarkable. Atlantic trade for the first time in French history challenged trade to the Mediterranean as the dominant sector in French commerce. But the costs of catering to the growth of Saint-Domingue were considerable. As William Doyle notes, it is not "a complete exaggeration to suggest that the costs of upholding a colonial system that could only work through slavery were what ultimately brought down the ancien régime in France."[63]

The remarkable economic success of these two colonies was acknowledged, eagerly welcomed, and sometimes resented in Europe because of the relentless materialism of both societies. The last point became more salient over time. Metropolitan observers increasingly felt that the white residents of these societies were morally deficient in their attachment to material gain, their indifference and sometimes hostility to traditional values, and, most of all, their shortsighted and seemingly un-European exploitation of their servile laborers. It was not the economics of slavery, but the cultural practices that were associated with the institution, that caused most concern in France and Britain. We can see these practices in a close examination of society and economy in the two colonies. We turn to this examination now.

CHAPTER 2

The Plantation World

Eighteenth-century Saint-Domingue and Jamaica were difficult societies for contemporaries to comprehend. Amazingly profitable and extraordinarily brutal, they were afflicted by horrific rates of disease and death, not only among enslaved peoples but also among their ruling elites. Death defined the peculiar cultural ambience of each place better than anything else.[1] Free residents seemed addicted to the fervent pursuit of pleasure as well as money in ways that fascinated but also unnerved observers coming from North America or Europe. J. Hector St. John de Crèvecoeur, for example, the author of the iconic text on American identity, *Letters from an American Farmer*, found Jamaica impossible to place in his schema of American development. Crèvecoeur, who appears to have visited Jamaica in the 1780s, understood the cruelties inherent in the plantation slavery that sustained societies from Virginia to northern Brazil and believed this system was deeply irrational.[2] Nevertheless, Jamaica disturbed him in ways that Spanish America—which he saw as lethargic, indolent, superstitious, and backward—did not. Jamaica's essential characteristics, he thought, were restless wandering, corruption, and pervasive dishonesty.

He noted Jamaica's "great Glare of Richesses." He was "shocked at that perpetual Collision & Combination of Crimes & Profligacy which I observed there." He noted the "severity Exercised agt ye Negroes" and lamented the illicit sexuality that raised some black females to a "Pomp" from "which the rest were reduced" and that was derived from "a perversion of appetites." In a place with no religion "save few Temples," everything was sacrificed to business and sensuality. He lamented that "a perpetual pursuit of Gain & Pleasures seem'd to be the idol of the Island." Jamaica was "a Chaos of Men Negroes & things which made my Young American head Giddy." Fighting his way "through this obnoxious Crowd," the innocent narrator reflected that "the Island itself looked like a Great Gulph, perpetually absorbing Men by the power of Elementary Heat, of Intemperance by the force of every Excess" so that "Life resembled a Delirium Inspired by the warmth of the sun urging every Passion & desire to some premature Extreme." His only response to these extremes, to the "Exhausting Climate," and to "the perpetual struggle subsisting between the two great factions which Inhabit this Island," was to take leave of Jamaica and not think about it again.[3]

Travelers to Saint-Domingue were similarly dislocated. Colonists there were highly materialistic, and had few communal institutions or spaces that were reminiscent of France. Isolated on their estates, many colonists adopted the customs of the buccaneers that preceded them and the enslaved Africans that surrounded them. It was a place that seemed at one and the same time to be very French and also the least French place imaginable. The colony inspired both desire and disgust in Alexandre-Stanislas de Wimpffen, a minor nobleman who sought his fortune there in the late 1780s. He was entranced, on the one hand, by the colonial landscape and by beautiful "mulâtresses" who "combine the explosiveness of saltpeter with an exuberance of desire, that scorning all, drives them to pursue, acquire and nourish pleasure." But he saw Saint-Domingue as a world turned upside down, where morality was absent, mainly owing to slavery. It was a society based on pursuit of profit, not religious virtue, social cohesion, or imperial loyalty. He proclaimed, "The Commerce of France is the true owner of Saint-Domingue."[4] Lieutenant Colonel Desdorides, stationed in Saint-Domingue in 1779, was also struck by colonists' pursuit of profit and pleasure. He lamented that "it seems that in Saint-Domingue violent agitations of the heart take the place of principles; except for illusions of love, dreams of pleasure, extravagances of luxury and greed, the heart knows no other adorations."[5] Desdorides believed that an obsession with money had fundamentally transformed the character of French colonists: "In Saint-Domingue, men and women behave in a manner totally opposed to what I have described [of the French character]. Men there reduce everything to financial gain."[6]

In this chapter, and the next, we look at a variety of images that help us understand these places—even if they lead us to the same feelings of displacement, dislocation, and unease felt by contemporary visitors. We start by examining large-scale maps, which show the geography and diversity of environment in each colony and which also give a glimpse into some of the underlying ideologies animating the West Indian ruling elite. We proceed to a treatment of the sugar plantation (the quintessential institution in Jamaica and Saint-Domingue) through two allied but very different portraits—one an idealized pastoral scene that hides the reality of enslavement and the other an estate plan that makes abundantly clear the underlying violence of the plantation machine, as well as the luxury it allowed owners. From an extensive survey of rural life in the two colonies in Chapter 2, we move, in Chapter 3, to examine the vibrant culture of colonial towns.

In 1763, the year that the Peace of Paris confirmed Britain's victories in the Seven Years' War, Thomas Craskell and James Simpson published an elaborate map of Jamaica (Figure 1). Encompassing twelve sheets, it was the culmination of the first detailed survey of the whole of the island, done under the

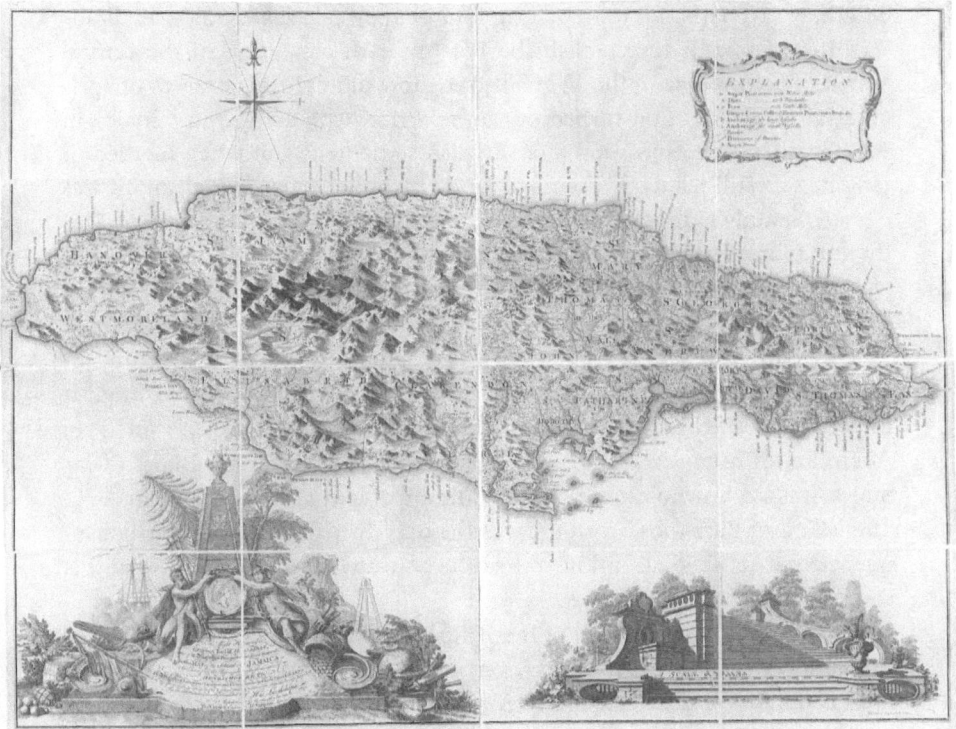

Figure 1. "To the Right Honourable George, Earl of Halifax . . . This map of the island of Jamaica . . . humbly inscribed by Thos. Craskell, Engineer, and Jas. Simpson, Surveyor." London: Fournier, 1763. © Courtesy of the John Carter Brown Library, Providence, Rhode Island.

direction of Governor Henry Moore between 1756 and 1761. This map stood for the next forty years as the most detailed image of Jamaica, identifying sugar plantations, some with water mills, and some with wind and cattle mills. It also located ginger, cotton, and pimento estates, livestock pens, military barracks, principal anchorage points, and Jamaica's road system.[7] By the 1760s the English had been developing Jamaica for a century.[8] Though it was roughly one-third the size of Saint-Domingue, it was much bigger than any of Britain's Lesser Antilles islands, more geographically varied, and consequently more capable of future development. Indeed, at the start of the Seven Years' War much good land was still uncultivated.[9]

Unlike the smaller colonies, Jamaica had mountains, swamps, and arid sections, as well as fertile coastal plains. This variety made it far more difficult than, say, Barbados to transform into a large-scale sugar producer. Francis Price's attempts to establish the Worthy Park plantation in the central parish of St. Thomas in the Vale illustrate how difficult this process of agricultural transformation turned out to be. Price, who arrived in Jamaica in 1655, acquired the land soon after English settlement, but when he died in 1689 he was still relatively poor. His land was only partly cleared, and it was devoted mainly to food crops and pasture rather than sugar. Price's farm was the sort of modest pioneer property that might have been found in the backwoods of seventeenth-century Virginia. It was Price's grandson Charles, who died as Speaker of the House of Assembly in 1772, who transformed Worthy Park into a large plantation with hundreds of slaves and significant amounts of cane land.[10] Jamaica's slow development disappointed English imperialists. Celebrating British American growth and expansion in 1776, Adam Smith reminded his readers that a century before, "the island of Jamaica was an unwholesome desert; little inhabited and less cultivated.... The island of Barbados, in short, was the only British colony of any consequence of which the condition at the time bore any resemblance to what it is at present."[11]

Jamaica became the jewel of the British Empire because of the ability of its planters to extract great wealth from sugar. But the risks involved in starting a plantation were considerable. A remarkable set of planter's records from Jamaica in the 1670s details the frustrations of the process. Cary Helyar was an aspiring but impecunious younger son from a genteel background. He came to Jamaica in the early 1660s, made some money in slave trading, and developed connections with leading politicians. He bought prime sugar land. He boasted to his brother that his Bybrook plantation of 1,236 acres was

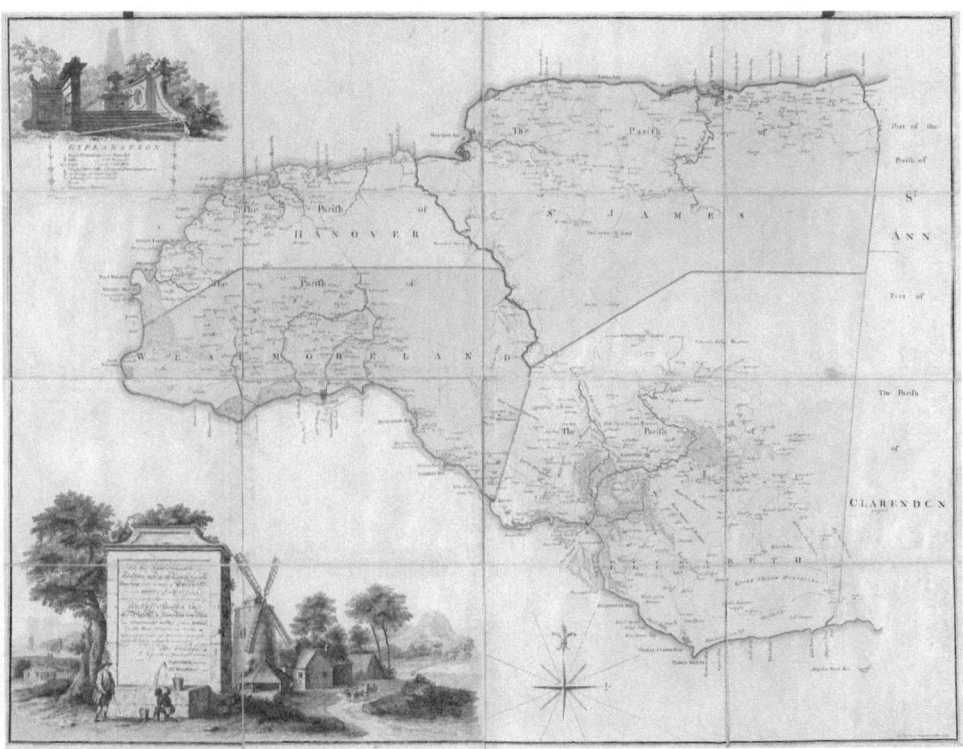

Figure 2. "To the Right Honourable Robert, Earl of Holdernesse, this map of the county of Cornwall, in the island of Jamaica . . . humbly inscribed by Thos. Craskell, Engineer, Jas. Simpson, Surveyor." London: Fournier, 1763. © Courtesy of the John Carter Brown Library, Providence, Rhode Island.

"almost square, of as good land and as well-watered as any in the island." Yet to make sugar at Bybrook required a large capital investment in slaves, equipment, and livestock.[12] Only in the late 1680s did Bybrook generate significant profits, and those disappeared in the 1690s, after Helyar's son, who had inherited the estate, returned to England. The Helyars could not find honest and skilled managers to supervise their overseers. Sugar production dropped, dead slaves were not replaced, and equipment wore out. By 1713, when the family sold it, the property was close to worthless. As Helyar's experience suggests, sugar planting was a tricky business.

Small planters preferred other options to the risky enterprise of

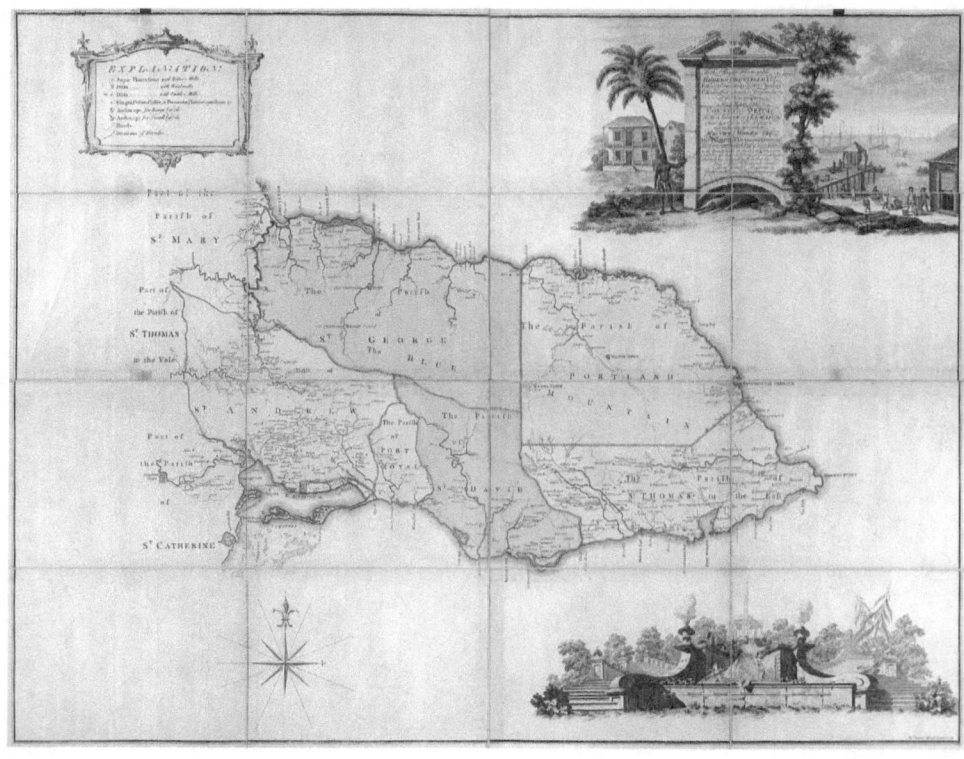

Figure 3. "To the Right Honourable George Grenville, Esq., First Lord Commissioner of Surrey in the Island of Jamaica ... humbly inscribed by Thos. Craskell, Engineer, Jas. Simpson, Surveyor." London: Fournier, 1763. © Courtesy of the John Carter Brown Library, Providence, Rhode Island.

large-scale planting. Piracy was one option.[13] For ordinary white men an enjoyable life could be had roistering in the narrow streets of Port Royal, seeking fame and fortune through plundering expeditions against the Spanish. They could supplement their income and ensure a measure of landed independence by growing small quantities of crops on a few acres of land in nearby parishes. A British migrant, John Taylor, gives us a lively picture of Port Royal at its zenith. It was not only Port Royal's "merchants and gentry" who "live here to the hights of splendor," but "all sorts of mechanicks and tradesmen ... all of which live here verey well, earning thrice the wages given in England, by which means they are enabled to maintain their famallies much better than in England, by which tradesmen 'tis much advance both in strength and wealth, still becoming more formidable."[14]

Port Royal was badly damaged in a 1692 earthquake, effectively ending privateering. Around this time, between the 1680s and the 1720s, wealthy colonists boosted their profits by adopting the integrated plantation model. A massive influx of enslaved Africans ensured that this model would be successful. While the black population was relatively equal to the white population in the 1670s, by 1700 Jamaica was populated mainly by enslaved blacks. In 1710, the Crown allowed private traders into the African slave trade previously monopolized by the Royal African Company. This greatly increased the number of captives arriving in Jamaica. Large sugar estates bought most of those Africans so that by the 1720s, most slaves lived and worked in labor forces of one hundred or more. The dominance of large plantations remained virtually unchanged until the end of slavery. Simultaneously, white servitude declined dramatically. Until the 1680s, most planters with medium to large slave forces also had indentured white servants as part of their labor force. In 1690 Dalby Thomas described a sugar plantation with fifty slaves as needing seven servants.[15] From the 1690s, however, the numbers of white servants noted in inventories faded away until by the 1720s a white servant in an inventory was a rarity. By this time, land prices were soaring as the best territory in the settled parishes of the southern coast and in central Jamaica to planters had already been distributed. The hard work of slaves and servants transformed these properties into fertile sugarcane fields.

The rise of the large integrated plantation had two significant consequences for Jamaica. First, it meant that the colony's population consisted primarily of enslaved African slaves sold to Jamaican planters to labor as sugar workers. Second, their labor produced a planter class with wealth and influence unprecedented in the eighteenth-century British Atlantic. Elite planters remained the dominant social and political force on the island, despite demographic disaster and metropolitan opposition, until after the end of slavery in 1838. This new plantation world was reflected in the Craskell and Simpson maps of 1763. The engraver, Daniel Fournier of London, included ornate title cartouches that showed just how far Jamaica had developed since perilous times in the 1690s and 1700s. The three scenes showed a confident planter class at work and at play. In the left-hand bottom corner, a planter posed next to an ornate stone on which the names of the governor commissioning the map and the two surveyors were engraved. A kneeling and subservient black man carrying pails of water accompanied the planter. Behind him was a well-ordered plantation with a working windmill, cattle trudging along a pathway, and a harbor with ships docked in the distance. Notably absent were the mass of enslaved people necessary for a functioning sugar plantation.

Counterposed to this scene of rural productivity was an urban scene in the right-hand top corner. Likely a composite picture of British West Indian ports rather than Kingston itself, the cartouche depicts a harbor full of merchant vessels and one military gunboat, showing the welcome protection of the Royal Navy for West Indian commerce. The crowded commercial scene is full of dockside workers ferrying sugar, rum, molasses, and provisions overland and across the harbor in small skiffs. Two white men occupy prominent positions in the image, one a gentleman planter and the other a merchant, each with puncheons of rum and hogsheads of sugar at their feet. An open account book sits on a cabriole-legged desk near the two men. To the left is a substantial and elegant merchant's house with a handsome portico. And in front of another stone engraved with the names of the prime minister, the Jamaican governor, and the two surveyors are two black men, a bag of coffee, and a stack of tropical lumber. The scene advertised Jamaica as a lush and bountiful land where Britons made money and engaged in genteel pursuits, assisted by African workers.

The third cartouche illustrated those pursuits. It showed a white man on foot, with dogs, cornering a large boar in a heavily forested countryside. The three scenes, overall, showed to English spectators an idealized Jamaica: a land of flourishing plantations and bustling towns, with abundant and quiescent black laborers serving wealthy aristocratic Europeans. Sugar and industry anchored the whole tableau. And women and free people of color were invisible.

Maps drawn in the 1720s of Saint-Domingue show not only that it was far larger than Jamaica, but that it was an even wilder place, at least in the first half of the eighteenth century. Jamaica became an English possession in one fell swoop in 1655, when an English invading force drove out a small Spanish colony. In contrast, Saint-Domingue became a French possession incrementally, as hunters and pirates living there began to accept the authority of French governors only around 1665. After this date the colony, or pieces of it, developed slowly under the control of various royal monopoly companies, coming under full royal governance only in the 1720s. In 1700 Jamaica had about seven thousand whites and forty thousand blacks; Saint-Domingue had 4,560 and 9,082 respectively.[16] Although early eighteenth-century French maps depict a network of overland roads, the colony's mountainous interior made shipping the most practical way to transport goods and people. Nevertheless, the colony's complex coast was difficult to sail around, and local pirates preyed on Saint-Domingue's coastal traffic until the 1730s.[17]

There were at least four factors besides geography and buccaneers that kept Saint-Domingue economically two or three decades behind Jamaica. First, Spain recognized French possession of the western coast of Hispaniola only in 1697, in the Treaty of Ryswick. Second, the French navy was ill equipped to protect the kingdom's transatlantic commerce from foreign enemies or pirates. From a high point of one hundred ships of the line in 1680, it

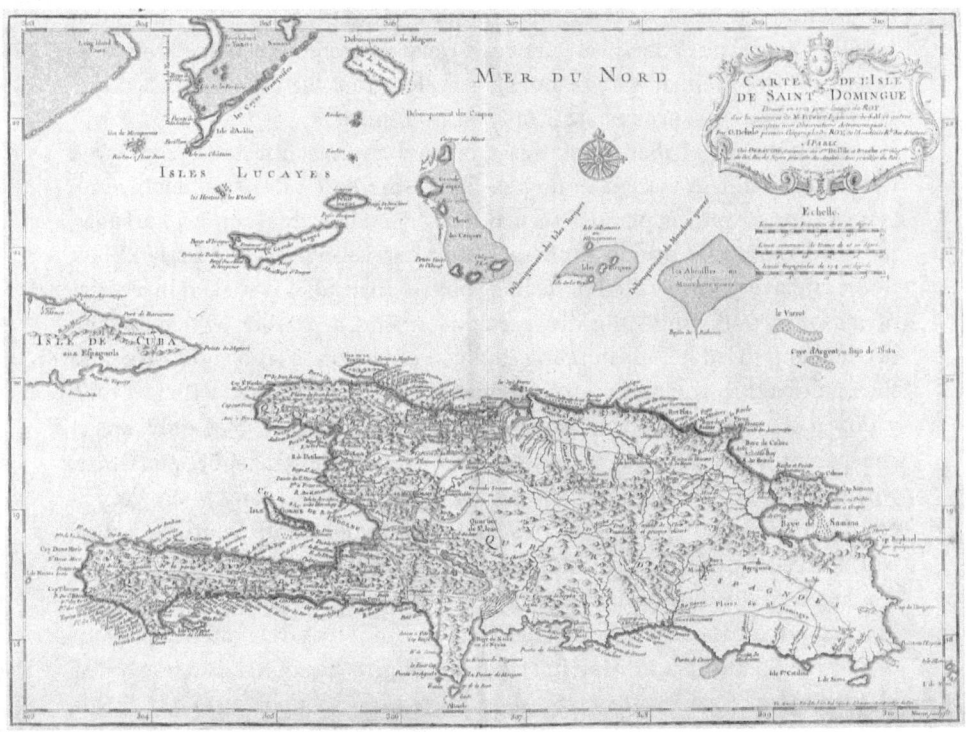

Figure 4. "Carte de l'isle de Saint Domingue. dressée en 1722 pour l'usage du roy, sur les mémoires de Mr. Frezier, ingénieur de S. M. et autres, assujetis aux observations astronomiques, Guillaume Delisle." Paris, 1780. © Courtesy of the John Carter Brown Library, Providence, Rhode Island.

shrank to forty-nine in 1725.[18] Third, the French slave trade started slowly. It was only in 1725, after bitter complaints from wealthy colonists about the Compagnie des Indes, which controlled France's trade with West Africa, that Versailles opened the slave trade to all private merchants.

The fourth factor was the amount of investment sugar required. Father Labat, a Dominican priest who managed his order's plantation in Martinique, described the technological and human workings of sugar plantations at the beginning of the eighteenth century in his widely read *Nouveau voyage aux isles de l'Amérique* (1724). In a long and detailed chapter that was copied in commercial handbooks, he advocated a workforce of 120 slaves, which he believed in 1696 would produce net annual revenue of 38,030 livres for over a century, with "a little thriftiness." But when Labat compared such an estate to the cacao walks he observed in Martinique around 1702, he noted that sugar required three times the investment as a cacao estate that produced the same revenues. This comparison, he noted, revealed "that the cacao walk is a rich gold mine, while a sugar estate is only an iron mine."[19]

About the time Labat's book was published, a fungal infestation destroyed the cacao sector in Saint-Domingue. But in the first half of the eighteenth century, most aspiring planters turned to indigo rather than sugar. Like sugar, indigo dye must be heavily processed after harvesting. But the indigo plant does not have to be crushed or its juice boiled. Instead, it is soaked in a series of masonry tanks until the dye precipitates into a powder, which is then drained and dried before shipping. Indigo cultivation is labor intensive, and the manufacturing process can be subtle, but it can be handled with less than a dozen slaves, a few water basins, and a skilled refiner. Not only was it cheaper to make than sugar, but indigo was easier to transport and store, which made it better suited for smuggling in parts of the colony where French commercial shipping was rare. Up to the mid-eighteenth century, indigo works in Saint-Domingue outnumbered sugar mills ten to one.[20]

Sugar cultivation began in Saint-Domingue in the 1690s. The colony's first sugar exports to France arrived in 1697. But it was only after the end of the War of Spanish Succession in 1713 that colonists began to build integrated plantations and to create large slave forces. Only after 1740 did Saint-Domingue reach the level of development Jamaica had attained by 1720. In 1730 it had just over seventy-nine thousand slaves, while in 1734 Jamaica had roughly eighty-six thousand slaves. By 1744, the enslaved population of the larger French colony had surpassed that of its neighbor, with 118,000 slaves compared to roughly 112,000 slaves in Jamaica.[21] Besides buying African workers, each of whom cost roughly as much as a French silk weaver made in an entire year, Saint-Domingue's largest sugar planters also invested in new plantation refineries to produce more valuable clayed, rather than brown or muscovado sugar.[22] In 1730 the colony had eighteen *sucreries en blanc*, amounting to only 5 percent of its sugar estates. By 1753 there were 235 of these more elaborate factories, making up 42 percent of all Saint-Domingue sugar estates. In the region around Cap Français, 189 out of 290 sugar refineries (65 percent) produced clayed sugar.[23]

Irrigation was another central element of Saint-Domingue's transformation after 1730. Before the Revolution, very few irrigation projects were completed in France itself. Most of them ended because of litigation caused by competing or overlapping jurisdictions and property rights. In both Saint-Domingue and France after 1760 the royal state eventually worked to promote irrigation schemes, but before 1760 irrigation works were private projects, involving dozens or even hundreds of planters.[24] The first irrigation works were built in 1731 for two plantations in the Cul-de-Sac plain that would one day become the hinterland of Port-au-Prince. In 1737 another system in this region brought water to twenty-four sugar plantations, and after 1739 four plantations shared an irrigation works at Aquin on the southern coast. Other early sugar areas, like the plains around Léogane and Petit Goâve, followed with private irrigation schemes in the 1740s.[25] These investments expanded sugar planting into less fertile regions of the colony. They also provided power for water-driven sugar mills, helping make Saint-Domingue into the most efficient producer of tropical commodities in the late eighteenth-century Atlantic World. In the 1740s, as Saint-Domingue's slave imports and sugar estates continued to grow, a new plantation crop emerged. Coffee, first planted in Saint-Domingue in 1738, thrived on hillsides that would not support indigo or sugarcane. In 1753 administrators counted nearly thirteen million coffee bushes. Moreover, the colony's overall population nearly doubled, increasing 86 percent from 1730 to 1753, almost entirely because of the accelerating importation of enslaved Africans. In 1753 Saint-Domingue had 161,859 slaves, a number well above Jamaica's slave population of 106,592 in 1752.[26]

The nature of port records and other primary sources in this prestatistical age make it difficult to assess what Saint-Domingue's total plantation exports might have been at midcentury. But it is clear that production from France's largest Caribbean colony had caught and probably surpassed that of Jamaica and perhaps that of the entire British West Indies. Charles Frostin estimates that in 1740 Saint-Domingue's sugar production of forty thousand tons surpassed that of the whole British Caribbean production of thirty-five thousand tons the same year.[27]

Because the sugar plantation was the emblematic feature of the landscape in both Jamaica and Saint-Domingue, by the last decades of the eighteenth century the wealthiest owners of these estates wanted to see images of the properties that supported them. In the 1770s, William Beckford, the wealthiest man in Jamaica, commissioned George Robertson (1735–1821) to paint views of Beckford's Westmoreland sugar estates. Four of Robertson's landscapes were popular enough that John Boydell made them into copperplate engravings in 1778. All four of the paintings featured tropical vegetation with rushing rivers in the foreground, with black men or women washing, fishing, or tending animals. One showed Beckford's Fort William estate in the

Figure 5. "A View in the Island of Jamaica, of Fort William Estate, with Part of Roaring River, belonging to William Beckford, Esqr. Near Savannah La Marr, Westmoreland, 1778," drawn on the spot by George Robertson, engraved by Thomas Vivares. © Courtesy of the John Carter Brown Library, Providence, Rhode Island.

background with the Roaring River in the foreground (Figure 5). Robertson drew Beckford's sugar works, and above it, on a hill, the planter's residence. Smoke rises from the boiling house, suggesting a proto-industrial estate, but, in the twilight scene, the enslaved blacks with their baskets and livestock look more like exotic peasants than mill workers. In the center of the image, Beckford's large white house dominated. Robertson thus suggests that the planter has successfully tamed not only the wild Jamaican landscape but also alien African workers.

All mid-eighteenth-century British America plantation societies entered into a prolonged period of affluence that lasted from around the 1730s until the end of the Seven Years' War, as they also did in the French Antilles.[28] Boydell's engraving of Robertson's painting reflects that prosperity. Profits in the West Indies, however, were far greater than those in the southern colonies of British North America. From the 1740s through to the mid-1770s, British West Indian returns on capital were over 10 percent per annum, reflecting the efficiencies of the integrated plantation model.[29] Jamaica became the leading exporter of sugar in the eighteenth-century British Empire. Productivity was extraordinarily high. Between 1750, when per capita productivity per annum based on both the white and the black population was around £8 sterling, and 1800, when per capita productivity per annum was £29.2, or 675 livres, Jamaican productivity expanded to reach probably its natural limits. The Jamaican economy performed strongly not only in comparison with other plantation economies but also relative to emerging industrial nations. On the eve of the American Revolution, when individual wealth (if not productivity) probably peaked, Jamaica was as important to Britain in terms of wealth creation as a large British county.[30] Saint-Domingue may have exceeded it in average wealth, but we do not have the figures to justify this assertion. Jamaican sugar planters were among the most accomplished capitalists of their time. Today we see them as ruthlessly using and discarding hundreds of thousands of men and women of African descent. But contemporaries viewed them as entrepreneurs who successfully manipulated a complex agro-industrial technology, supplying their compatriots with a highly desirable product at an ever cheaper price. Their activities were the lynchpin of an Atlantic trade network, linking Africa and Europe with the Caribbean.[31]

Jamaica's sugar plantations increased in number from approximately 150 in 1700 to 775 in 1774. Sugar output increased from five thousand tons in 1700, to sixteen thousand tons by 1734, twenty thousand tons in 1754 and to forty thousand tons by 1774. At the same time total exports, of which somewhat over half were sugar exports, increased nearly eight-fold between 1700 and 1774, from £325,000 in 1700 to £1,025,000 in 1750 and £2,400,000 (or 55 million livres) in 1774.[32] Sugar was produced by industrial-sized plantation units, customarily containing between 150 and 300 enslaved persons. Such labor forces were the largest in British America, far larger than the crews of around fifty slaves that worked large tobacco plantations in the Chesapeake and the slightly larger forces that labored on rice plantations in South Carolina.[33]

From the mid-1740s, Jamaica entered into an explosive period of growth. The establishment of peace with the Maroons (independent communities of people of African descent living in the Jamaican interior who from the early eighteenth century fought a twenty-year war against the British government) in 1739 allowed sugar planting to expand into the northwestern parishes of St. James and Hanover as well as into the eastern parish of St. Thomas-in-the-East. It is noteworthy that these years saw only very slow white population growth. The number of whites barely increased in the disease-ridden years between the earthquake at Port Royal in 1690s and the start of the American Revolution. The only firm figures we have are that there were 7,768 whites in 1673 and 8,230 in 1730 and probably no more than ten thousand by the start of the Seven Years' War. In 1774, a census, most of the details of which are lost, suggested a white population of 12,737. Thus, the number of whites on the island, despite considerable migration from Britain, had grown by only 4,969 between 1673 and 1774 or by less than fifty people per annum.[34] At the same time the black population increased rapidly, entirely owing to an ever expanding slave trade. The number in slaves in Jamaica catapulted in the eighteenth century to 74,523 in 1730 and to 192,787 in 1774.[35] Outnumbered eighteen to one, whites understood that this influx of workers was the source of their rising prosperity.[36]

The wealth of Jamaica was extracted out of the bodies of enslaved people. Apart from those in Saint-Domingue and perhaps Dutch Guiana, the lives of the enslaved population in Jamaica were the most miserable in the Atlantic World, especially in the first half of the eighteenth century, when planters were carving out plantations from frontier land and when the great majority of slaves were traumatized, brutalized, and alienated migrants from Africa. The sources for this period, nearly all produced by whites, militate against any true appreciation of what slaves went through. But it seems certain that the first half of the eighteenth century marked the nadir of black life in Jamaica. Slavery had always been brutal in British America, but the level of violence exercised against Africans dramatically increased as the slave population grew.[37]

British Jamaica was always an incredibly violent place, even before the sugar planters arrived. In the seventeenth century, royal authorities subjected pirates to gruesome executions while masters commonly whipped and chained their white servants. Africans, however, got the worst of the treatment. John Taylor, writing in 1688, dwelt almost lovingly on the barbaric tortures that planters forced on slaves caught in rebellions. He was convinced that it was only through terror that Africans could be controlled. He told a lengthy story about how "Collonel Ivey" discovered a plot against him and how he cornered his slaves and read out details from Jamaica's slave code (a book that "is hated by those slaves, and they still say 'tis the divile's book"). Ivey acted as judge and jury of his frightened slaves. He "caused all his slaves to be bound and fetter'd with irons, ... and then caus'd them to be severely whip't, caused some to be roasted alive, and others to be torn to peices with dogs, others he cutt off their ears, feets and codds, and caused them to eat 'em; then he putt them all in iron feters, and soe with severe whipping every day forced them to work, and soe in time they became obedient and quiet, and have never since offer'd to rebel. Thus did God bring to nothing their damnable disigne, and prevent the horrid rebellion and murther they intended."[38]

Matters got worse for slaves in the eighteenth century. Even planters' defenders admitted that they used abnormal levels of cruelty against slaves. Indeed, they reverted to punishments, such as castration and burning by slow fire, which had lost favor in Britain since medieval times. The rector of St. Catherine parish noted in 1751 that when planters sought "to deprive [negroes] of their funeral rites by burning their dead Bodies, [it] seems to Negroes a greater punishment than Death itself."[39] He wrote this just a year after Thomas Thistlewood had arrived in Westmoreland parish. Thistlewood's diary entries prove the truth of what the minister wrote to the Bishop of London. Twelve days after arriving in Westmorland, on 12 April 1750, he watched his employer, William Dorrill, apply "justice" to runaway slaves by whipping them severely and rubbing lime juice, salt, and pepper into their wounds. Not long after he saw the chief slave on the estate getting "300 lashes for his many crimes and negligences." On 1 October 1750, visiting the St. Elizabeth town of Lacovia, he "Saw a negroe fellow named English ... Tried [in] Court and hang'd upon ye 1st tree immediately ([for] drawing his knife upon a White Man) his hand cut off, Body left unbury'd."

This unbridled cruelty included forms of psychological debasement. On 19 March 1752, for example, Thistlewood related with amusement some stories that speak volumes about the character of life on the Jamaican frontier. "This morning," he noted, "Old Tom Williams," the patriarch of a major planting family, "Call'd and made his observations as usual." Thistlewood recorded two stories that he evidently found funny. Williams told him first how he had once killed a young girl with diarrhea by "Stopping her Anus with a Corn Stick." Then he told a story about how when he felt a domestic was doing a poor job cleaning his house, "he Shitt in it and told her there was Something for her to clean." With attitudes like this, it is little wonder that Charles Leslie declared in 1740 that "No Country excels them in barbarous Treatment of Slaves, or in the cruel Methods they put them to death."[40]

In the end, however, it may have been planters' interest in increasing workers' productivity, not their commitment to brutality, which affected slaves most. Caribbean planters were relentless innovators, anxious to try new methods or tinker with existing technology to increase their profits. The efficient use of slave labor was the key to the profitability of the integrated sugar estate. As Henry Drax of Barbados explained in the mid-seventeenth century "There must be especial Care taken in working the Hands, that every negro doth his part according to his Ability. . . . The best way that I know of to prevent Idleness, and to make the Negroes to their work properly, will be upon the change of work, constantly to Gang all the Negroes in the plantations in the Time of Planting."[41] Sugar planters saw themselves as enlightened improvers and from the mid-eighteenth century began to publish books outlining their plantation methods.[42] Many of these ideas came out of the Lesser Antilles, smaller islands where geography placed a premium on efficiency, but Jamaica also participated in such reforms. The Jamaican historian Edward Long declared in 1774, "A spirit of experiment has of late appeared which, by quitting the old beaten track, promises to strike out continual improvement."[43]

The key part of what Antiguan planter Samuel Martin called a "well-constructed machine" was its human component. Planters devoted enormous attention to improving their human capital so that enslaved people could work more efficiently and, more important, could increase in value over time.[44] One indication of the hardheadedness and calculating business sense of Jamaican slave owners was their determination to get as much value out of enslaved women's labor as they could. West Indian planters divided slaves by physical capacity rather than by sex. They insisted that the "stoutest and most able slaves . . . without any regard being had to their sex" should do the hardest work, such as digging cane holes, dunging, and cutting and harvesting cane. By the early eighteenth century, women were the majority of field hands on Jamaican sugar plantations. Plantations that needed extra labor hired women and men at the same rate, suggesting that there was little distinction in what they did as ordinary field laborers. This lack of gender differentiation in field work persisted until the end of slavery. David Collins in 1803 insisted that there are "many women who are capable of as much labour as man, and some men, of constitutions so delicate, as to be incapable of toil as the weakest women."[45] Planters were so determined to utilize women's labor capacities that they gave little attention to alleviating field work for sick or pregnant women. Planters were reluctant to lose women's labor. Slave reproduction came very low on the list of their priorities. Most women worked in the fields up to close to the time they gave birth and were returned to the field soon after. Able female field hands probably had no more than five weeks before and after birth where they were released from arduous field work.[46]

There is some evidence that planters worked women even harder than men. They were more likely than men to be field laborers, and they did extremely physically demanding tasks, such as cane holing and dunging, in greater proportions than did men. Accordingly, women tended to have very high rates of sickness. By the late eighteenth century planters began to admit that the demands they placed on women, especially young women at the start of their childbearing years, were too great. Jamaica planter Gilbert Mathison conceded in the early nineteenth century that "perhaps young females are too much subjected to hard labour at an early and critical period of their lives." Nevertheless, slave women proved demographically tougher than slave men. Slave labor patterns on Mesopotamia Estate in the second half of the eighteenth century suggest that women had longer life expectancies than men and enjoyed lower death rates.[47]

Despite planters' Enlightenment ethos of rational improvement, they were largely unwilling to apply those ideals to improve the lives of their enslaved laborers. The hallmarks of the large integrated plantation were discipline, coordination, and coercion. Planters were determined to make slaves work as hard as they could. They concerned themselves little about the deaths such demands produced. Our best data on the workload of Jamaican slaves comes from late in the eighteenth century. At Prospect Estate in the developing parish of Portland, slaves worked twelve hours a day for an average of 272 days, with sixty days off. Illness or other problems stopped work for thirty-three days.[48] During their days off, enslaved people produced their own food, constructed their own housing, and attended to the needs of themselves and their children. Like the cotton mills of early industrial England, which depended on fresh inputs of laborers from the countryside to replace worn-out or dead workers, the sugar plantation depended on a functioning slave trade to maintain numbers. It was a consuming industry to cater to the growing consumer markets of western Europe.

In the French Caribbean, early eighteenth-century missionary accounts provide some indication of colonists' attitudes toward the enslaved men and women whose labor was transforming Saint-Domingue. When the Jesuit Pierre-François-Xavier de Charlevoix described Saint-Domingue sugar slavery in the 1720s, he was struck by the workers' misery, the rags they wore, and their lack of food. Yet he contrasted their "perfect health" with the "infinity of illness" that plagued their French masters.[49] Frustrated by the apparent inability of some Africans to learn even the Lord's Prayer, Charlevoix used a mechanical metaphor to describe their "extremely limited intelligence," describing Africans as "machines whose springs must be rewound each time that you want them to move." Though he called the whip a necessary disciplinary tool, Charlevoix believed that the French were kinder masters than the English. If Saint-Domingue bordered an English colony, he claimed, most of the slaves would decamp to the French side.[50]

Writing a few decades earlier in Martinique, Jean-Baptiste Labat, who was both a missionary and a planter, also believed the English were crueler than the French. In 1696, after warning his readers about the frequency of deadly accidents at the sugar mill, where careless workers might be dragged into the rollers, he claimed that the English used their sugar mills to execute slaves who committed major crimes, gradually lowering their living bodies into the machine.[51] Yet Labat highlighted the industrial character of plantations like the one he directed in Martinique: "Whatever you say about the work in an iron forge or glass factory or others, it is clear that there is nothing harder than that of a sugar mill, since in the former there are no more than 12 hours of work, while in a sugar mill there are 18 hours of work a day." Because of mill workers' lack of sleep, Labat advised planters to force them either to sing or to smoke so they would not doze off and fall into the cauldrons of boiling sugar.[52]

There are almost no such missionary accounts from the French Antilles after the 1730s, but census reports show the rapid expansion of the plantation sector in Saint-Domingue after this date. The number of sugar estates expanded from 450 in 1739 to 636 in 1771. These were large operations, though still notably smaller at this date than estates in Jamaica. For the 1770s, David Geggus calculates that the average size of a sugar estate slave force was roughly 160. Exports from the Antilles to France, of which Saint-Domingue's products constituted the majority, increased from nearly 60 million livres tournois in 1749 to over 178 million in 1778 (£7.7 million). This output was largely the result of the fact that the colony's enslaved population more than doubled, from just fewer than 109,000 in 1739 to just under 220,000 in 1771. The free population of color expanded at an even faster rate over the same period, from 2,545 to 6,480. During the same time the white population grew only 60 percent, from 11,600 to 18,400. Where whites were roughly 9 percent of the total population in 1739, in 1771 they were only 7 percent.[53]

By the 1780s, the twin themes of profit and misery were more central than ever to French Caribbean colonization. An unusual and disturbing map and vignettes of Saint-Domingue's Févret de Saint-Mésmin sugar estate captures these overlapping realities for an absentee investor. On the one hand, the water-colored estate plan drawn by the royal engineer Louis de Beauvernet conforms to the Enlightenment ideal of rational planning (Figure 6). Trained in the latest surveying and cartographic techniques Beauvernet likely had seen the widely circulated prints of the royal salt works conceived by Claude-Nicholas Ledoux between 1775 and 1778, years when Beauvernet was in France. Ledoux planned a large preindustrial workplace, the Salines de Chaux, at Arc-et-Senans about eighty-nine kilometers from Beauvernet's native city of Dijon.[54] Like Ledoux's celebrated image, Beauvernet's sugar plantation map appears, at first glance, to celebrate the idea of harmony between the forces of nature and man's organizing genius. Tropical savannah land and cane fields dominate the central panel of the image, but they are traversed by rectilinear roads, as well as by a meandering stream. At the center are the plantation's buildings, the most distinctive of which are the large sugar mill and, nearby, twenty tiny rectangular slave huts aligned in two close columns. One key lists the thirteen buildings and another key names the thirteen parcels of land, mostly cane fields, that surround the manufacturing center. A

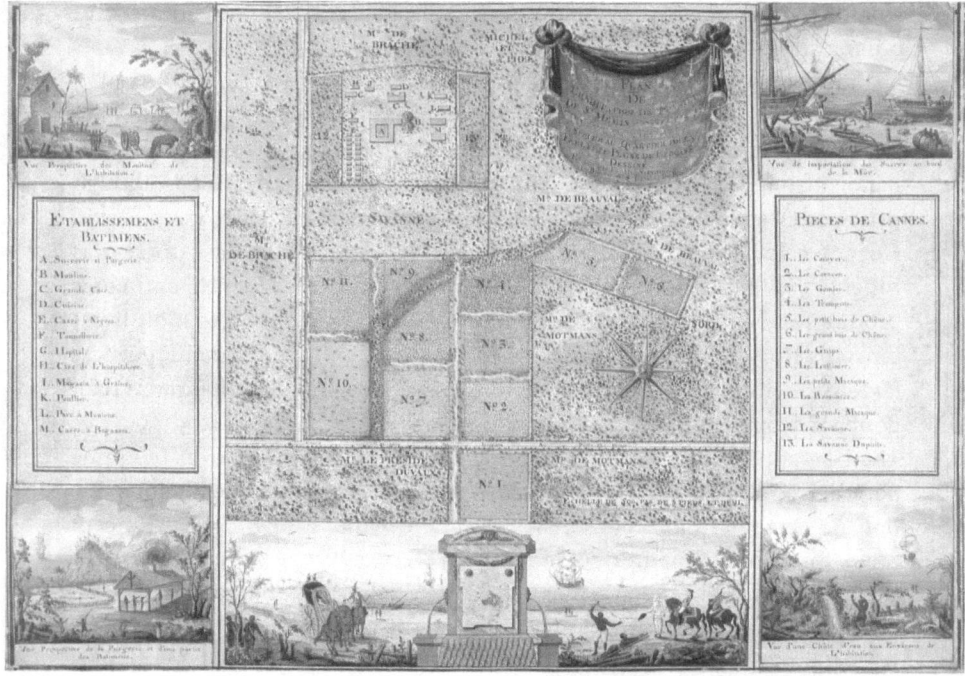

Figure 6. "Plan de l'habitation de Févret de Saint-Mésmin, à Saint-Domingue, Louis de Beauvernet." © RMN-Blérancourt, Musée franco-américain du château de Blérancourt.

compass rose and an elaborate cartouche identifying the estate, its owner and location, and the artist provide the aesthetic trappings of an elaborate map.

Beauvernet drew the plan for the estate's absentee owner, Févret de Saint-Mésmin, a parlementary judge in Dijon who had inherited the estate from his creole mother and army officer father. He kept decades of plantation registers and letter books about this estate.[55] Yet the six vignettes that Beauvernet painted around the central map made this document more an artistic object than a management tool. The Févret family had a notable painting collection, and, during the French Revolution, the judge's son Charles became a well-known portraitist and engraver in New York, until returning to Dijon to serve as director of the city museum. Descended from a Dijon noble family himself, Beauvernet likely knew of the Févrets' artistic interests.[56] Given the modest size of Beauvernet's estate plan (17 inches by 25.5 inches; 44 cm by 62 cm), Févret likely kept it in a portfolio or album, perhaps for showing to guests in the family's Dijon mansion.

Beauvernet's six vignettes of plantation life are somewhat realistic. For example, the plantation had the single refinery and two animal-driven sugar mills he portrayed.[57] Nevertheless, the engineer took liberties that made his scenes strikingly metaphorical. All but one vignette shows ships on the sea, although the Févret plantation was completely landlocked. The managers of Févret's plantation probably used the wharf of the adjoining Motmans' estate, as Févret's wife was a member of the Motmans family.[58] But Beauvernet's fascination with shipping, as well as with the machinery of the sugar plantation, reveals his vision of Saint-Domingue as a world of commerce and manufacturing, as well as agriculture. The ships linked the estate to metropolitan consumers, and to the Févret family, who consumed the profits from the estate. Beauvernet understood that his patrons were likely preoccupied with the moment when their barrels of sugar would be boarded onto a ship. In 1776 their instructions pressed their estate manager to send them 120 barrels of sugar a year.[59] Beauvernet devoted one vignette to this moment. In it, white men repair a beached ship, and men wave to another vessel out at sea. Two or three white children sit on timbers as a smaller boat arrives at the shore. This is the sole painting in which whites outnumber blacks, who are depicted here only as three diminutive figures in the foreground rolling a large cask toward the sea.

Although he was a royal engineer, Beauvernet was not a military man but rather came to Saint-Domingue to make his fortune.[60] There is no evidence that he saw plantation slavery as pernicious. Indeed, when he painted this image he was probably involved in a scheme to open up new sugar land in the undeveloped Cayes de Jacmel parish. Beauvernet's wife was one of at least six investors who received a royal land grant as part of a group that planned to construct a joint irrigation works on this virgin territory.[61] This project was precisely the kind of technical investment that Beauvernet was trained to design and implement. As an aspiring sugar planter, Beauvernet likely believed that slavery was even more essential to tropical agriculture than irrigation.

In nearly all his images, naked and suffering black bodies outnumbered white figures, which were always clothed, often luxuriously. In one vignette, two white men on horseback arrive at the beach where they are greeted by a third white man. They peer at ships on the sea, apparently oblivious to the fact that several yards away a nearly naked black woman, tied to a ladder, is being whipped by a black man, who wears only white breeches. Beauvernet's

Figure 7. "Vue Perspective de la Purgerie et d'une partie des Batimens, Plan de l'habitation de Févret de Saint-Mésmin, à Saint-Domingue, Louis de Beauvernet." © RMN-Blérancourt, Musée franco-américain du Château de Blérancourt.

vignettes depict a violent plantation world in which whites ignore the pain that surrounds—and indeed supports—them. His view of a waterfall near the plantation shows a large alligator or caiman that has surprised two black bathers. One of them, in the center of the image, is about to be devoured by the creature, whose jagged teeth and large eye make him much more of an individual than his victim.

Two of Beauvernet's vignettes focus on the productivity of Févret's estate. Figure 7 shows the sugar refinery in full operation. Here, as in figure 8, naked suffering slaves are shown as part of the tropical landscape. The refinery building has no walls, allowing Beauvernet to depict five black figures with

Figure 8. [Untitled cartouche of woman in carriage with slaves] "Plan de l'habitation de Févret de Saint-Mésmin, à Saint-Domingue, Louis de Beauvernet." © RMN-Blérancourt, Musée franco-américain du Château de Blérancourt.

long-handled tools, stirring the sugar syrup as it boils, while a ship approaches in the distance. In the foreground a black man holds a whip high above his head as he chases a woman who raises her arms in distress. Another vignette, not shown here, shows the plantation's two animal-driven cane mills and contains eight human and three animal figures. Two or perhaps three of the humans are holding whips in midstroke, above the animals in two cases and perhaps above a human in the third. One trio consists of two persons standing with a third kneeling between them.

Beauvernet not only showed Févret his estate's productive assets—its processing centers and their coerced human and animal workers—he also illustrated the wealth these machines generated. In figure 8 two horses with ostentatiously studded harnesses draw a handsome carriage along a coastal road. Inside the carriage, at the very center of the image, sits a richly dressed white woman. This may have been Judge Févret's creole mother-in-law, Mme. Motmans, who was criticized for her extravagant domestic service.[62] She is not driving the vehicle, for its reins are being held by a black man riding alongside on a third horse, carrying the omnipresent whip. Two black postilions stand on the back of the carriage, wearing the same livery as the rider—white turbans with a single red feather, red jackets trimmed in white, and a white shirt.

In the left foreground, Beauvernet depicts a black couple whose naked suffering stand in stark contrast to the comfortable passenger. The man and woman wear heavy chains around their wrists. The larger of the two figures is also wearing a special iron collar, a device designed to humiliate and to isolate slaves found eating sugarcane or committing other minor violations of plantation discipline. The iron arms of the collar, with their barbed ends, stick out at a forty-five-degree angle, extending far over the man's shoulders. The composition reinforces the marked contrast between the grandeur of the men and horses accompanying Mme. Motmans and the abject misery and animalistic treatment of the chained pair. Moreover Beauvernet depicts Mme. Motmans and her attendants as unaware of this scene of degradation. The cruelty of plantation life, Beauvernet suggests, is what keeps the wheels of commerce turning.

In 1779 the army officer Desdorides confirmed this when he described colonists' casual attitude about cruelty in Saint-Domingue: "They do not always take care to remove the children when the slaves are punished. Those who frequently witness these punishments become hardened to it; they run as to a game to see an unhappy slave be whipped. . . . I have heard a mother boast of her son that at the age of ten he was strong enough to 'cut' a slave, that is to remove his skin with the stroke of a whip."[63] The Févret images illustrated an omnipresent reality in the slave colonies of the Greater Antilles. Violence was at the base of all the wealth and social and political power in slave societies in the region.

In his influential study of the English West Indies in the seventeenth century, Richard Dunn put forward a thesis on the social character of the West Indies that has proven very enduring. Looking at the development of the British West Indies as a student of Puritan New England and early modern European history, Dunn discounted the economic success of places like mid-seventeenth-century Barbados. He concluded that these societies were social failures, monstrous societies that were moral indictments of the process of European colonization in the Americas.[64]

That the colonial experience of the West Indies was socially and morally repugnant has been a strong theme in writings on the period. The Trinidadian writer V. S. Naipaul famously described them as places of mimicry where nothing original was created.[65] This characterization of the West Indies as a cultural and political wasteland has, of course, been extensively critiqued, from Aimé Césaire to Kamau Braithwaite and beyond. These opponents of Naipaul's depressing vision, however, have generally seen the Afro-Caribbean struggle against oppression as the fundamental source of Caribbean creativity, while describing the plantation as the essential colonial institution that creative Caribbean peoples struggled against. Vincent Brown has recently reinforced these arguments, mainly as a way of trying to overcome what he sees as the nihilistic assumptions embedded in the notion that Caribbean slavery was a form of "social death," in which the slave self was irrevocably harmed through the process of being ripped away in the Atlantic slave trade from communities and familiar landscapes in Africa. He believes that slaves "must have found some way to turn the disorganization, instability, and chaos of slavery into collective forms of belonging and striving, making connections when confronted with alienation and finding dignity in the face of dishonor."[66]

This literature on colonization, creolization, and cultural mimesis in the Greater Antilles is rich, confrontational, and intellectually challenging. What we want to suggest, however, is that the plantation world of the eighteenth-century Greater Antilles should not be viewed only in terms of horror.[67] Death and despair were abundantly in evidence on the plantation, as the Févret map and vignettes show. That neither the white nor the black populations of eighteenth-century Saint-Domingue and Jamaica could naturally reproduce their population is one more testimony to the destructive character of the plantation system. Yet the plantation system was also a place of dramatic vitality. Enslaved persons' creative attempts to overcome or evade slavery are reason enough to see slavery's power as both destructive and productive; in other words, enslaved people's fear of social death was not incapacitating but generative.[68] It was more than just a machine for the accumulation of wealth. We turn in the next chapter to look at some manifestations of that world in the urban life of both colonies and in the startlingly untraditional relationships between men and women of all races between the Seven Years' War and the start of the French Revolution.

CHAPTER 3

Urban Life

The sugar plantation was the most remarkable institution in Jamaica and Saint-Domingue. Nevertheless, both colonies were also home to lively cities in which a much greater percentage of the population lived than was usual for British and French North America. The "generative fusion," or creative adaptation, that Philip Morgan argues was characteristic of eighteenth-century Caribbean culture was most apparent in these dynamic urban places. It was in cities and towns that some of the transformations in race, and indeed gender, that we will be concerned with in this text were most obvious.[1]

Towns in Jamaica and Saint-Domingue were not mere appendages to the plantation world. They were "electric transformers," to employ Fernand Braudel's metaphor for early modern towns, places that "increase[d] tension, accelerate[d] the rhythm of exchange and constantly recharge[d] human life."[2] Colonists in these cities found few if any of the institutions that controlled the sexual, economic, or religious behavior of European urban residents. Urban spaces were also liberating places for free and enslaved people of color, who often found more autonomy, a wider set of social relationships, and greater earning power than in rural areas. Free women, both black and white, found it easier in towns to re-create themselves as people of cultural and economic importance.

These New World towns were part of a general urbanizing trend in the eighteenth-century British and French Atlantic Worlds.[3] Like provincial towns in Britain and France, they had shops and theaters, markets and commercial exchanges, and new housing developments. Yet these places were even more dynamic than their European counterparts. They had high mortality rates and a constant flux of migrants, merchants, soldiers, and slaves. The pulse of the Atlantic trade made the towns of Saint-Domingue and

Jamaica into sites of continual reinvention, as new arrivals struggled to adapt to a new and sometimes bewildering environment.

Late eighteenth-century maps of Saint-Domingue's chief cities, like René Phélipeau's 1785 map of Cap Français and its surroundings, reveal an extraordinary vision of rectilinear order and social sophistication (Figure 9).[4] Such images make it easy to remember that Port-au-Prince, Saint-Domingue's capital, was a critical stop for Charles Marie de La Condamine's 1735 expedition to determine the shape of the earth, and that the first hot air balloon ascension in the Americas occurred in Cap Français, the colony's major port city. Descending closer, the image of enlightened urban sociality becomes more impressive. Cap Français had a theater that could seat fifteen hundred spectators. Cap Français was, moreover, just one of the eight towns in Saint-Domingue with theaters. Actors and musicians from Europe regularly toured

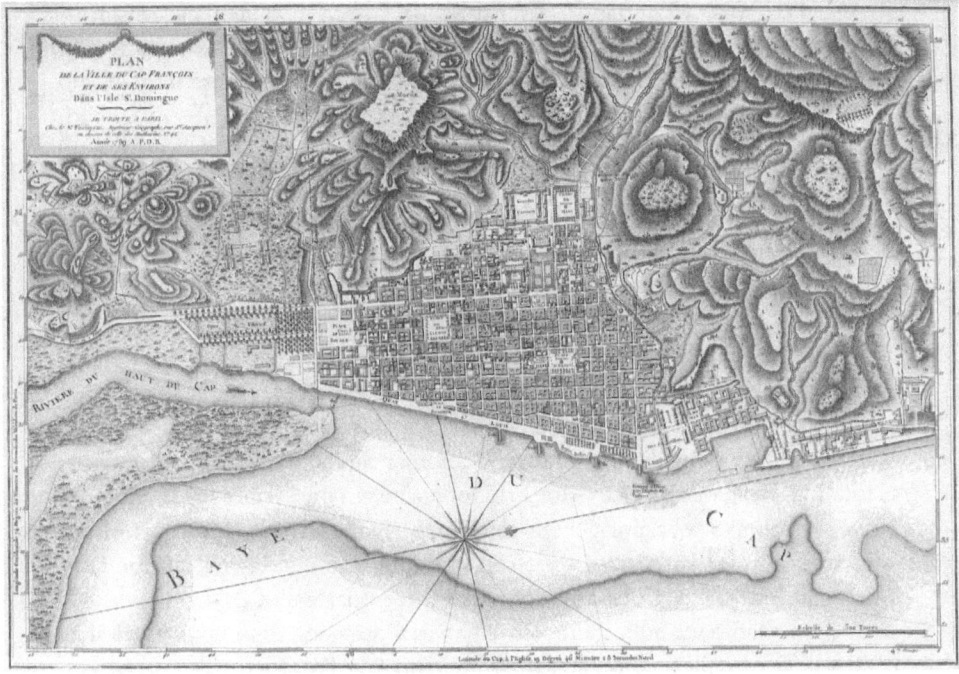

Figure 9. "Plan de la Ville du Cap François et de ses Environs Dans l'Isle St. Domingue, Nicolas Poncé." © Courtesy of the John Carter Brown Library, Providence, Rhode Island.

the colony, performing the latest plays from Paris. Cap Français and Port-au-Prince had full-time professional troupes and orchestras. Saint-Domingue was also blessed with many booksellers, forty-four Masonic lodges, several public parks and squares, five subscription-based reading clubs, three Vauxhalls (or pleasure gardens), and a scientific academy that included corresponding members like Benjamin Franklin.[5]

But this portrait of the colony as a center of Enlightenment activity is misleading. For one thing, urban sociability came late to Saint-Domingue. Many of these institutions were established in Saint-Domingue as an attempt by French authorities after 1763 to create "civilized" public spaces binding Saint-Domingue's colonists closer to France. Second, the colony's towns became sizeable only in the 1770s and 1780s. Cap Français, for example, had around forty-four hundred residents in 1771, including slaves, and tripled to fifteen thousand by 1788, still counting slaves.[6] Third, although Saint-Domingue had three regional capitals—Cap Français, Port-au-Prince, and Les Cayes, on the southern coast—it was no more extensively urbanized than Jamaica. This proliferation of capitals was largely a function of how difficult it was to travel around the colony and in particular reflected the military vulnerability of the northern coast. Counting the population of these regional capitals, about 5 to 6 percent of Saint-Domingue lived in cities, a number about half of the 10 to 12 percent of Jamaica's population that lived in towns by the middle of the 1780s. When Saint-Domingue's roughly fifty parish towns are counted, with populations of about three hundred apiece, however, the two colonies are roughly equivalent in people in urban or semiurban settings, with about 10 to 12 percent of people living in urban or semiurban places.[7]

It was only in the middle of the eighteenth century that France began to spend large sums mapping and planning its colonial cities.[8] The case of Port-au-Prince illustrates this process. In 1749, when the Crown officially designated it as the site for Saint-Domingue's new colonial capital, Port-au-Prince was a plantation. Religious services were held for a while in the estate's sugar refinery, while the colony's governor-general lived in the main plantation house. In November 1751 royal authorities were still granting land and laying out the city when an earthquake struck and destroyed three-quarters of the one hundred houses that had been built there. The city was a cesspool, until 1770, when the residents began filling in the rutted-out and overly wide streets between houses.[9] In 1770 another earthquake hit. After this, administrators decreed that all houses were to be built in wood, or masonry between posts, rather than stone, to minimize earthquake damage. A decade later Port-au-Prince had few buildings with a second story. Even in the late 1780s, Moreau de Saint-Méry observed that "nothing about it is reminiscent of a large city."[10] On the north coast, by contrast, Cap Français was a larger, older, and more architecturally sophisticated city than Port-au-Prince, with many two-story stone and masonry buildings. In 1756, however, it had only few of the dance halls, public gardens, fountains, coffee houses, print shops, and bookstores that would later lead travelers to describe it as the Paris of the Antilles. These buildings and public spaces were mostly erected in the 1770s and 1780s. The theater, discussed below, was an exception, having first made its appearance in 1740.[11]

At midcentury, colonists in Saint-Domingue and Jamaica were already beginning to argue that these places had moved past their buccaneering origins.[12] In 1750, for example, Emilien Petit, a judge and planter born in the former pirate town of Léogane, argued that "this country is no longer, as in its origin, inhabited almost entirely by crude, unknown, undistinguished people ... seeking refuge in another world from the consequences of their crimes."[13] This conviction that Saint-Domingue had attained a new stage of development grew stronger after the Seven Years' War, as royal authorities began to increase the size of the colonial bureaucracy, allowed the establishment of a printing office, and established regular transatlantic and intracolonial mail service.[14]

These post-1763 changes in the colony's urban culture were so noticeable that in 1769 the planter G. Lerond proposed that Saint-Domingue have its own literary academy. Sophisticated gentlemen, he claimed, had replaced illiterate buccaneers: "All fashions are found in the colony today: plays, concerts, libraries, sumptuous parties where gaiety and wit oppose irksome boredom. ... Pirates have given way to dandies with embroidered velvet jackets. ... A love of learning accompanies this love of luxury. Those who previously could not read or write are today poets, orators, and scientists." But an anonymous critic pointed out that the new breed of colonists were obsessed with making money, not polite conversation: "many intelligent people will be found in Saint-Domingue but I repeat that they will justly be too jealous of their time to attend literary conferences."[15]

It was in the cities that European visitors were most struck by colonists' relentless materialism, for they expected that the Church would be at the center of urban life. This was true only in a spatial and legal sense. Saint-Domingue's towns were built, as in Europe, around parish churches. The Catholic Church was an important part of French colonial ideology, and parishes served as the fundamental unit of local government. As the Code Noir of 1685 proclaimed, all religions besides Catholicism were illegal, though there were a small number of Protestant and Jewish families in the colony. The code specified that all slaves were to be baptized and instructed in the Christian faith. Few planters followed this aspect of the slave law, but the Jesuit order did baptize and catechize slaves.[16] The powerful symbolism of the church meant that all towns and cities had sanctuaries of some kind.

Yet colonists were notoriously irreligious. In 1768, Lieutenant Desdorides, an artillery officer recently arrived in the colony, wrote his father: "Religion, which everywhere consoles the just and slows the wicked, enflames or restrains almost no one in Saint-Domingue. Priests, by their bad conduct . . . lose much of the merit [by which] they will persuade [congregants] of the maxims they are charged with teaching. They assemble a very small number of faithful; thus one sees deserted churches. . . . Without religion . . . one sees the continual victory of error and disorder."[17]

In the 1780s, Cap Français was the largest French city in the Americas, with fourteen hundred houses and seventy-nine public buildings. In 1748 the Jesuits built a two-story masonry residence there, but in 1763 the town's only church could still be described as "wooden shack ready to collapse." A proper stone church was not in place until 1774. Few residents attended services, except during major feast days.[18] Moreau de Saint-Méry illustrated the religious climate by describing how the parish of Cap Français raised funds. Every year the parish, just as in France, named two sextons to administer parish property and other financial matters. In France local notables pulled strings and made donations to be nominated to this prestigious office. In Cap Français, however, the parish deliberately nominated men who had no interest in church affairs, then informed them that they could escape the duty by making a large donation to the church. The nominee usually complied, and the committee went on to another victim. In some years they tapped two or three of these reluctant sextons in a single day.[19]

Royal government, like the Church, was another metropolitan institution that looked more important on late eighteenth-century maps than was actually the case. In theory, Saint-Domingue was under absolutist rule from the 1660s. But it had little in the way of centralized royal administration until 1763. The Superior Council of Cap Français met in the royal storehouse for twenty-four years.[20] Not until December 1763 did the Crown convert the former headquarters of the Jesuit order into a proper "Government House," with meeting rooms and offices for the Council, the lower royal courts, and naval administrators.[21]

The colonial capital, Port-au-Prince, was the official residence of Saint-Domingue's governors and their staffs. The end of the Seven Years' War marked the beginning of a bureaucratic and mercantile influx that doubled the size of Port-au-Prince to 683 houses and then to 895 in 1788. This made it about the size of Kingston twenty-five years earlier.[22] In the 1780s, Port-au-Prince had a total population of ninety-four hundred, but it was still smaller than Cap, which had between twelve thousand and fifteen thousand people. Despite its status as capital, it had few free residents able to participate fully in civic life. Most of its population consisted of soldiers and sailors (thirty-two hundred) and slaves (four thousand), leaving eighteen hundred whites and four hundred free people of color.[23] The city had a theater and, after 1782, a fashionable Vauxhall, though it eventually dissolved because of gambling disputes. Indeed, Moreau claimed that in Port-au-Prince, unlike Cap Français, male colonists gathered only to gamble. In this respect it resembled St. Jago de la Vega, the small inland capital of Jamaica.[24] Cap Français had three charity hospitals, but Port-au-Prince did not get such an institution until 1787, when the governor-general brought state financing to the project.[25]

Saint-Domingue's towns were also the site of its military garrisons, which added both a leavening of high status officers to complement the existing elite of merchants and planters and also a larger number of men of lower status. European troops, however, died quickly in the Greater Antilles. Records from Saint-Domingue in 1765 show a minimal annual death rate of 21 percent among rank-and-file soldiers.[26] It did not help that rank-and-file soldiers in the colony were given poor rations and shoddy clothing and were forced to do manual labor outside their military function.[27] Slaves and free people of color referred to them as "white slaves" (*nègres blancs*). Many deserted, often in groups of five or six at a time. In 1766, for example, two weeks after a contingent of 647 soldiers debarked at Cap Français, forty-four of them had already deserted, swelling the growing *petit blanc* class.[28]

Towns were also important in Jamaica, although the English colony had only one large urban area, Kingston. Even the capital, St. Jago de la Vega, also called Spanish Town, was a hamlet, not a substantial town. Kingston was the island's major port, and trading was so important in Jamaica that in the mid-1750s that Governor Charles Knowles tried unsuccessfully to move the capital there, arguing it was illogical to make ship captains and merchants trek thirteen miles to Spanish Town to conduct business. Knowles also hoped to build a wedge of merchant support in the Assembly against politically powerful planters, who wanted the capital to remain inland. He lost that political battle, but Kingston's economic and cultural importance remained intact.[29]

Kingston may have been an important place of business, but it was, like Port-au-Prince, an unprepossessing place. It was built around a Spanish plan, which placed the formal center of the town—a large square with the parish church at its center and public buildings dotted around it—at some distance from the commercial district. That was Kingston's real heart, full of bustling streets lined with shops and warehouses, located around a magnificent natural harbor. A map produced in 1745 by Michael Hay shows Kingston when it was probably at the height of its influence (Figure 10). Hay showed a town,

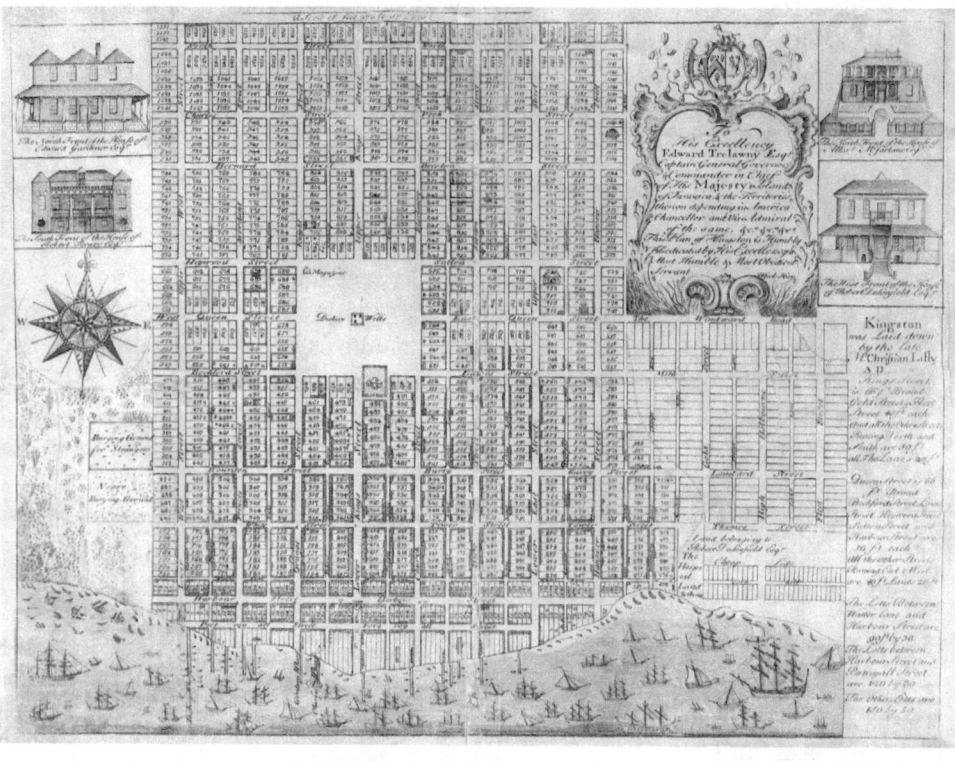

Figure 10. "[Plan of Kingston] to His Excellency Edward Trelawny, esqr., captain, general governor & commander in chief of His Majesty's island of Jamaica & the territories . . . ; this plan of Kingston is humbly dedicated by His Excellency's most humble & most obedient servant, Mich. Hay." [Kingston, n.d. (1745)]. © Library of Congress, Geography and Maps Collection.

dominated by merchants and mercantile houses, which was beginning to expand east of the harbor. The cartographer highlighted that commercial identity in his map by featuring sketches of the houses of four prominent merchants—Edward Gardiner, Robert Turner, Alexander Macfarlane, and Robert Duckinfield. He also outlined, to the southwest of the town, the location where "strangers" and "negroes" were buried and showed the location of the hospital, in "Cheapside," in the southeast.[30]

Kingston was a practical town, built for business rather than aesthetics, but parts of it were pleasing. In his history of 1774, Long praised the houses in Kingston for their construction and sturdiness, noting they were "mainly of brick, raised two or three stories, conveniently disposed, and in general well-furnished; their roofs are all shingled; the fronts of most of them are shaded with a piazza below, and a covered gallery above."[31] Lord Adam Gordon on a trip to British America in 1764 thought the town "very considerable, being large and well Inhabited, the Streets spacious and regularly laid out," as befitted "the most . . . trading Town on the island."[32] But against this were less positive comments. Gordon also noted that Kingston was "a very unwholesome place," "often visited by sickness." Long blamed the mortality on the "loathsome practice" of using human excrement to pave the roads and on burying people in too shallow graves near the middle of the town.[33] Thomas Hibbert, the town's greatest merchant, gives some credence to Long's views. In his notably deist will of 1780 he instructed his executors that "in order to show my detestation and abhorrence of the prevailing superstitious custom of Interning dead bodies in Churches and Church yards and to prevent mine being added to the noxious Mass that is daily Corrupting in the Centre of the Town, I desire that my executors will see it placed in the deep Vault which I have provided for that purpose in the Garden belonging to my House in Kingston with the least expense consistent with decency."[34]

We don't have a census for 1745 for Kingston, but extrapolating from a census of 1730, when it had 4,461 residents (1,468 whites, 269 free people of color, and 2,724 slaves) and assuming modest growth per annum suggests that by 1745 the town had somewhere between five thousand and six thousand people. There were 844 householders in the town in 1745, suggesting that the average household size was very small—probably not much more than two whites or six free people and slaves per household. Nevertheless, Kingston accounted for 7 percent of Jamaica's population in 1774. By this time it was the third largest town in British America, with around 14,200 people, more than twice the size of contemporary Cap Français, which had only 6,143 people in 1775.[35] More important, just under one-third of white Jamaicans dwelt in Kingston, and 40 percent of free people of color lived there. In contrast, 25 percent of Saint-Domingue's whites lived in the three largest cities, and less than 10 percent of the free people of color were urban residents.[36] Kingston's population increased rapidly after the American Revolution. By 1788, its 26,478 people meant that it had 12 percent of Jamaica's population, including 6,539 whites and 3,280 free people of color. Most slaves lived in the countryside, but with 16,659 slaves, Kingston accounted for 7 percent of Jamaica's enslaved population. Kingston had more slaves and free people of color combined than in all the urban centers of British North America.[37]

It was a heterogeneous place. The cacophony of voices and jumble of complexions in Kingston struck observers forcefully. Curtis Brett, a Dubliner from a middling commercial background, who arrived in Kingston from London in 1748, wrote vividly about the heterogeneous character of the town's population, noting upon arrival, "Instead of the Morning London Cries, of Old Clothes, Sweep, &c. my Ears were saluted with Maha-a, Maha-a, the Cries of Goats, kept in Most Houses for their Milk. And presently I heard called the Names of Pompey, Scipio, Caesar &c. and again those of Yabba, Juba, Quasheba (Negro Boys and Girls, Slaves in the Family) which first raised an Idea of being in old Rome; & then again of my being transported suddenly to Africa." He thought "the Inhabitants very open, courteous, lively & very ready to serve and assist a stranger. But how different did everything appear to me. . . . People of almost all colours! White, black, yellow, in abundance. Many pale white, and great Variety in the Shades of Black and Yellow. Very few, not one in a Thousand, of a ruddy Complexion."[38]

Kingston's great merchants were very wealthy, with the largest leaving estates of more than £100,000 or 230,000 livres. They made their fortunes in several ways. Kingston's principal business was the importation of African slaves into the Americas, the largest and most complex international business of the eighteenth century.[39] Between 1700 and 1758 the city was the sole port of entry for Africans shipped into Jamaica and the major port for such shipments between 1758 and 1807. During this time nearly 830,000 slaves were imported into Jamaica. The total value of this trade amounted to nearly £25 million. Perhaps £200,000 per annum passed through the hands of Kingston merchants. Henry Bright, a Bristol factor resident in Kingston, called the trade to Africa the "chief motive of people venturing their fortunes abroad."[40]

Many of those African captives did not stay in Jamaica but instead were reexported from Kingston as part of the *Asiento*, or legal slave trade with Spanish America. Richard Sheridan estimates that between 1702 and 1808,

193,597 Africans landing in Jamaica, or 24.4 percent of all captives landed on the island, were immediately shipped onward to Spanish America. The peak years of exportation were between 1710 and 1740, when 42 percent of slaves were reexported. Saint-Domingue had no such reexport sector, and its local merchant class was consequently far less wealthy.

A well-informed account of Britain's American colonies in the mid-1740s thought that Jamaica was the most substantial financial contributor to the British Empire, with its trade to Spanish America worth £900,000 sterling per annum in 1745, making its total trade worth £1.5 million, compared to New England's £1 million.[41] James Abercromby, a Scottish imperial thinker who had lived in the colonies, argued in a similar vein in 1752. He ranked twenty-three colonies in British America in terms of their worth to the empire. Jamaica ranked thirteenth in fighting men but second in value of produce and first in value of the British goods it imported.[42]

As this data suggests, the wealth of Kingston's merchants continued after the Asiento ended in 1740. The most recent estimate of intercolonial slave departures from Jamaica (mainly from Kingston) suggests that between 1741 and 1790, 62,600 Africans were exported from Jamaica to foreign markets with a further 11,075 leaving in the same period for other British colonies. It was the most active such entrepôt in British America. Cuba was a popular destination, as was Cartagena, with the southern regions of Saint-Domingue also attracting considerable illegal smuggling. By the 1760s, Kingston merchants had begun to expand their trade in slaves and other goods to other places, such as the Bahamas and Honduras, as well as to familiar Spanish American and North American destinations. The American Revolution was a blow to such trade, but it quickly picked up again in the 1780s.[43]

Kingston merchants also made substantial sums from money lending. Planters approached them for credit to acquire slaves, livestock, land, and mill equipment as well as to fund conspicuous consumption and to pay out family inheritances. The biggest moneylenders provided large sums to their clients. Fifteen Jamaican colonists (eight of whom were Kingston merchants) left inventories showing they lent out more than £60,000 or 138,000 livres to creditors. To keep this in perspective, only one man in British North America—Charles Carroll of Annapolis—lent money on the Jamaican scale, with £57,400 out on loan in 1776.[44]

Located on the Atlantic coast, the "French Cape" was, like Kingston, a major location for the arrival of Africans in the New World, and a significant site for the transculturation that created a creole society in Saint-Domingue. Although it did not dominate colonial commerce like Kingston did, Cap Français received roughly half of Saint-Domingue's incoming African slave ships; Port-au-Prince, the colony's second slave port, received only 30 percent. While eighteenth-century Jamaica received slave ships coming from fifty-eight sites on the African coast, according to surviving records Saint-Domingue received human cargoes from seventy-six African ports. Even more than Kingston, therefore, Cap Français was a place where different African peoples became Americans.[45]

Unlike Kingston, however, Saint-Domingue did not reexport its African captives. Rather, the French colony received slaves from other empires, often illegally. Saint-Domingue, as one expert describes it, was continually in the grip of a terrible labor shortage. Planters blamed imperial restrictions on foreign trade, though the real culprit was the relentless work regime on many plantations, and the ongoing expansion of those plantations. David Geggus estimates that in one region of Saint-Domingue's southern peninsula, between 10 and 15 percent of slaves there had been purchased from British traders, probably sailing out of Kingston.[46]

By the end of the colonial period, the ten thousand enslaved people living in Cap Français made up 67 percent of the city's population. Port-au-Prince and Les Cayes had similar compositions. Thus slaves, who made up 90 percent of the overall colonial population, were underrepresented in the cities.[47] But their labor and actions determined cities' character. David Geggus has summarized what little we know about the 4 or 5 percent of slaves in Saint-Domingue who lived in towns, villages, or hamlets.[48] Their characteristics resembled, in exaggerated form, those of the slave population as a whole. They were mainly male, African, and overwhelmingly adult: 65 percent of slaves in Cap Français were adult males, and only 13 percent were children, compared to 41 and 23 percent on sugar estates in the northern province. Nearly two-thirds of urban slaves were not locally born, the great majority being born in Africa. A few were described as "foreign creoles," and they tended to be slaves transported for being troublesome in other places like Martinique, Guadeloupe, or the Mississippi. They continued to be troublesome in their new homes, being disproportionately represented in fugitive slave advertisements.[49]

The lives of urban slaves, while hard, were probably better than the lives of rural slaves. Colonists certainly believed they were better fed and healthier, and had more personal property. Indeed, their clothes, which seemed to some whites like gaudy finery, gave them away as relatively privileged, at least compared to slaves on distant plantations, who often went about nearly naked or dressed in rags.[50] As in Kingston, they had more opportunities than rural slaves to establish independent lives where they hired themselves out and, after paying a portion of their wages to their owners, lived separately. Colonists found such independence disturbing, seeing these slaves as "thieves and receivers of stolen goods."[51] They warned that these hired-out urban slaves encouraged rebellious thoughts and rebellious actions in the plantation workers who flocked to the cities every Sunday to sell their produce. Moreau, for example, was highly critical of the ribaldry, breaking of laws, potential for clandestine gathering, and unwelcome practice of subversive religion that went on at these markets, or elsewhere. He described how slaves gathered together for dances in an old cemetery at the edge of Cap Français contemptuously, as "a spectacle of fury and pleasure."[52]

A remarkable report by the Chamber of Agriculture in 1785 gave voice to white fears of slave depravity in Cap Français and to its potential consequences. Bemoaning that "the Negroes are so open in their insubordination that the line of demarcation between whites and slaves has almost vanished," the authors of the report predicted "grave events" if no one "put an end to this evil." They urged that the town authorities put a stop to their tolerance of slaves' "nightly gatherings and gambling dens, their nocturnal dances, associations, and brotherhoods." What particularly disturbed the chamber was how slaves insolently refused to give way to whites on the street—they gave several horrified accounts of such insolence. They were even more concerned at slaves' tendency to travel at all times "with a large stick." On holidays, they lamented that "you find 2,000 of them gathered at La Providence, La Fossette and Petit Carénage [neighborhoods on the edge of the city] all armed with sticks, drinking rum and doing the kalinda."[53] Nevertheless, despite these provocations and despite the potential for riot that large gatherings of armed and drunk young adult men provided, urban slaves were remarkably politically quiescent. There were no urban revolts in any of Saint-Domingue's towns before the start of the French Revolution and little involvement by urban slaves in the initial stages of the Haitian Revolution. David Geggus argues that while towns facilitated social flux and opportunities for subversive gatherings, they were also places full of whites, including a concentration of soldiers and sailors, thus making armed uprisings difficult.[54]

Another critical group in Saint-Domingue's cities, indeed in all Caribbean cities, was free people of color. On islands where whites controlled most arable land, ex-slaves and their descendants tended to gravitate toward the cities to find work. In late eighteenth-century Barbados, for example, free coloreds were about 6 percent of the residents of Bridgetown, but only 3 percent of the total colonial population.[55] In Jamaica and Saint-Domingue, however, vacant rural land was available. In Saint-Domingue specifically, a system of royal land grants made it possible for free coloreds to become peasants, market farmers, and ranchers. A small but significant minority became indigo, cotton, and coffee planters.[56] As we explore in Chapter 6, in Jamaica colonial elites took measures after Tacky's Revolt in 1760 to prevent whites from bequeathing land and slaves to free people of color. In Saint-Domingue such laws were discussed but never implemented, making it possible for free families of color to eventually accumulate enough land and enslaved workers to establish plantations.[57]

One consequence of the economic opportunities available in Saint-Domingue's countryside was that free people of color were not especially concentrated in the cities, unlike whites and unlike their counterparts in Jamaica. They were only 6 percent of the population of Port-au-Prince and Les Cayes, while they were 5 percent of the overall colonial population. Cap Français was unusual in that 10 percent of its population was free colored by the 1780s.[58] Seen another way, about 10 percent of Saint-Domingue's free population of color lived in the three capital cities, while 40 percent of Jamaican free coloreds lived in Kingston.

In the cities or in the country, most Dominguan free people of color were quite poor. But the vibrant urban economy of the 1770s and 1780s did allow the emergence of a small population of free colored merchants and property owners. The most prominent among these was the quadroon Vincent Ogé the younger, who claimed to have been worth 350,000 livres or £15,000 by the early 1780s. This was a large sum, two to three times as much as the net worth of a wealthy Parisian merchant at the same time.[59] Ogé's case was unusual, because he was from a mixed-race coffee-planting family, was educated in Bordeaux, and had an uncle who was a merchant in Cap Français. In the 1780s, Ogé sold French cargos around the colony, was part owner of a schooner, and leased and subleased apartments in Cap Français. Another man of one-quarter African descent whom contemporaries described as a large-scale merchant (*négociant*) was Joseph-Charles Haran, born in the town of Léogane around 1744. In 1785 he subleased a property in Port-au-Prince that included a shop, twenty-seven slaves, a flat-bottomed boat, eight carts, twenty-four mules, and three horses. Unfortunately the lease fell through, and by 1787 he owed his creditors 440,000 livres or £19,000 while his assets were worth 215,000 livres.[60] Such possibilities were also open to a few exceptional free women of color. Indeed, two-thirds of the clients of color who appeared before notaries in Cap Français or Port-au-Prince to buy or sell property between 1776 and 1789 were women. The most successful was Zabeau Bellanton of Cap Français, who bought and sold over 100,000 livres or £4,300 worth of slaves and real estate.[61] Dominique Rogers has found considerable numbers of women of color in Cap Français and especially in Port-au-Prince who lived on profits from rental properties.[62]

But it was unusual for free colored people to be wealthy, even in Saint-Domingue, which had the wealthiest free population of color in the Americas in the eighteenth century. The poverty of most urban free coloreds can be seen in the 1776 cadastral census of Cap Français. Free colored properties were on the outskirts of town and at the ends of streets that ran up into the hills in a region named Petit Guinée. Other cities and towns—Les Cayes, Saint-Marc, Port-de-Paix—had similar districts.[63] This segregation was based on economics, not formally on race. While free coloreds accounted for 10 percent of Cap Français's population in 1775 and owned 16 percent of its houses, their property was only 5 percent of the total value of city residences. The census shows that free women of color were far more active within their class than were white women; 42 percent of free colored proprietors were women; only 14 percent of white owners were women. The average value of property that these white women owned was equivalent to the average value of property owned by white men, a rental value of 2,600 livres per year, equal to the purchase price of an adult male slave. Free women of color, on the other hand, owned property that was on average valued far less than this, with an annual rental value of 636 livres, or £27.5, compared to 798 livres for free men of color. Free colored property like this was typical of those recorded in the Cap Français census: "one-fifth of a city lot on which there are several wooden shacks belonging to Pierre known as Beau Soleil, free black, and occupied by him."[64]

Saint-Domingue's whites, as in Jamaica, lived disproportionately in the cities. They made up 5 percent of the colony's overall population, but they were 22 percent of the population of Cap Français and nearly 30 percent of the population of Port-au-Prince.[65] Some of these were administrative personnel and local merchants, but many were involved in transatlantic commerce. The 1776 census of Cap Français shows that approximately two-thirds of the city's houses were leased to tenants.[66] While Jamaica's eighteenth-century sugar planters paid to ship their produce back to Britain to be sold by merchants, a practice some Dominguan planters followed, many Saint-Domingue planters expected metropolitan merchants to come to colonial ports and to buy their produce there. Because British sugar colonies imported their food from North America, ships sailing from Britain to the Caribbean often did not have full cargoes. But French merchants held tight to their monopoly system, which prohibited colonial planters from buying foreign supplies. French ships, therefore, sailed to the Saint-Domingue full of wheat, dried beef, and other provisions, which their captains expected to sell for a good profit to the colonial market. They then bought sugar, coffee, and other crops to carry back to Europe.[67]

As in Jamaica, therefore, Saint-Domingue had a class of wealthy urban merchants, though they were more likely than their Kingston counterparts to be affiliated with metropolitan firms. They occupied the most desirable real estate in Cap Français. The median rent for the city's nine hundred plus houses or plots was 2,000 livres per year, strikingly higher than in central Paris, where in the same period 500 livres would rent a ground-floor apartment of three or four rooms for a year.[68] A few streets between the sea and the cathedral square in Cap Français had rents of 6,000 to 9,000 livres per annum, the value of two to three skilled slaves. Some houses near the harbor rented for as much as 15,000 livres.[69] The profit that merchants reaped from the commercial monopoly also explains the deep tension between metropolitan merchants and colonial planters over the future of this system, as described in Chapters 7 and 8.

A key aspect of the mercantile life of Cap Français was the exchange of information in new kinds of public spaces. In addition to its various open-air markets, its theater, and the fountains, squares, and gardens that proliferated after 1763, the city had literary societies, book stores, and a biweekly broadside, the *Affiches américaines*.[70] From 1761 to 1778, it had a commercial exchange, a space devoted to the buying and selling of letters of change and other instruments of credit. Yet colonial commerce did not lend itself to sociability. Cap was full of men who came to make their fortune, or who were tied to merchant houses in France. As Martin Foäche informed a young friend due to arrive in Cap Français in 1760, "There is little or no custom [here] of going to eat [with colleagues] at the inn, even those with whom one is doing business. When mid-day strikes, everyone goes in his own direction."[71] He told his friend to seek out guest tables instead of eating at his inn during his first six months in the city. This would allow him to meet people and gain information about the colony. Meeting like-minded people was especially important, as travelers to Saint-Domingue stressed the lack of connection that many colonists felt to any country. In his 1754 *Essai sur les colonies françaises* Pierre-Louis de Saintard wrote, "The Europeans who live in the colonies, having become by voluntary transplantations outsiders everywhere, no longer pretend to have a fatherland."[72]

The pervasive notion of rootlessness and of the lack of connections among colonists also explains why Saint-Domingue was probably the most heavily "masonized" society in the eighteenth-century Atlantic World. In France in 1789, Freemasonry involved less than 1 percent of the eligible male population.[73] It was at least five times more popular in Saint-Domingue, where there were around one thousand Freemasons, in a population of roughly twenty thousand white men. Not all men had the money or education to be a Freemason. James McClellan estimates that about 25 percent of "sociologically eligible white men" in Saint-Domingue were Freemasons.[74]

In France, Freemasonry was imported from England in 1725, and this pattern was duplicated in Saint-Domingue. In Bordeaux and other French Atlantic ports, English or Irish merchants often organized early lodges.[75] Jamaica opened its first masonic lodge in 1739, and English merchants involved in contraband trading brought the new institution to southern Saint-Domingue sometime before 1747. The names of many of the Jewish trading families active around the southern port of Les Cayes can be found among lodge members in Jamaica.[76]

In Europe and in the Atlantic World, eighteenth-century Freemasons consciously thought of themselves as part of a diaspora.[77] A network of correspondence joined lodges on a national or geographic basis. Such a web of connections was especially useful for transatlantic merchants. Masons developed what have been called "management tools of mobility"—initiation certificates, interlodge affiliations, passwords, and maps of lodge locations, all designed to insure that a traveling mason could find a friendly lodge in a new city.[78] While most of Saint-Domingue's lodges were part of a French Atlantic network, some stretched farther. One of Cap Français's leading lodges was part of a network centered on the St. Jean d'Ecosse lodge of Marseilles, which also had an affiliated lodge in Martinique. But this network was primarily invested in creating fraternal and commercial connections in the Mediterranean, with affiliates in Naples, Sicily, and the Ottoman Empire.[79]

Jamaica also had a lively Freemasonic culture, with more than a dozen lodges listed in the Jamaica Almanac by 1789.[80] But we know more about Freemasonry in Saint-Domingue, where its extraordinary popularity reflected the colony's cosmopolitanism, its lack of established social institutions, its network of provincial cities, and the absence of a well-established

French religious culture. It is likely that for some colonists Freemasonry provided a way to forge new communities within a highly mobile population, and to structure those associations around something besides wealth. The popularity of so-called Scottish Rite Freemasonry in Saint-Domingue, with its elaborate hierarchy of over two dozen levels of initiation, is an example of that desire to negotiate new hierarchies. Established in Bordeaux in 1743, the Scottish Rite by 1762 had twenty-five degrees and was very complex in ways that seem to have appealed to colonists wanting new forms of distinction.[81] The Scottish Rite came to Saint-Domingue with the wine merchant Etienne Morin, who had been one of its founding members in Bordeaux. A traveling salesman who also sold religious books and Sèvres porcelain, Morin is said by some historians to have been in Saint-Domingue in the 1740s, perhaps making multiple trips establishing lodges.

At some point around midcentury, he returned to France, where the leaders of the Scottish Rite were in the process of forming a kingdom-wide organization. This group gave Morin some kind of credential to establish new lodges in the Americas. But during the Seven Years' War the British captured Morin on his return voyage to the Caribbean. Imprisoned in Great Britain, the salesman established contact with Freemasons in England and even is said to have visited Scotland. In 1763 he returned to Saint-Domingue. He landed in Jacmel, a port on the southern coast, which suggests he had come from Jamaica. Morin quickly went to work founding new "Scottish" lodges in Saint-Domingue, but by 1766 masonic authorities in France accused him of exceeding his authority to grant higher degrees. They especially criticized him for deputizing others to do the same. Although Bordeaux sent an inspector to Saint-Domingue to investigate his activities, Morin continued to establish lodges in Saint-Domingue and eventually returned to Jamaica, where he died around 1772.[82]

Morin's career illustrates the entrepreneurial and cosmopolitan side of Freemasonry, which was certainly part of its success in Saint-Domingue. But some of Freemasonry's colonial popularity stemmed from the reasons it was popular in France. Lodges provided a setting for elite sociability, and masonic doctrines dovetailed with enlightenment ideas of improvement, self-government, and "public" discussion. Despite its universalist ideals, Freemasonry was based on a dichotomy between the mystical brotherhood and the "profane" world, between "light" and "darkness," between civilization and barbarity. Despite historians' suggestions that Toussaint Louverture was a Freemason in Saint-Domingue, there is very little evidence that he or any other free man of color was permitted to become a masonic brother. The colony's lodges were fiercely discriminatory against people of African descent, even removing white members from leadership positions because they had married women of color.[83]

While the hierarchical and closed social aspects of Freemasonry were well suited to Saint-Domingue, this was also true in France's coastal cities, like the port of Le Havre in Normandy, where Freemasonry had a very different political and social profile from that of the freethinking lodges of Paris.[84] The relative cultural conservatism of Le Havre's Freemasons made it quite common, Eric Saunier finds, for them to also be members of religious confraternities, which were part of the rich associational life of many French cities.[85]

Given the institutional weakness of the Catholic Church in Saint-Domingue, it is not surprising that the colony appears to have had no confraternities. It seems likely that in Saint-Domingue some men were attracted to Freemasonry because it offered a spiritual context they remembered from France. One of the leading figures of French esoteric Freemasonry, Martinès de Pasqually, died in Saint-Domingue, where his followers had established at least two lodges. In France, Martinès had founded the Elus de Cohen order, which overlaid Christian mysticism and elements of the Jewish kabbala tradition over Scottish Rite Freemasonry. One of his leading followers, Bacon de la Chevalerie, spent much of his military career in Saint-Domingue. Martinès claimed his rites could bring forward angels or other spiritual beings who could guide men toward a reintegration with the Deity, essentially restoring them to spiritual status that Adam had enjoyed before falling from grace.[86] Indeed, some of the spiritual figures or *loas* in Haitian Vodou were masons, suggesting that there was an overlap of spiritual ideas from these different traditions.[87]

The Freemason's Hall was a gendered place, a center of a particular kind of male sociability. But West Indian towns were not exclusively male. Indeed, they were places in which enslaved women but also white women and free women of color were prominent. As in Jamaica and Saint-Domingue generally, the majority of the urban population was male, but the percentage of women in towns was much greater than the percentage of women in the countryside.

The principal determinant of gender relations in Jamaica and Saint-Domingue was the demographic lottery that governed white life. As in Saint-Domingue, Jamaica's high mortality rates made family life something of a game of chance. Multiple marriages were common, so much that some contemporaries described these relationships as a series of fleeting encounters.

The bewildering uncertainty of life in the tropics and the failure of Jamaicans to establish a settler society on the model of settler societies in British North America meant that Jamaica came to be seen by metropolitans as a vortex of social disorder.[88] Similar conditions existed in Saint-Domingue. For Moreau de Saint-Méry, "there is perhaps no country in which second marriages are as common as in Saint-Domingue, and there have been women there who have had seven husbands."[89]

Slavery and racial categories caused another set of problems. White men and women lived irregularly, with concubinage as common as marriage. In Jamaica, marriage between whites and free people of color was always regarded as illegal. Nevertheless, colonial authorities seem to have allowed the relatively small number of free colored women who were sought as brides by white men to "pass" as whites.[90] After 1761, as part of new racial laws examined in Chapter 6, the Jamaica Assembly made passing much more difficult and thus interracial marriage became close to impossible. In early eighteenth-century Saint-Domingue passing was also possible: before the 1760s notaries and clergy often did not describe the color or ancestry of wealthy free women of color in official documents but did give them courtesy titles like "Madame" or "Demoiselle," suggesting that they were white.[91] Even at the end of the century, when stricter laws governed racial categories, unions between white men and free colored women were never outlawed. The percentage of religious marriages celebrated between white men and free women of color reached 17 percent in some parishes.[92]

Before midcentury, in both colonies the policing of boundaries between whites and free coloreds was done very loosely, if at all. In Jamaica, for example, it was not that marriages between whites and free coloreds were banned as that social convention stopped people from even thinking that such unions were possible. Jamaican patriots proudly declared that they maintained their Britishness by preserving a "marked distinction between the white inhabitants and the people of colour and free blacks." But sexual contact between white men and women of color was very common in Jamaican households.[93] Unlike British North America, where interracial sex was frowned on and kept out of the public sphere, in Jamaica such practices were highly public. Gossip about rich married men and their mulatto mistresses was commonplace, and white bachelors lived openly with their "housekeepers." The purported lines of division between whites and blacks, however enshrined in law and ideology, were nevertheless violated every day in the household.[94]

The Jamaican family was not the British family, just as the

Saint-Dominguan family was very different from the French family. It was customary for white men to cohabit outside marriage, both with white women and also with black or colored women. Few white families lasted long or produced surviving children. As the prevalence of venereal disease in the white male population graphically showed, social constraints against the free exercise of white male sexual power were virtually nil.[95] The inability, or unwillingness, of white Jamaicans to establish flourishing family lives and regular marriages was a serious reputational problem for colonists within the empire well before the Seven Years' War.[96] The same was true of Saint-Domingue, where since the late 1600s the French government had tried in vain to bring in European women to marry male colonists. In 1750 Emilien Petit still believed the problem of creating stable white households in Saint-Domingue was the lack of European women.[97] Metropolitan governments were very concerned about observing and regulating familial relationships and sexual behavior. They saw population, security questions, and racial purity in colonial settings as indelibly linked together. As Kathleen Wilson asserts, "the fate of nation, colony, and empire was tied to individual sexual choice; the well-governed colony and the self-governing individual went hand in hand."[98]

Contemporary commentators in both societies described the black or mulatto mistress as morally subversive.[99] British West Indians were occasionally moved to write verse celebrating the beauty of black women; the most famous being the "Ode to the Sable Venus," composed by the Reverend Isaac Teale in 1765 and published in Bryan Edwards's 1793 history. For Teale, the black woman represented forbidden but easily accessible sensual pleasures. In his heavily eroticized prose, the white man sought the "sable queen's . . . gentle reign . . . where meeting love, sincere delight, fond pleasure, ready joys invite, and unbrought rapture meet." Other writers saw that beauty in more threatening terms. Edward Long, for example, recognized that many white men found black women intrinsically erotic, but this recognition filled him with disgust, given his shrill belief that there was something essentially animalistic about black women. Famously, he opined that black women were attracted to white men for the same reasons that he believed orangutans supposedly lusted after black men. In this reading, Long argued that black women and female orangutans sought to improve themselves by attaching themselves to a superior species. Clearly in Long's mind the gap between the most advanced animals—great apes—and the least advanced humans—sub-Saharan Africans—was not at all that great, placing him—as a potential believer in polygenesis rather than the conventional Christian belief in monogenesis—as both a progenitor of early forms of pseudoscientific racism and also beyond the pale for most thinkers wedded to ideas of a single human origin.[100]

For all their differences, Long and Teale shared a conviction with nearly all contemporary observers that sexual relations between black women and white men were inevitable, regardless of slavery. When white writers condemned Jamaican men for not marrying and for attaching themselves to colored mistresses, they couched their reproaches in terms of white male weakness and black female aggressiveness. Edward Long's argument was typical: when a white man, through weakness of flesh, succumbed to feminine charms, he became an "abject, passive slave" to his black mistress's "insults, thefts, and infidelities."[101] White men, masterful everywhere else, were powerless when in the clutches of conniving mulatto and black women. Black women were scheming Jezebels, "hot constitution'd Ladies" possessed of a "temper hot and lascivious, making no scruple to prostitute themselves to Europeans for a very slender profit, so great is their inclination to white men."[102]

White colonists and travelers in Saint-Domingue described women of color in the same hypersexualized images and just as often portrayed themselves as under the erotic power of these women.[103] Girod-Chantrans observed that "these women, naturally more lustful than Europeans, and flattered by their power over white men, have, in order to keep that power, gathered all pleasures to which they are susceptible. Sensual pleasure is for them the subject of a special study."[104] Hilliard d'Auberteuil, moreover, claimed that "mulatto women are much less docile then mulatto men, because they have acquired a dominion over white men based on debauchery."[105]

Just as in Saint-Domingue, Jamaican white men blamed their indiscretions on white women's deficiencies. White men had a schizophrenic attitude to white women. They alternated between praising them for their fidelity, attractiveness, and devotion to maternal duties and lambasting them for their lack of education, poor manners, unpleasing appearance, and violent temper toward their slaves. John Taylor in 1688 denigrated white women as "vile strumpets and common prostitutes" who so "infect" Port Royal "that it is almost impossible to civilise it" and who "trampass about their streets, in this their warlike posture and thus arrayed they will sip a cup of punch rum with anyone." William Pittis in 1720 proclaimed that the climate so affected people that "if a woman land there as chaste as a Vestal she becomes in forty eight hours a perfect Messalina and that it is as impossible for a woman to live at Jamaica and preserve her Virtue as for a Man to make a Voyage to Ireland and bring back his Honesty." By the late eighteenth century even the agency that was afforded by being a whore or a petty criminal was denied to ordinary white women, who were regarded as constant and affectionate but also as inordinately lazy and small-minded. William Beckford of Hertford Pen wrote in his 1788 history of Jamaica that white women were a sex that "suffers much, submits too much and leads a life of toil and misery."[106]

The same progression of images also was something that occurred in Saint-Domingue. The practice of sending women from French poor houses to the Antilles did not succeed in balancing the gender ratio in seventeenth-century Saint-Domingue as it did in Martinique.[107] Throughout its colonial history, Saint-Domingue had far more white men than white women, despite royal efforts to provide additional female colonists. Recent historians have argued that these women were from respectable households, but the contemporary image of these *filles du roi* in the Caribbean was overwhelmingly negative: they were troublemakers, prostitutes, and poorly suited for marriage.[108] In the 1780s the Baron de Wimpffen relayed this common image of the first female migrants, noting that "they sent whores from the Salpêtrière [prison], sluts picked up in the gutter, cheeky tramps."[109] Other women may have been put off coming to Saint-Domingue by the negative representations made about women who were already there. In 1713 the colony's administrators informed Versailles, "We need at least 150 girls, but we ask you not to take any from the bad parts of Paris as usual; their bodies are as corrupted as their morals, they only infect the colony and are not at all good for reproduction."[110] The complaints about the morals and health of white female colonists continued in 1743 when Saint-Domingue's governor complained that France was sending women "whose aptitude for reproduction is for the most part destroyed by too much use." As he put it a few months later, "Real colonists are only made in bed," referring to the need for an island-born population.[111] In the same vein, in 1750 Emilien Petit, who was himself a creole, advocated that the Crown send more single women to Saint-Domingue to attach male colonists to the colony permanently.[112]

After midcentury, however, French officials and visitors gave up on the idea of using marriage between white men and women to repopulate and stabilize the colony. In 1779, Desdorides described relations between men and women in Saint-Domingue as completely different from France. He commented that "men here pay all their attention to their financial interests. Their passion for wealth weakens their desire to be loved. Therefore between them and women there are none of those sweet sentimental emotions that bind two honest hearts. . . . Those wives who come to America . . . have little to do and are relegated to their plantations."[113] Baron de Wimpffen described white women as living in decadence and boredom and agreed with

Girod-Chantrans that they were crueler to their slaves than were most men.[114] Thus, the same assumptions and stereotypes that operated in Jamaica were also working in Saint-Domingue.

Moreau de Saint-Méry agreed with de Wimpffen: "The state of idleness in which creole women are raised, the heat they are accustomed to experiencing, the indulgence perpetually extended to them; the effects of a vivid imagination & an early development, all produces an extreme sensitivity in their nervous system. It is this very sensitivity that produces their indolence which pairs with their vivacity to create a temperament that is fundamentally a little melancholy." He continued, "Who would not be disgusted to see a delicate woman who cries over the story of the slightest misfortune preside over a punishment she ordered! Nothing can equal the anger of a creole woman who punishes the slave that her spouse may have forced to soil the marriage bed."[115]

This succession of stereotypes from both colonies illustrates that white men did not see white women clearly. They tended to describe women's differences in terms of race rather than social class, a tendency that has been replicated by most modern historians. Edward Long and Moreau de Saint-Méry, among others, divided women in their societies among consuming white women, producing black women, and parasitical brown women. Long condemned white women for their "constant intercourse from their birth with Negroe domestics whose drawling, dissonant gibberish they insensibly adopt," meaning that their "ideas are narrowed to . . . the business of the plantation, the tittle-tattle of the parish, the tricks, superstitions, diversions and profligate discourses of black servants." Moreau followed a similar line but with more analytical reasoning and a more interesting conclusion. He argued that white women were removed from the processes of production almost completely by black women and that they had lost their sexual role within the planter household to brown mistresses. All that was left for them, he suggested, was a largely symbolic role as the keepers of racial purity, achieved through their reproductive function as the producers of white children.[116] Yet white women were not merely present in Jamaica and Saint-Domingue. They were active agents in urban life.[117] There is no reason why we need to accept as true the gendered inaccuracies put forward by contemporary male observers in both colonies who saw women as dangerous strumpets in the late seventeenth and early eighteenth centuries and as redundant ornaments by the late eighteenth century.

The best example of a white woman as active agent in mid-eighteenth-century Jamaica was Teresia Constantia Phillips.[118] A famed beauty, as her portrait by

Joseph Highmore in 1748 reveals, she cut a swathe through fashionable society in Britain as a participant in the demimonde of Augustan London, first as the mistress (shamefully abandoned, she argued) of the future fourth Earl of Chesterfield and then as the wife and lover of numerous other rich and fashionable men. A colossal spendthrift, a lover of theatre and social assemblies, she ended

Figure 11. *Teresia Constantia Phillips* by John Faber Jr., after Joseph Highmore, mezzotint, 1748. © National Portrait Gallery.

up cutting her losses in love and money and moving to Jamaica around the year 1751 in order to be with her wealthy Jamaican lover, the Clarendon planter Henry Needham. There, she continued her scandalous life as a courtesan, arbiter of social life in the capital, St. Jago de la Vega, and devoted self-fashioner. She was singular as a woman in Jamaica in having an official government post, as Mistress of the Revels, a largely invented position given to her by Needham's friend, Governor Henry Moore. She received a small government stipend for this role and presided over Jamaica's small but flourishing dramatic scene. In her official capacity, she oversaw and orchestrated all events involving the governor, such as the balls, assemblies, and entertainments held in the governor's honor. More significantly, she gave official approval to all theatrical productions, earning herself 200 guineas for this task.[119] It was a perfect position for her, as her life was as theatrical as a life could be in a mid-eighteenth-century British colony.

Phillips's theatrical and ceremonial responsibilities made her a figure to be reckoned with.[120] To an extent, she demonstrated the "cultural heteroglossia" of mid-eighteenth-century Jamaica, where multiple displaced, avaricious, and exiled people created a syncretic culture, although Phillips did not show any interest in non-British theater or revelry. But she demonstrated how Jamaica manifested a particular kind of Englishness, an aspect of metropolitan culture that disturbed contemporaries with its materialism, and acceptance of all kinds of transgressions—financial, ethical, and sexual.[121] As Kathleen Wilson notes, Phillips "appreciated her cachet as an émigré who could perform the role of the poised and witty English lady with considerable aplomb." She mixed with leading planters and denigrated creole women, whom she thought immodest, dull, and crass with tongues that expressed "the meanest satire." She made a great deal of her Englishness and her patriotism, even writing an anonymous letter published in the *Kingston Journal* that condemned the overly exuberant celebrations of planters in Spanish Town following their defeat of Governor Knowles in 1756 when he attempted to transfer the capital to Kingston. The planters, in their enthusiasm, burned not just an effigy of Knowles but also an effigy of his royal vessel, including its flag. Phillips (ungallantly outed as the author by the printer when he was taken to task by an outraged House of Assembly) declared this action "a most atrocious mark of their ingratitude to his majesty, as well as a very impudent insult upon the gentlemen of the Navy." [122]

Thus, Phillips adopted the role of grand English lady, albeit one of dubious reputation, bringing English culture and English standards of behavior to less culturally advanced creoles. Her promotion of theatre was part of her educative mission in Jamaica: to transform the colony into a place where English values could flourish and where unlearned creoles could be exposed to English cosmopolitan manners. That seems to have been her intention as Mistress of the Revels. In particular, her oversight of elaborate gubernatorial "performances" was designed to enhance the authority of executive office by implanting in Jamaica the symbols and practices that linked politics with theatre. She was not averse to lecturing Jamaicans about their lack of patriotism and to pronounce against Scots and Irish merchants, whom she thought were in a perpetual power struggle to dispossess English landed gentlemen from their rightful place as rulers of the island.[123]

In England, Phillips was a symbol of the dangerous and rebellious woman, as willing to sleep with Catholics as with Protestants and entirely removed from the maternal realm. In Jamaica, she was a symbol of a different kind, one usually associated with free women of color in the historical literature, rather than with white women. She was especially resourceful, even though reliant on female charm rather than inherited advantage to make her way in the world, declaring in her memoir that "My Beauty, while it lasted, amply supplied the Deficiencies of my Fortune."[124] As Wilson comments, "her extravagant lifestyle, proclivity for vulgar display, recurrent overwhelming debts and lavish attention to her own natural resources made it clear that she had taken the laws of imperial mercantile capitalism to heart . . . [she was] an excessively consuming female" with a "taste for the sensual, the sensational and the luxurious."[125] These were the types of terms in which writers like Long and Moreau deplored the lifestyles of free women of color. The towns of Jamaica, like towns and cities in France and Britain, offered opportunities, social and economic, for those women enterprising enough to challenge social conventions and willing to accept the risks that independence brought. That was as true for white women as it was a fact for free women of color. White women in Jamaica did not necessarily lead as cloistered lives as Long and other defenders of patriarchal order imagined.[126]

Despite Phillips's rackety private life and unusual public career, some aspects of her sui generis Jamaican experience illuminate the workings of gender in the colony. In this respect, the most salient fact about her was that she was childless. White women in Jamaica were not defined by maternity. Few women had children, and even fewer had surviving children. Most marriages were short and were interrupted by the sudden death of one partner. For many women, such childlessness and the experience of fragile marriages was undoubtedly a tragedy, especially if they needed to eke out an uncertain living in fickle urban economies. But some women, such as Teresia Phillips, developed personas that fit well with the frenetic hedonism and risk-taking character of Jamaican and Dominguan society.[127]

Phillips was no simpering, uneducated, and passive white woman in thrall to African vices like those Edward Long denigrated. She was indeed a "consuming" woman, but her consumption—of goods, ideas, and new modes of behavior—was active rather than passive and marked her out as a principal agent in fashioning slave societies into new and disturbing places where traditional gender roles came under considerable stress and sometimes alteration. What Teresia Phillips resembled most was the popular characterization of the alluring free colored woman—the mulâtresse, usually depicted by travel writers and colonial authors as avaricious, alluring, sensuous, and highly disruptive.[128]

If the reality of white women's lives in Jamaica and Saint-Domingue was more varied than the stereotype put forward by commentators like Long and Moreau, so too was the stereotype of free colored women at variance with the lives they actually lived. As we have seen, for observers in both societies, the hypersexuality of women of color, free and enslaved, was one of the most enduring tropes of colonial life. This trope was especially well expressed in accounts of the urban environment, where the overwhelmingly male colonial population plus the thousands of European sailors and soldiers based in its cities created a lively sexual marketplace. Enslaved prostitutes earned money for their masters and mistresses—and male colonists sought free or enslaved women of color as housekeepers, a position that was widely considered to involve sexual services. In 1776 Hilliard d'Auberteuil estimated that of 23,100 free people living in Saint-Domingue's cities, there were two thousand married white women and one thousand married free mulatto or black women. He also noted a slightly larger population of thirty-two hundred "prostitutes or women living as concubines" comprising twelve hundred whites and two thousand free mulattos or blacks.[129] These numbers may not be accurate, but they reveal the highly charged sexual atmosphere of Saint-Domingue's towns.

Yet this sexualized image of free women of color is incorrect, for they occupied a variety of economic niches besides sex work. Surviving leases, receipts, and inventories reveal that free women managed slaves and business interests; they built networks of patronage and affection with whites that did not involve sex, as business clients, neighbors, landlords, tenants, employers, and employees. Because many free women of color never married, especially those in the cities, some were able to escape male control and direct their own business interests.[130] Being a housekeeper or concubine to a male colonial was often just one stage of a woman's life. Many women used these positions to acquire real estate and slaves, which they then used in their own businesses. Although white male colonists created a narrative in which they used and discarded women of color as objects of pleasure, there is ample evidence that such men formed valuable partnerships and emotional relationships with concubines and mistresses, as well as with free colored neighbors, business partners, and friends.[131]

Saint-Domingue's lively theater scene, the area of colonial life in which Phillips made her mark in Jamaica, illustrates how free women of color operated in spaces in which they were a sometimes conspicuous minority. Besides Freemasonry, theater was Saint-Domingue's other distinctive urban institution. In Bordeaux and Paris in the 1780s, the ratio of theater seats to city residents was one to forty-six. In Saint-Domingue, Lauren Clay calculates, the comparable ratio was one to seventeen, including whites and free people of color. She also notes that colonials in the parterre paid five times what Frenchmen in the provinces would have paid; free colored theatergoers paid twice the price they would have paid in French provincial cities for equivalent seats.[132] Cap Français and Port-au-Prince alone had more than twelve hundred theatrical and musical performances in the 1780s, according to surviving newspaper accounts.[133] Actors and touring groups from Europe arrived regularly. Like theaters in provincial French cities, these colonial playhouses received many of the most popular plays from Paris soon after their premiers. In 1765, for example, after the Seven Years' War, Saint-Domingue's governor arranged for the patriotic play *The Siege of Calais* to be performed in Saint-Domingue, just four months after it debuted at the Comédie Française.[134] Although there were performances in creole, and companies sometimes adapted European scenes to colonial settings, performances in Cap Français remained closely aligned with French metropolitan styles.[135]

Like Freemasonry, theatrical performances provided an occasion for socializing in an urban society marked by individualism and a scramble for wealth. For Moreau de Saint-Méry, "one cannot miss a show at Cap Français, especially since [attending] has become the custom. There is little social life in this city and [at the theater] we are at least assembled if not united."[136] Unsurprisingly, the board of directors of the Cap Français Comédie touted the business advantages this sociability created for the colony: "A harmony of minds and an agreeable ease of conducting business, the custom of seeing and talking to one another prevents and dissipates personality [clashes] and often gives rise to new projects and new business which favors the growth of the country and the advantage of its residents."[137]

The theater was such a central institution in the social lives of colonial cities that after 1770 it became a prominent site for attempts to segregate whites and free people of color, a phenomenon whose political context is described in Chapter 6. The Cap Français theater admitted free men and women of mixed race, who sat in the upper lodges. After 1775 free black women obtained a legal ruling allowing them to attend the theater as well. Mixed-race women refused to sit with them, however, so the petitioners were assigned their own separate section.[138] Most theater attendees, however, were white men. Moreau described as extraordinary a performance during the carnival season attended by as many as 130 women, this in a room with 1,500 seats. Moreover at Cap Français there were only 60–80 places out of 1,500 for people of color, compared with 90–120 out of 750 in Port-au-Prince and 7 out of 400 in Léogane.[139] Racial borders were thus drawn carefully in this later period.

The appearance of women in the theater audience, especially, was very sexualized, as was fitting for the town where the *redoutes des filles de couleur* were invented. These dances, where white men could enjoy the charms of seductive women of color, were later translated to New Orleans, becoming the famous "quadroon balls" of the nineteenth century.[140] Moreau emphasized how the public display of feminine beauty at the theater overcame even the color line that had become so important by the 1780s. Writing about both white and mixed-race women, he observed, "They go to the show to parade their charms and their suitors; one notes that nearly all women dress with the same elegance, which shows that in the colony the charming sex is divided into only two classes, those who are pretty and those who are not."[141] Indeed, the Comédie was just one of those public spaces in which the city's role as a sexual marketplace was on full display: "Publicity, I will repeat, is one of their greatest pleasures and it is to this pleasure that we owe the custom that every night at bedtime, one sees the girls of color leave their homes, often lit by a lantern carried by a slave, to go to spend the night with the man they love the best or who pays them the most."[142] Yet Moreau condemned the sexuality of street life in Cap Français and praised the public decorum of women of color in the much smaller port city of Saint-Marc. He observed that "the small number of colored courtesans and the way they dress show that Saint-Marc has not arrived at that excess of civilization where there is a kind of pleasure in offending public decency."¹⁴³

Because women of color were on display for white males in the audiences of colonial theaters, as well as on the streets, it is interesting that in the 1780s an actress of color emerged as one of Saint-Domingue's best-known performers, mentioned some forty times in the colonial press during in a nine-year career. Minette, whose real name was Elizabeth Alexandrine Louise, was born in 1767. Although most free people of color in Saint-Domingue's cities were poor, Minette's grandmother was a propertied free woman of color, and her mother, described sometimes as a free quadroon, sometimes as a "mestive," meaning her ancestry was mostly European, had a widely acknowledged concubinage relationship with Minette's father, a white naval accountant in Port-au-Prince. The couple had two girls, both of whom had elite white godparents. In 1770 the colonial governor and intendant granted permission for Minette's mother to accompany her father back to France, though it isn't clear if the couple also took their daughters there.¹⁴⁴

Trained to sing and act by a white actress named Mme. Acquaire, Minette first appeared on stage in 1780 at the age of fourteen to sing arias in a Christmas performance in Port-au-Prince. Two months later, the director of the city's theater hired her for 8,000 livres or £347 per year, a sizeable sum for a beginning performer in this colony where singers regularly earned between 3,000 and 12,000 livres annually. Although free colored and enslaved musicians and singers did perform in colonial theaters, Minette's leading roles on the stage were still controversial. Moreau de Saint-Méry, despite his opinions about the dangers free colored courtesans posed to the colonial public, celebrated Minette's "triumph" over colonial prejudices.¹⁴⁵ We know little about Minette's private life. Like most free women of color she did not marry. In 1786 a French mapmaker in Port-au-Prince gave her two enslaved girls in his will, specifying that "she was the only person who helped me in this country where I had no family [and] where I would have certainly died given the nature of my illness."¹⁴⁶ But there is no other suggestion that this was a concubinage relationship like that between her mother and father.

Like Constantia Phillips in Jamaica, Minette called on Saint-Domingue's public to embrace its European identity. In a 1782 newspaper advertisement she decried "those ephemeral productions that bastardize and degrade the lyric stage, which are only local and which very often only address the everyday events of private society."¹⁴⁷ This statement was not merely a reference to the widespread use of tropical sets, costumes, and creole dialog and lyrics.[148] Minette was also criticizing the racial stereotypes that increasingly defined even freeborn people of color like her as part of the enslaved population. For example, in 1786 the husband of her former acting teacher produced a pantomime entitled *Arlequin mulâtresse sauvé par Macandal* (Arlequin, Mulâtresse Saved by Macandal).[149] The text no longer exists, but Macandal was an escaped slave said to be the mastermind of a poisoning conspiracy, an episode we examine in Chapter 5. Minette's sister Lise appears to have performed in such plays, for example, as in the 1786 performance of *Les Amours de Mirebalais* (The Loves of Mirebalais),[150] a colonial adaption of Rousseau's opera *Le Devin du village* (The Village Soothsayer).[151] Minette's public defense of the French theatrical tradition against ephemeral colonial productions shows that she rejected being defined solely by her race or sexuality, even if she worked in an occupation and in a space that was highly sexualized. Minette seems to have wanted to highlight the theater, as Lauren Clay puts it, as "an imaginary... that represented... a world in which free people of color could be fully and equally French."[152]

The lives of Teresia Phillips and Minette show that we need to be skeptical about descriptions of women in these societies. Their admittedly extraordinary careers belie the sharp contrasts that Long and Moreau made between passive but consuming white women and passionate but corrupting free women of color. These were working women, who defined themselves by what they did as actresses and devotees of the theater, rather than by their maternal role or even by their relation to white men. The distance between what white women and free women of color actually did was not that great in mid-eighteenth-century Jamaica and Saint-Domingue. Colonial ideologies stressed the distance between the two, especially after the Seven Years' War and the American Revolution, as efforts to make racial categories more distinct took hold in both colonies. What the lives of both women outlined in this chapter show is that, for freeborn people, urban society was a place of opportunity and, to some extent, liberation. While the mix of people and cultures, the lack of conventional religious institutions, and the prevailing materialism bewildered commentators, this "chaos of men" or "excess of civilization" provided a way for women like Phillips and Minette to construct identities that defied stereotypes.

By the start of the Seven Years' War, both Saint-Domingue and Jamaica had moved past the frontier stage of their development. They had established themselves as conspicuously successful plantation societies. They were culturally vibrant, dynamic, and economically valuable imperial possessions. Commentators new to West Indian mores and West Indian slavery found them disturbing places, where much of quotidian existence seemed immoral or at least dysfunctional. The harshness of slavery contrasted strongly with white peoples' devotion to pleasures of the flesh; the colonies had become vital imperial possessions, but colonists were only partially attached to European norms. White elites rejected religion, paid little attention to social rank, and embraced money as the measure of all things.

In the next forty years, these characteristics became more firmly entrenched as Jamaica and Saint-Domingue entered into the period of their greatest prosperity. It was also a period in which Jamaica and Saint-Domingue became important within the geopolitics of the French and British Atlantic empires. That geopolitical importance became clear during the Seven Years' War, despite the fact that it was largely fought outside the Greater Antilles. It is to this conflict we now turn.

CHAPTER 4

The Seven Years' War in the West Indies

The Seven Years' War in the West Indies was important in two ways, besides including the Caribbean in what is sometimes called the first global war. First, Britain's capture of Canada, Guadeloupe, Martinique, and Havana threatened the cohesiveness of France's New World empire. Although British negotiators refrained from a peace treaty that would allow George III to dominate the Caribbean, the war showed that London's power in the region was as much commercial as military. French planters in Guadeloupe and Martinique, like their counterparts in Cuba, profited from the British occupation, while colonists in Saint-Domingue chafed under what they saw as an overly harsh wartime regime. The postwar history of Saint-Domingue was devoted to government reforms designed to strengthen colonists' loyalty to the French Empire. Second, the conclusion of the Seven Years' War marked the beginning of a period of prolonged economic prosperity in the plantation economies of Jamaica and Saint-Domingue. Planters were never more powerful than in the three decades following the Peace of Paris in 1763.

Nevertheless, colonists faced challenges from metropolitan authorities. London and Versailles recognized the importance of each colony to imperial prosperity and insisted that they could control sometimes recalcitrant colonists. In Jamaica, moreover, planters faced the most violent slave uprising that had ever occurred in the Caribbean to that date. Their reaction to that challenge was to establish a new racial regime on the island, as Chapter 5 explains.

After establishing their first Antillean colonies on the smaller islands of the eastern Caribbean in the 1620s and 1630s, Britain and France found footholds in the Spanish-dominated Greater Antilles several decades later. For the next century, the two nations tangled with each other in the western Caribbean in a series of intense but short-lived local conflicts, while building up their plantations and transforming their colonies into profitable but brutal slave societies. Conflict heated up in the 1740s, starting in 1744 when France entered an Anglo-Spanish conflict that had begun in 1739 (the War of Jenkins's Ear), thereby joining it to the ongoing European War of the Austrian Succession. It is instructive to dwell a little on this conflict because it highlights, in its limited aims, how different the Seven Years' War was from previous Caribbean wars. France began the war in 1744 with only twenty-seven ships of the line, compared to the British navy, which had seventy-seven ships of the line.[1] This naval imbalance allowed England to attack Spanish ports in Peru and Chile and capture a Spanish treasure galleon in the Philippines.[2] Despite its naval superiority, Britain gained no new Caribbean territory in this conflict, thus preserving the balance of power it had established with France and Spain in the 1650s. It captured Spanish Porto Bello in 1739 and made an abortive attack on Cartagena in 1741.[3] But these conquests were relinquished at the peace, for the primary aim of these assaults had been to harass the Spanish, not to gain new territory. Significantly, in the War of Austrian Succession (1744–48) Britain did not attack France's Caribbean colonies. This war's major theaters were in North America, India, and Europe as British colonists did not want to gain new sugar lands, since expanding sugar production might reduce its overall price. Perhaps more important, Caribbean warfare was amazingly destructive of manpower. Even a short occupation, as in Cartagena in 1741, resulted in thousands of deaths of British regular and colonial troops, mostly from disease.[4]

Consequently, the only clamor for annexation of French and Spanish West Indian islands came from a couple of British newspapers unaware of the dread demography of the region. A correspondent in the *Newcastle Courant* wondered "whether Quebec, St. Augustine, the Havannah, St. Domingo, or the fortress of Martinico, be not of more importance to us" than land in Flanders.[5] In any event, British statesmen were impervious to such calls for action. The Peace of Aix-la-Chapelle of 1748, which ended the war, left Caribbean territory unchanged. Britain returned the fortress at Louisburg, in modern Nova Scotia, to France, which saw this installation as essential to its status as a maritime power.[6] In return France withdrew from the Austrian Netherlands, an ideal staging ground for an invasion of Britain.[7]

The War of Austrian Succession illustrates how commercial shipping, more than naval combat, was at the heart of British/French rivalry in the Caribbean. During the conflict, in the years 1745–48, only seven French slave ships went to Saint-Domingue, unloading 1,741 Africans.[8] Meanwhile, 109 British slave ships docked at Jamaica, carrying 29,786 slaves.[9] In the 1740s conflict, unlike the Seven Years' War, the French navy mounted an effective system of convoys between France and the Caribbean. But in 1747 the British learned how to intercept these expeditions, cutting Bordeaux colonial sugar imports in half in 1748.[10] British naval superiority ensured that Jamaica's average annual exports to Britain were more affected by drought than by the war, falling only 5 percent in value during the conflict. The island's sugar exports actually increased slightly, from £311,680 per annum to £317,390 in the same period.[11]

The Peace of Aix-la-Chapelle in 1748 did not change French and British fears about each other's intentions. British West Indians were especially concerned about Saint-Domingue's growing prosperity. Admiral Edward Vernon proclaimed in Parliament that this expansion would soon lead to the French being in control of all the sugar islands.[12] Governor Edward Trelawny of Jamaica was especially fierce on the matter, declaring "unless French Hispaniola is ruined during the war, they will, upon a peace, ruin our sugar colonies by the quantity they will make and the low price they afford to sell it at."[13] Such thinking led to the last act of the war, a daring raid by Admiral Charles Knowles, supported by Trelawny, on the French fort of St. Louis, on Saint-Domingue's southern coast.[14] This was one of Saint-Domingue's best defended harbors, but militia forces there laid down their arms after only eighty-five minutes of British cannon fire. Local indigo planters then concluded a massive sale to the attackers, transferring the dye onto British warships. Later, in Europe, Admiral Knowles signed and remitted a bill of exchange from St. Louis planters. To top it off, down the coast at Tiburon, the Jamaica governor, Trelawny, accepted planters' invitation to come ashore for tea.[15] As this anecdote suggests, French planters were more interested in trade than in war. That attitude would be a major theme in the Seven Years' War, to the dismay of Versailles.

The Seven Years' War began deep in the North American interior. In the 1750s, financial exigencies and a lack of stable leadership in the Naval Ministry weakened France's Indian alliances and claims over the territories south and west of the Saint Lawrence River.[16] At the same time, the rapid population growth of British colonies like Pennsylvania and Virginia put new pressure on those claims. After Versailles began to build forts in the Ohio valley to strengthen its position, conflict broke out in July 1754 when a troop of French soldiers and Canadian militia defeated a militia unit commanded by

Colonel George Washington in modern-day Pennsylvania. A more serious incident occurred a year later when French, Canadian, and Indian troops defeated General Edward Braddock on the Monongahela River.[17]

These imperial disputes fed into a balance-of-power struggle in Europe, as Austria, with French and Russian backing, sought to defend itself from the growing power of Frederick II's Prussia, allied with Britain. By the time war was formally declared on 15 May 1756, British and French navies had already clashed off the North American coast. In the autumn of 1755, the British captured hundreds of ships and thousands of French sailors.[18] In spite of the fact that France had built 34 new ships of the line between 1749 and 1754, its navy was still far behind the British. In 1755 they had just 57 battleships and 31 cruisers compared to 117 and 74 respectively for the British. By 1760, the gap was even larger, with the British having 135 battleships and 115 cruisers compared to France's 54 and 27.[19] Moreover, France's new ships, designed to accompany convoys, were lighter and with fewer guns than their counterparts.[20]

France survived the first half of the Seven Years' War without disastrous losses in North America, largely because Britain did not devote enough troops or ships to the area. Versailles's strategy was to use the overwhelming size of its European armies, now allied with Austria, to exert pressure on the British either through victories in the Austrian Netherlands, or in George II's native Hanover. The West Indies became important only after William Pitt became Britain's leading minister in late 1757. Of course, Britain had long noted with concern the growing wealth of the French sugar islands, especially Saint-Domingue. One correspondent of the Duke of Newcastle believed that "France was pushing for Universal Commerce" and that this push made him "more afraid . . . of French commerce than of French fleets and French armies."[21] The British were also concerned that the French colonists, in violation of the Treaty of Aix-la-Chapelle, had established themselves on St. Lucia, St. Vincent, Dominica, and Tobago, and "neutral" islands in the Lesser Antilles,[22]

As imperial tensions rose in North America in 1755, British and French Caribbean colonists had every reason to assume that they would be attacked. Jeremiah Meyler, a merchant in Savanna-la-Mar in western Jamaica, described how the threat of war "certainly has thrown things into confusion for the present & occasions people to desist from purchasing any quantity of sugar, rum &tc. the market being so uncertain." Meyler's partner, Robert Whatley, moved Meyler's account books from Savanna la Mar to Kingston. He made copies of all the entries in case there was an invasion. Jamaican fears about invasion persisted during the war. When the French sent fleets into the Caribbean in 1756, 1757, 1759, and 1762, white Jamaicans were sure they were about to be invaded.[23] French colonists in Saint-Domingue experienced similar fears about possible British invasions. Their awareness of British naval superiority deepened their hostility and cynicism regarding the naval officers who governed their colony. After French ships lost a naval skirmish off Saint-Domingue's northern coast in 1760, a critic of Saint-Domingue's government wrote that "the poor navy, which furnishes us with officers who govern the colonies so well, is, they say, reduced to a very bad state. Happily . . . according to the accounts we have heard, there were few casualties. Thus, though we lack ships we still have officers and governors from this illustrious corps."[24]

The war also disrupted trade. As Robert Stanton, a Kingston merchant, wrote to Henry Bright, in June 1757, "This damned wicked French war as it 'tis carried on will be the ruin of many, both planters and merchants, because no goods shipt from hence." Stanton believed that no one could "bear the heavy charges of freight and insurance at the extravagant rate they now are." The war also meant that trade with Spanish America, "by which means there was introduced a great deal of money, and large quantities of European goods taken off," had virtually halted. At the same time it hindered the delivery of slaves to the island as ships that might have been sent from Bristol to Africa were instead turned into privateering vessels. Such complaints were carping, however, given what British American colonists in North America were being forced to deal with in the first two years of war, when the war went very badly for them.[25] Indeed, once the war turned in favor of the British, Jamaican merchants began to profit from the disruption of trade in the Caribbean. As British and Anglo-American privateers seized enemy ships, they brought them to West Indian ports, especially to Jamaica. By 1763–64, 123 former prize ships were noted as entering Jamaican ports and clearing customs. Of these, seventy-seven were registered in Jamaica. Thus, the war expanded the island's commercial fleet, providing ships for Kingston merchants to use in regional trade.[26]

Before Pitt's ascension to power in 1757, the British experience of the war was one dismal event after another, culminating in the capitulation of the Duke of Cumberland to the French at Klosterzeven on 8 September 1757. This defeat allowed Pitt to take control of the navy, army, and diplomatic corps, giving him close to total direction of the war. He immediately initiated what he called his "system," a pragmatic, fluid mixture of strategies that allowed Britain to attack France where it was weakest—in North America, India,

West Africa, and the West Indies—rather than where France was strongest—in Europe. Leaving the European war mainly to his Prussian ally, Frederick, he turned his attention to the imperial periphery, where for the rest of the war Britain fought most of its battles. It was a bold and original strategy, for no one prior to Pitt had seen the war as a means of attacking the sources of French wealth.[27] Neither were the French able to take advantage of Britain's lack of focus before Pitt's ascension. Until late 1758, when the Duc de Choiseul rose to power, the French war effort was bogged down by a Council of State whose ministers were divided over war aims, by generals who were unable to take advantage of their victories in Germany, and by financial problems that prevented the navy from building a fleet as large as that of the British.[28]

Pitt's first overseas victories came in West Africa in 1758 when expeditions seized French trade counters at Senegal and Gorée and on the River Gambia.[29] Encouraged by this success, Pitt turned to the West Indies, where he hoped to knock out French power and destroy French colonial wealth. William Beckford, the London-based head of Jamaica's most politically important family, urged him to take an ill-guarded Martinique. This easy conquest, Beckford argued, would acquire for Britain an island with slaves and property worth more than £4 million or 92.1 million livres. He exhorted Pitt: "For God's sake, attempt it without delay."[30]

The minister was well aware of other strategic reasons for such a campaign. Not only did France produce more sugar at cheaper prices than Britain, but in the struggle between the two empires geography and wind patterns favored the French. Privateers sailing out of Martinique posed a serious threat to Barbados, which had no capacity for a naval harbor. Antigua, which did have a British naval station, was downwind from both Martinique and Barbados, so its ships found it impossible to shut down French privateers.[31] Attacking Martinique would strike a major blow at the privateers and at the French sugar industry. Pitt did not, however, intend to keep his West Indian conquests. As he saw it, Martinique would be an ideal chip with which to gain back the Mediterranean island of Minorca, which the French had taken from Britain in 1756.[32]

The French were keenly aware of their naval inferiority, but Britain's 1759 campaign in the eastern Caribbean drove home a more unsettling realization: French planters would not sacrifice themselves or their property in order to repel a British attack. In the years following the Seven Years' War, this insight had a profound effect on politics in Saint-Domingue, as Versailles sought to build imperial patriotism in its most valuable colony. It was the example of Guadeloupe and Martinique in 1759 and 1762 that created this reaction. On 12 November 1758, Pitt sent off a large force of nine thousand men in seventy-three ships to the West Indies. The expedition was hindered by the medical complications that European troops always faced in the Caribbean. Within a month of arrival in Barbados, disease had reduced the number of men fit for service to five thousand. Within another month this force had nearly halved again to just fewer than three thousand. The primary killer was, as always, yellow fever.[33]

Nevertheless, the fleet arrived at Martinique on 16 January 1759. Intimidated by the guns of Fort Royal and unaware that the island was low on food, within a week the attackers moved on. They sailed seventy-five miles north to Guadeloupe, where the British took the capital city and fort of Basse-Terre on 23 January, after a single day of bombardment. Guadeloupe's defenders withdrew into the steep wooded hills behind the city, waiting for disease to ravage the invaders. The British established a new base at what is today the city of Pointe-à-Pitre and began to work their way down the southeastern coast, destroying as many as eighty-four sugar estates. In the third week of April, when the attackers reached the valuable plantations of Capesterre parish, Guadeloupe's planters forced Governor Nadau de Treil, who owned a plantation himself and was married to a creole, to sign a capitulation agreement with the British. Although disease had reduced the initial British force to fewer than one thousand, Guadeloupe's colonists were done with fighting.[34]

The final surrender agreement was highly favorable to French planters. The British promised that colonists would retain their property, their French laws, and their current administrators. No Englishmen would be allowed to settle in the colony during the war. The new government would not requisition planters' slaves for construction nor billet troops in their homes. The residents were not required to fight against the French or against anyone else. Of great importance was that the British allowed Guadeloupian goods to be sold in British markets—or other foreign markets, if London was closed. Moreover, they would permit British and British American traders to sell slaves, provisions, and goods in the captured territory. Guadeloupe's merchants would consequently have the same commercial rights as any member of the empire.[35] The benefits for Guadeloupian planters of British protection were immediately apparent. The slave trade expanded dramatically. Between 1713 and 1755, Guadeloupe had received only fifteen slave ships and 2,998 slaves.[36] French trade law forced planters to buy almost all their workers in

Martinique. Between 1759 and 1763, however, seventy-eight British slave ships with 23,834 Africans docked at Basse-Terre.[37] The number of sugar plantations in Guadeloupe increased from 185 to 447. Not only did its production quickly surpass that of Martinique, but in 1761 it led the British Empire in exports of sugar, cotton, rum, and coffee, as well as in its purchase of British and American products.[38] Guadeloupe's surrender, specifically the way its planters quickly shifted their allegiance to France's mortal enemy, shaped French policy in the Antilles for years afterward. Versailles quickly realized that its West Indian subjects, including colonists in Saint-Domingue, were not as loyal as they proclaimed. What motivated them most was profit.

Despite these British victories in the Caribbean and the capture of Québec and the rest of French Canada in late 1759, Versailles refused to accept British terms in peace negotiations in 1760 and early 1761.[39] Therefore, in the second half of 1761 and the first few months of 1762, the British struck again in the French Caribbean. In June 1761 British troops from Guadeloupe captured the island of Dominica, located between Martinique and Guadeloupe. Dominica was technically neutral but had been occupied by French colonists.[40] Pitt's goal, as before, was to capture Martinique in order to force Versailles into additional concessions at the peace negotiations. By the time Pitt secured cabinet permission to route soldiers from New York and Europe to the Caribbean, the hurricane season had already begun. The thirteen thousand British troops gathered at Barbados had to wait until 5 January 1762 to sail to Martinique.

Martinique's planters had observed how Guadeloupe had profited from its surrender in 1759. In 1760 Martinique's intendant, Pierre-Paul Mercier de la Rivière, wrote, "The over-riding notion in this country is that the colony is seen as being foreign to the metropole and the metropole is regarded as foreign to the colony."[41] François de Beauharnais, Martinique's governor from 1757 to early 1761, warned France that he needed more European troops to ensure the loyalty of the colony's elite. As part of a series of reforms undertaken in 1761–63 to assure creoles that Versailles was responsive to their needs, Choiseul replaced Beauharnais with a governor from an old creole family, Louis-Charles Levassor de la Touche, overruling a regulation against governors serving in colonies in which they had family.[42] But the appointment did little to rally the island to the metropolis. On 17 January 1762 the fleet from Barbados anchored before Fort Royal, Martinique's capital. A British blockade prevented a French squadron from reaching the island in time to repel the invasion. With only one thousand regular troops, and eight militia battalions, the colony capitulated on 13 February 1762.[43]

At this point it appeared likely that Saint-Domingue, France's one remaining American colony, would be Pitt's next target. In naval terms, it stood virtually undefended during much of the war. Not a single naval vessel from France had reached the colony between July 1757 and 13 January 1758, during which time Jamaica had an average of a dozen warships on station. The disruption of France's Caribbean trade allowed British sugar growers temporary access to European markets, at French expense.[44] Saint-Domingue's garrison, however, had approximately fifteen hundred troops. In March 1762 a squadron of seven ships originally destined to defend Martinique disembarked another fifty-five hundred French soldiers in Saint-Domingue, as well as the colony's new governor-general, Gabriel de Bory. Because Bory was more of a scientist and administrative reformer than a soldier, on 31 July the Naval Ministry entrusted all military preparations in Saint-Domingue to the Vicomte de Belzunce.[45] Belzunce threw himself into preparations for a British attack, building a network of roads, creating military camps in the mountains behind Cap Français, scheduling weekly militia musters, requisitioning planters' slaves for construction projects, and requiring them all to plant food crops. He also took command of the maréchaussée, the mounted constabulary used to hunt maroon slaves. Faced with the reluctance of colonists to sacrifice their time on militia training, Belzunce created a new unit of free colored troops. His enthusiasm for the military aptitude and self-sacrifice of these men shaped the debate over race and patriotism in Saint-Domingue in the aftermath of the war.[46]

Even as Saint-Domingue prepared for an invasion from Jamaica, the arrival of so many French troops in the region caused Jamaicans to send for help to the Leeward Islands. Moreover, in January 1762 Spain joined the war on France's side. The Jamaican overseer, Thomas Thistlewood, who had hardly mentioned the Seven Years' War in his copious diaries before this, became agitated by news of Britain declaring war against Spain. He noted in his diary for 29 January 1762 what he read in a Kingston newspaper, that there was "Some Conjecture, by Letters found on Board a French Prize lately took that the French and Spaniards are going to Join their Forces, and invade this Island." It was a worrying prospect. Thistlewood noted, "Last Night I dreamed that a Spanish Fleet, Come and took the Shipping in Savanna la Mar Road, and Sa: la Mar, without any opposition! only a dream I thank god."

As an active militia man, called for duty "under Arms" at least eight times in February and March, Thistlewood knew how underprepared Jamaica was for invasion. He noted on 5 March 1762 that "Kingston Can raise about 900 Armed Militia, this parish about 400, and the Whole Island about 5000, only exclusive off the Officers, and reformades; and about 14 hund:d Regulars, all together a poor strength indeed." Although Jamaica's militia was reasonably well armed and mustered regularly, he was uncertain about how they would fare against regular troops. After all, as he noted, many militiamen were unable to cope with a muster on a hot day. On 1 March 1762, there was a "general Muster off horse and Foot," but during the review by "Brigadier General Norwood Witter, Colo: Lewis, &c. &c" it was "very hott," and many men "fainted, and Were obliged to quit their Ranks." Moreover, the ranks of the militia were becoming depleted as people like his underling, Richard Lloyd, decided to try their luck at privateering.[47]

In fact, while Saint-Domingue prepared for a British attack, Choiseul proposed to Madrid a joint French and Spanish invasion of Jamaica.[48] That attack never occurred, but some Frenchmen and Spaniards raided Jamaica in order to steal slaves. Thistlewood noted on 4 December 1761 that "the French have been at Negril guard house [on the southwestern coast] last Wednesday night, and Carryed off Some Negroes." A year later, a large force of seventy-two Spaniards, accompanied by five Frenchman and an Italian, "Went up to Colo: Barclay's Polink, on the other Side the hope 3 or 4 miles, and about a Mile and Quarter Within land, and took 15 Negroes (Some Say 20 others' 27.), men Women and Children Clear off, besides Some Fishermen," out of a gang of nineteen slaves. Thus, "only 4 Escaped, old Coffee one off them, Who I Saw today about Noon." The raid met with a fierce response. The militia and a body of free black rangers quickly apprehended the raiders and took them to Savannah-la-Mar. Thistlewood was struck by the poignancy of the capture, noting on 7 December 1762 "that yesterday Evening, dr: gorse, Mr: Randolph donaldson, Mr: Abraham lopez, &c, Conducted the Spanish prizoners thro' Egypt towards Savanna la Mar." Thistlewood observed, "The Captain and another rode in the Kitterine, the Common Man Walk'd." He thought that "the Captain looked very Thoughtffull," and the Jamaican "Could not help pittying them. . . . The girl he kept even made him help Carry his own things down to their Canoe, Stripp'd him Naked except an old dirty Check Shirt they gave him."[49]

Fortunately for Thistlewood and Jamaica, it was Britain who struck at Spanish possessions rather than the other way around. Instead of attacking Cap Français, Britain laid siege to Havana on 6 June, capturing the city on 12

August 1762. It was a hard-won victory. As usual in the West Indies, deaths from disease far outweighed battlefield casualties; the British lost over 40 percent (forty-seven hundred of twelve thousand soldiers) of their invading army to yellow fever and other tropical diseases. Indeed, the extent of the losses through disease leads John McNeill to consider Britain's triumph at Havana a fluke that could have been easily prevented if the Spanish commanders had not surrendered to the British so precipitously. Moreover, the British commander, George Keppel, Earl of Albemarle, was fortunate in that he could supplement his forces with four thousand troops from North America. These troops allowed him to mine under the major defenses of Morro Castle and take the heavily defended fort by storm. Disease prevented him from taking the entire island. If Cuban planters, like their counterparts in Guadeloupe and Martinique, had not been so eager to trade with the enemy, the British would have been unable to hold Havana.[50] British colonists thought the conquest of Havana one of the great events of the war, if also one of the mostly costly. As Benjamin Franklin wrote, "It has been however the dearest conquest by far that we have purchas'd this war when we consider the terrible Havock made by the sickness in that brave Army of Veterans, now almost totally ruined." Reflecting back on the event in 1771, Samuel Johnson lamented the cost of the victory: "May my country never be cursed with another such conquest."[51]

On 10 February 1763, France and England signed the Treaty of Paris, ending the war and reconfiguring France's overseas empire. Pitt's strategy in the West Indies from 1757 can be judged an enormous success.[52] All French Canada went to the British, with the exception of Grand Banks fishing rights and two fishing bases in Newfoundland. In the Caribbean the British kept the islands of Grenada, St. Vincent, Tobago, and Dominica but returned Martinique and Guadeloupe to France. St. Lucia remained in French control.[53] In India, though France had committed more troops there than Britain during the war, the French had lost all their territories. Britain returned France's five Indian trading counters, but from this period on, France's ambition to extend itself militarily in India vanished.[54] In West Africa, France lost its base at Senegal, though the treaty did return its station at the Ile de Gorée.[55]

The nature of French versus British naval power in the Caribbean by the mid-eighteenth century meant that, on the surface, Saint-Domingue and Jamaica experienced very different Seven Years' Wars. For example, although both colonies feared an enemy attack, each had a distinctive military system. Saint-Domingue was administered by military officials, from the governor-general to the parish commanders. This centralized system meant that the naval ministry had authority over coastal guards, siege preparations, the construction of forts, and the allocation of funds to pay for these operations. In Jamaica, on the other hand, the Colonial Assembly funded defense works, leading to peacetime lulls in construction, and then overbuilding when war broke out.[56] The French sent large numbers of men to the West Indies, with squadrons leaving in 1757 and 1759 and two detachments, taken away from the Atlantic fleet, in 1758 and 1761. That the French government was willing to send ships from the Atlantic fleet, vital for the protection of France itself, shows the significance they attached to defending the region.[57] Nevertheless, the war stretched France very thin. Worried about losing more colonies to Britain, Versailles could not protect shipping between France and the Caribbean. From 1756, Versailles made the defense of Québec its top priority in the New World, directing convoys there while leaving Caribbean trade unprotected. After the British took Québec in 1759, France did not mount convoys to the Antilles. Saint-Domingue remained under the worst commercial blockade it had ever known, with massive increases in maritime insurance.[58]

For most Jamaican colonists, however, the events of the war itself were relatively minor. Unlike North American colonists, most West Indian colonials saw little or no military action, and Jamaican business affairs were only marginally affected. Thistlewood viewed the war as evidence of the superior qualities of British arms and the British system of government. In his diary he rejoiced over British victories at Cape Breton in 1758, in West Africa in 1759, and at Pondicherry in India. Hearing of the last success, he celebrated "what Glorious Success we are blest Withall."[59]

Britain decisively won the war. But Pitt was unhappy with the terms of the Treaty of Paris. He made a three-hour speech in the House of Commons on 9 December 1762, denouncing the agreement as a betrayal of all he had worked for as chief minister of the Crown. He argued that by restoring to France "all the valuable West Indian islands . . . we have given her the means of recovering her prodigious losses and of becoming more formidable to us at sea."[60] Nevertheless, the great majority of the House of Commons voted to approve the peace terms, convinced that it was a phenomenal diplomatic coup.[61] Britain's chief negotiator, the Duke of Bedford, was determined not to entirely destroy France's New World empire. This, he believed, would "excite all the naval powers of Europe to enter into a confederacy against us, as adopting . . .

a monopoly of all naval power, which would be at least as dangerous to the liberties of Europe, as that of Louis XIV." Pitt, however, played the long game. He hoped to prevent France from ever reemerging as a naval threat to Britain. The Duc de Choiseul privately made it clear that Pitt, not Bedford, was right: "If I were in charge . . . England would be reduced and destroyed by us within thirty years . . . our navy is in bad shape, but it will recover; not everything is done in a day."[62]

As this quote suggests, France and Spain were surprised by the Canada decision, for they believed that Britain's ultimate aim was control of Mexico. Choiseul predicted dire consequences for France if the British became "masters of Mexico," arguing that "if the English took it [Mexico], they would furnish it exclusively with their own manufactures, which would occasion a loss of trade . . . and in consequence ruin the greater part of our industry."[63] Spain had joined the French war effort because of a mutual sense of "fear and impotence" about Britain's power in the Americas. In order to counter British intrusions into the Caribbean, Madrid spent vast sums shoring up Havana's defenses. The capture of the city in 1762 was therefore an enormous shock, precipitating a decade of Spanish colonial reforms. The fall of Havana illustrated that New Spain was vulnerable to a similar attack on Veracruz, leading to an attack on Mexico City. It was for this reason that Spain insisted so strongly on keeping Havana in the Peace of Paris. Even if it meant losing West Florida, the Honduras coast, and their right to fish off Newfoundland, Spain had to regain Havana so as to block Britain from Mexico.[64]

Britain's decision to keep Canada also perplexed the multiple authors (marshaled by Guillaume-Thomas Raynal) of the *Histoire des deux Indes*, a best-selling six-volume interpretation of Europe's impact on the New World and the New World's impact on Europe, first published in 1770. British actions in 1763 meant that "from this time England lost the opportunity . . . of seizing all the avenues, and making itself master of the sources of all the wealth of the New World." Britain, they thought, could have easily taken Mexico and from there the rest of the South American continent. The authors of the *Histoire* dreamed that British penetration of the Spanish Main might lead to a creole revolution in Spanish America, thus changing, for the better, "the whole face of America," as "the English, more free and more equitable than other monarchical powers, could not but be benefited by rescuing the human race from the oppressions they have suffered in the New World." Indeed, British advances in the Caribbean and the establishment of a large tropical empire would lead to an empire for liberty, to where all sorts of "distressed and unfortunate persons," notably "the peasants of the north, [probably Russian] slaves to the nobility, who trample upon them," would flock into "these regions, which require only just and civilised inhabitants to render them happy."[65]

A few Britons also saw the different actions taken by the French and the British following the Seven Years' War and pointed out the mistakes that they thought that the British were making in their future planning for the West Indies. William Burke, cousin of Edmund, was a colonial official in Guadeloupe in 1760. He was convinced that it was folly not to capitalize on temporary French weakness. For Burke, the North American colonies were so less valuable to Britain than the West Indies that their interests had to be subordinated to those of that region. After all, he argued, "it is by Means of the *West-Indian* Trade that a great Part of *North America* is at all enabled to trade with us.... [I]n Reality the Trade of these *North American* Provinces ... is, as well as that of *Africa*, to be regarded as a dependent Member, and subordinate Department of the *West-Indian* Trade; it must rise and fall exactly as the *West-Indies* flourish or decay."[66] If this was the case, then it was folly of the highest order, he believed, to allow the French West Indies to recover. Burke echoed the opinions of Pitt, who wanted to squeeze France so hard in the peace negotiations that it would be unable to challenge British dominance overseas ever again.[67]

This folly, as Pitt and Burke saw it, was compounded by the extensive economic development that the British had fostered during the period of occupation in both Guadeloupe and Cuba. The conquest of Guadeloupe in 1759 ushered in a period of great profitability for the island with most of the profits, ironically, going to established French planters, who were allowed to surrender under such favorable conditions that the conquest became one of the best things that could have happened to them. The large profits that had been made in other parts of the British Caribbean encouraged British merchants. Soon they were supplying French planters with all that they could possibly need. Liverpool traders alone claimed to have imported 12,437 slaves into Guadeloupe during the occupation. Modern estimates back up these figures; the *Transatlantic Slave Trade Data Base* shows that the British disembarked 23,834 slaves onto the island during four years. This was a massive influx in an island that in 1755 had only 41,140 slaves.[68] As a result, sugar production increased rapidly, and exports to Britain from both Guadeloupe and Martinique were so munificent that the price of sugar declined by nearly a third, from 45 shillings 9 pence in 1759 to 32 shillings 6 pence in 1762.

The number of sugar plantations on the island shot up, with the number increasing from 339 to 420 in 1761–62. British merchants were instrumental in developing a new town, Pointe-à-Pitre, which quickly replaced Basse-Terre as Guadeloupe's most significant commercial center.[69] Britain's brief occupation of Cuba in 1762–63 was also instrumental in launching that island's transformation from a relative backwater into the major sugar-producing region of the Americas by the early nineteenth century.[70] The explosion of British commercial involvement in Guadeloupe and Havana in the early 1760s showed that British merchants were desperate to invest in West Indian plantation agriculture and in slavery. In November 1762, for example, 145 Liverpool merchants petitioned the secretary of state asking that Britain keep Guadeloupe because "the possession of that island has increased their trade beyond all comparison with its former state, in the demand of British manufactures for slaves."[71] Humiliation in the Seven Years' War and fear of Britain's long-term aims galvanized both Spain and France into reforms that would make their empires more profitable and more defensible. While Spain's so-called Bourbon Reforms are the subject of a large literature, historians have paid less attention to the ways that France restructured its West Indian colonies after 1763. Those reforms—political, military, economic, and social—are the subject of Chapters 6, 7, and 8.

While the British Empire was trying to avoid the appearance of monopolizing the Caribbean, the French were determined to restore their power there after the Seven Years' War. They would partly achieve this by building a more formidable navy, but Choiseul believed nothing could be done without addressing the problem of colonial loyalty. For whites in Saint-Domingue, the experience of being largely undefended by imperial forces had raised important questions about the colony's place in the empire. Already unhappy with the way French mercantilism restricted their trade and chafing under a military rule that leading planters described—without irony—as "slavery," Saint-Domingue had now discovered that the metropolis could not adequately defend it from the British. Instead of warships and troops, Versailles supplied administrators. These men dragooned colonists and their slaves into costly and humiliating preparations for an attack from Jamaica that planters believed would inevitably destroy their sugar estates. Choiseul's aim was to make Saint-Domingue more attached to the French Crown than their counterparts in Martinique and Guadeloupe had been.

His reform project was supported by a major wave of new European immigration onto the island. In contrast to the British West Indies, where white population growth remained mostly stagnant in the years between midcentury and the American Revolution, the European population in Saint-Domingue increased steadily after 1763. Historians estimate that more than a thousand French migrants moved onto the island every year. These new migrants pushed the white population from 14,258 in 1754 to over 27,000 by the 1780s.[72] The size of this influx was dwarfed by the extraordinary importation of Africans after the mid-1760s and by the increasing size of the free population of color. By 1788 the proportion of the population that was white had shrunk from 7 percent in 1764 to 5.8 percent. However, Saint-Domingue had nearly twice as many whites as Jamaica by the 1780s and over half as many as the total combined white population of all the British West Indies. This in-migration gave the colony a greater buffer against foreign invasion than before and strengthened its cultural connection to France. But the flood of striving European immigrants—dismissively labeled *petit blancs* locally—added a new tension within the white population.

Increasing the white population was essential to Choiseul's plans to create loyal and self-defending Caribbean colonies. As he noted, "the English were only able to make their conquests during the last war by means of their northern colonies, which are almost entirely populated by whites."[73] French officials understood that sugar planters were not typical "settlers." Their heavy investment in their plantations made their loyalty vulnerable to a damaging enemy attack. New arrivals deeply resented being forced to muster and drill in a society where free and enslaved blacks performed virtually all manual work. Officials might force colonists' slaves to build fortifications, but they could not use those bondsmen as reliable soldiers. To create a loyal colony of white settler-soldiers, Choiseul's grand and quixotic plan after 1763 was to develop white settlement in the Kourou river region of French Guiana in northeastern South America. In this barely settled region with just 575 white residents and 6,996 slaves in 1762, he planned a new American colony, a tropical Canada or Massachusetts, full of European farmers. The Kourou colony might also provide Saint-Domingue with cheap food and lumber, thereby discouraging smuggling from British North America. To ensure that Kourou would not become another plantation colony, Choiseul instructed its governor that black slaves should be introduced into the colony "only in the last extremity." He diverted massive resources to the project, leading fourteen thousand Alsatians, Rhinelanders, and Venetians, along with a few Frenchmen to embark for Guyana in 1763–64.[74]

The result was disaster on a colossal scale. The organization was abysmal, and promised food supplies never arrived. Nearly two-thirds of the settlers—some ten thousand people—died within a few months of arrival, and most of the survivors returned to France. In 1770, the white population of French Guiana was no more than 1,178. Although the scale of the disaster was magnified by chaotic mismanagement and wide-scale corruption, the major contributor to this debacle was tropical disease. For the authors of the *Histoire des deux Indes*, however, the disaster proved that "in order to obtain rich productions, one will have to have recourse, necessarily, to the sinewy arms of negroes."[75] Saint-Domingue and its exploitative plantation system were proving to be the only viable model for colonization in hot climates.

If the war led the French to seek a new kind of tropical colony, it led the British investors to look for ways to extend their plantation profits. Wealthy London merchants were involved with all aspects of the plantation enterprise, from supplying the manufactures and slaves that kept the plantations going, to receiving the produce of those plantations and owning the plantations themselves. The most dynamic areas of British commerce were tropical America, West Africa, and India. David Hancock shows how successfully one group of merchants integrated these regions through its business activities. Investors in all parts of empire, including the West African slave trade, they concentrated in particular on the British Caribbean, first in Jamaica during the Seven Years' War, and then, in the boom years of the mid-1760s, into the recently acquired Ceded Islands, where colonists in Grenada, Dominica, and St. Vincent were developing plantation societies at a phenomenal speed. These investors found little in former French Canada to rival the prospects of these new Caribbean lands. Hancock notes that British merchants moved quickly out of the northern and Chesapeake colonies immediately following the Peace of Paris into the rice and sugar economies of the Lower South and the West Indies.[76]

The amount of money Britain invested in these areas in the years after the Seven Years' War was astounding. Most of it was private money, but state involvement was crucial, especially in the large-scale Crown-supported settlement programs in the Ceded Islands.[77] This investment fever was directly inspired by Jamaica's extraordinary economic performance. Saint-Domingue may have eventually overshadowed Jamaica's plantation economy, but Jamaica's expansion after the Seven Years' War was remarkable in its own way. Moreover, that wealth was matched by the influence of Jamaican planters on imperial policy making in the 1760s. West Indians formed the most powerful colonial lobby in London. Their opposition to the Stamp Act, for example, was influential in encouraging British opponents of that act, like the Marquess of Rockingham, to increase their agitation against government policy. They were also able to plead successfully for preferential treatment from the home government. During the 1760s, Andrew O'Shaughnessy notes, "British colonial policy increasingly discriminated against the North American colonies in favor of the British West Indies," with, for example, the islands being exempted from some clauses of the Townshend Acts of 1767–70 and being reprieved from the jurisdiction of the new American Board of Commissioners of Customs.[78] Lord North used to say of West Indians that "they were the only masters he ever had," while Benjamin Franklin complained that the "West Indies vastly outweigh us of the Northern Colonies."[79]

In particular, West Indians received metropolitan respect for the wealth that sugar and slavery were bringing into Britain. Looking at thriving British ports full of ships going to Africa for slaves or carrying goods to the West Indies, British politicians found it easy to believe that Jamaica's riches stimulated British wealth creation. Not only were they wealthier than North American colonists, but Jamaicans were better prepared politically to cope with tougher imperial reforms following 1763. Their connections in London were strong, especially in the capital's commercial community. In the third quarter of the eighteenth century, for example, they developed a powerful and increasingly well-organized lobby in the London Society of West India Planters and Merchants.[80]

Yet clouds were gathering on the horizon. The British recognized the value of the sugar colonies, especially Jamaica, within its American empire. But they became increasingly disturbed by white Jamaicans' way of life and by their proclivity toward horrific violence against enslaved people. The Seven Years' War marked an important transitional moment in how Britain viewed both its empire and also its white subjects within that empire.[81] At the same time as Britons were developing new attitudes to nonwhites, in part brought on by the large accession of territories and new peoples into the empire after 1763, they were beginning to reject the insistent plea of white Americans and West Indians that they were British and entitled to all the rights and privileges that Britons could have in the metropolis. As Stephen Conway notes, "the inhabitants of the old colonies lost out primarily because they came to be viewed by at least some Britons less as compatriots and more as just another set of people to be ruled."[82]

The same process also occurred in Saint-Domingue, although metropolitan disdain for West Indian colonists there was less obvious than in Britain.

Indeed, the authors of the *Histoire des deux Indes* fantasized that if slavery were abolished, then the French West Indies could become the home of liberty. They denied climate theory, which blamed tropical heat for the languor and hot-temperedness of West Indian whites. Instead, they attributed such failings to the undeserved feelings of mastery that slave owners experienced when they had total power over unhappy slaves. Creoles, they argued, "were the most astonishing people that ever appeared on earth." If slavery was eliminated, "the spirit of liberty which they [white creoles] would imbibe from their earliest infancy; the understanding and abilities which they would inherit from Europe; ... the rich commerce they would have to form on an immense cultivation; ... and the maxims, laws, and manners they would have to establish on the principles of reason: ... would, perhaps, make of an equivocal and miscellaneous race of people, the most flourishing nation that philosophy and humanity could wish for the happiness of the world."[83]

What was important in both colonies was that the heights of colonial prosperity were reached at precisely the time when new understandings of racial difference were emerging. British and French thinkers in the 1760s and 1770s started to believe that it would be best if their growing empires were rid of the curse of slavery.[84] Such a prospect was unimaginable for West Indians who knew that their economic survival rested on the backs of enslaved African laborers. Their ambition was to continually increase the slave population to produce even more tropical commodities. The problem was, however, that in planters' "rage to push on their estates," they were "playing with edgetools," as a perceptive observer of Jamaican slavery warned in 1746.[85] It was thus during the Seven Years' War that Jamaican and Saint-Domingue colonists began to see the consequences of living in a society where they had enslaved over 90 percent of the population. While imperial navies battled in 1757 and 1758, Saint-Domingue experienced a frightening epidemic of poisoning attributed to a maroon slave called François Macandal. Two years later Jamaican masters were traumatized, in turn, by the biggest event to occur during the Seven Years' War in the British Caribbean, the large-scale series of slave rebellions in 1760 that go under the name of Tacky's Revolt. The nature of these challenges to slavery in Saint-Domingue and Jamaica are the subject of the next chapter.

CHAPTER 5

Dangerous Internal Enemies

Jamaica and Saint-Domingue were relatively unaffected by the events of the Seven Years' War, at least in comparison to mainland North America, where fighting was fierce and constant for seven years. Nevertheless, the conflict realized the worst fears of Caribbean colonists. Slaves in Jamaica rose up against their masters in 1760 in the most sustained rebellion in the eighteenth-century British West Indies. About the same time, Saint-Domingue experienced one of the most frightening episodes in its history, a wave of poisonings in 1757 and 1758 that authorities eventually attributed to a maroon known as Macandal. Neither of these events is usually placed within the context of the Seven Years' War—historians who write about Tacky's Revolt and about the Macandal poisoning scare seldom note that they occurred during wartime. Yet viewing Tacky's Revolt and the Macandal *affaire* within the wartime context dramatically transforms our understanding of them.

For colonists in both Jamaica and Saint-Domingue, these internal challenges were the most important events of the whole war. They forced both colonies to reevaluate whether they could become European-style settler societies or whether they should accept their own emerging hybridity and admit wealthy free people of color into the local elite. Just as important, these assaults shocked masters out of any complacency they might have had about enslaved peoples' attitudes toward them. Colonists in both societies witnessed how African-derived religions could be a source of political and cultural power for enslaved people. Yet the two colonies dealt differently with black religion after the shocks of 1757 and 1760. Both legal systems criminalized black spirituality in order to counter the power of enslaved spiritual leaders. In Saint-Domingue, however, these laws were less important than colonists' response, which was to claim untrammeled authority to investigate and discipline their slaves. The ambiguous nature of the poison threat and planters' violent responses on their private estates raised deep questions about whether masters' cruelty was in fact causing slave unrest. This issue remained highly controversial for the next thirty years, up to beginning of the Haitian Revolution.

White West Indians had long known that the enslaved African majority could drive them from the islands. Jamaica and Saint-Domingue each had a history of slave conspiracies and violent attacks by maroon bands. Michael Craton counts nine episodes of organized slave resistance before 1760 in Jamaica, while Jean Fouchard lists more than seventeen similar events in French Saint-Domingue before 1757.[1] Moreover, Jamaica's first Maroon War (1733–39) was well known in French colonial circles.[2] Colonists understood that importing thousands of new Africans each year increased the risk of revolt. In 1740 a well-informed commentator on Jamaican slavery foresaw destruction because of "the too great Number of Negroes in Proportion to white Persons."[3]

Nevertheless, there was no slave rebellion of any consequence in either colony in the first half of the eighteenth century. Thus, the two wartime episodes—the Macandal poisoning episode in 1757–58 and Tacky's Revolt in 1760—were the greatest threat either colony had ever faced from rebellious slaves. Tacky's Rebellion caused over five hundred deaths, making it the most violent slave revolt to occur in either the French or British West Indies before the Haitian Revolution of 1791. In Saint-Domingue, authorities ascribed over six thousand deaths to Macandal's poisons.

In this chapter we devote more attention to Macandal than we do to Tacky. This may seem perverse, as Tacky was undoubtedly a more serious shock to the British imperial system than Macandal was to the French imperial system. Yet we have less to add to what historians have already written about why Tacky's Revolt developed than we do about the Macandal affair.[4] The most conspicuous feature of Tacky's Revolt was how it developed in utmost secrecy. Authorities in Jamaica, unlike their counterparts in Saint-Domingue, appear not to have made any serious investigation into the causes of the revolt through interrogations of the leaders of the rebellion. It seems that colonial authorities' failure to interrogate rebels was part of a deliberate campaign to show white ruthlessness. John Hamilton, a planter in St. Elizabeth, wrote home to a relative in Scotland that when British troops seized rebels in June 1760, "they killed them with great Slaughter of the Prisoners they took, they hanged up without ceremony or Judge or Jury," thus hoping to "leave a Terror on the Minds of all the other Negroes for the future." [5]

Where we can add new material on Tacky is on the meaning and importance of its repression. The repressive measures used to counter slave rebellion after 1760, and the changes they evoked both in Jamaican politics and in the metropolitan response to white Jamaican brutality, are covered extensively in Chapter 6.

By contrast, we have considerable new material on Macandal. We argue that contemporaries and historians have missed three critical dimensions of the Macandal affair. The first slave poisoner to be interrogated described a secret plan to use magic powders, not toxins, against the whites. His idea was that one by one slaves would use magic to make their owners manumit them until the colony had enough free blacks to stand up to the whites. This confession, however, occurred at a time when thousands of slaves and perhaps hundreds of whites were dying mysteriously during a food crisis caused by a naval blockade at the beginning of the Seven Years' War. Consequently, we argue that the poisoning victims were probably killed by spoiled flour, not by a network of African poisoners. Finally, the deaths and the admission of a "secret among the slaves" occurred at the moment when West Central Africans—a group that colonists broadly described as "Congos"—became a majority of new captives arriving in Saint-Domingue. We believe a key reason Macandal frightened colonists so badly was that his spiritual practices differed radically from those of other accused Africans, by including Christian symbols.[6]

Tacky's Revolt and the Macandal affair illustrate the powerful effects of interimperial war on enslaved people. Both events show that slave rebellion, whether imagined or real, was always at the forefront of white anxiety about life in the tropics. It is important to note, however, that although white colonists were fearful, they were not overcome by their fears. Slave rebellions, they felt, were part and parcel of living in the tropics and if put down with maximum coercive powers could result in little lasting damage to the plantation economies that were vital to continued prosperity.

Let's deal first with the conspiracy that colonists misunderstood. The idea that slaves might use poison to attack masters fed deeply into the colonial imagination, especially in the French West Indies.[7] Saint-Domingue, Guadeloupe, and especially Martinique experienced serious poison scares in the late seventeenth century, and Saint-Domingue executed its first recorded slave poisoner in 1723.[8] The most famous case, however, is that of François Macandal, who was burned at the stake in the central square of Cap Français on 20 January 1758. Although half a dozen men and women had already died in this way in the previous six months, executed as convicted poisoners, colonists believed that Macandal was the mastermind of what they described as "a plot to exterminate all who were not black."[9] For white contemporaries and for the historians who have relied on their written accounts, the Macandal episode was the most serious threat by slaves to the rulers of Saint-Domingue's plantation society before the great slave uprising of August 1791. After his execution Macandal's memory lived on in the *macandals*—small bound-up objects with spiritual power—that slaves continued to make and sell. But the word "macandal" also became synonymous with "poisoner," "sorcerer," or even charismatic rebel, as French colonists used it. During the Haitian Revolution the French general Kerversau described Toussaint Louverture as "a kind of Macandal," suggesting a link between the charismatic black revolutionary and his presumed forebear.[10]

Before he escaped from slavery on the LeNormand de Mézy plantation in Limbé parish, Macandal was known as François. Most contemporaries said he was born in Africa, but his ethnicity has never been clearly established. Using linguistic clues and the notion that he could read and write Arabic, some historians identify him as being from the Islamic parts of West Africa, such as Senegal or Mali.[11] His name, however, which was also applied to the objects he created, likely derived from *makunda/makwanda*, meaning amulet or charm in the Mayombe dialect of West Central Africa.[12] Various pieces of evidence suggest he was from the region colonists broadly identified as the Congo and that modern scholars identify as West Central Africa. He lost his hand in a sugar mill accident and then worked with the plantation's livestock before escaping slavery sometime in the 1740s. Rather than join maroon bands in the mountains, Macandal lived in the plantation hinterland of Cap Français, the same area where he had been enslaved. In January 1758, he was arrested at a *calenda*, or slave dance, in Limbé parish.[13]

This arrest was the culmination of nine months of investigation, torture, and colonial hysteria. It began the previous May, when a slave named Médor from an estate in the mountains outside Fort Dauphin, more than thirty miles east of Limbé, confessed to a series of poisonings. He informed interrogators that there was a "secret that reigned among the blacks," and he named roughly a dozen enslaved and free people who trafficked in "poisons and spells."[14] Administrative correspondence before this date says nothing about mysterious deaths, but as colonists questioned people Médor named they began to see a network dedicated to producing, selling, and using poison.

Then they began to tally the deaths. As one colonist explained, planters had seen "three-quarters of our laborers perish from sicknesses of a cause unknown even to doctors."[15] Further interrogations produced more poisoners, so that by early September 1757 there were twenty-one accused men and women in the Cap Français prison.[16]

In January 1758, therefore, when Macandal was captured, authorities were predisposed to view him as a poisoner at the center of a conspiracy against the colony. They understood that he had extraordinary charisma. On 20 January 1758, Macandal ensured his cultural immortality by breaking away from the stake where he was being burned to death before a large audience of masters and slaves. The executioner and nearby soldiers quickly recaptured him and the execution proceeded.[17] But that brief escape, and Macandal's claim that the whites could never kill him, lived on as legend. It was retold in a pamphlet published in Paris later that year and then again thirty years later in 1787 in a highly romanticized account in the *Mercure de France* that compared Macandal's abilities to those of "ambitious fanatics" like Mohammad or Cromwell."[18]

The most puzzling aspect of the Macandal episode is the death count. Every tally of poisoning victims in 1757 and 1758 reported that livestock and slaves died in far greater numbers than whites. Some contemporaries estimated that the poison conspiracy had killed several dozen whites versus hundreds of slaves, but the colony's top administrators wrote Versailles twice that between six and seven thousand slaves had died, compared to "a very great number of white persons." Although it is possible the panic of the moment led officials to inflate this number, six thousand slaves amounted to 10 percent of the 63,859 enslaved people counted in the 1753 census of the North Province. The same census tallied fifty-nine hundred whites. If a "very great number" meant three hundred to six hundred whites, that would mean that approximately 5 to 10 percent of the white population fell victim to the mysterious poison.[19] Because Saint-Domingue's enslaved workers were poorly nourished and systematically overworked, these numbers suggest a poison that was more environmental than political, affecting slaves and masters roughly equally.

Most colonists shared Moreau de Saint-Méry's later expressed belief that Macandal "in his vast plan had conceived the infernal project to eliminate from the surface of Saint-Domingue all the men who were not black."[20] But there is only one contemporary report of such a conspiracy, an alleged plan to poison the water of the Cap Français garrison in the month before

Macandal's capture. That account is secondhand, and none of the official correspondence about the poisoning issue mentions it.[21] The governor and intendant, while insisting that Macandal was the "chief" of the poisoners, did not believe there was a conspiracy: "They are led only by the hope of promised liberty, by jealousy, resentment, and most often still by the simple desire to do harm." They noted that whites feared a plot, but they could not explain how a possible slave plot would lead to the deaths of "more than 6,000 slaves dead by poison at Cap Français and in the surrounding area in about three years."[22]

While many eighteenth-century colonists accepted virtually unprovable hypotheses to explain the deaths they attributed to Macandal, another theory may be more convincing. It is probable that when slaves admitted to poisoning their masters and fellow slaves, they were referring to spiritual practices that they believed would shape the actions and attitudes of others. Slaves may have understood the word "poison" to refer to spiritual and medicinal objects, and after seeing whites and blacks die mysteriously, they may have believed that they killed them. Moreover, the chronology of poisoning cases shows a progression of deaths, or at least arrests of poisoners, along the road that led from the smuggling center of Monte Cristo to Cap Français. This chronology in turn suggests that it was not Macandal who was killing thousands of laborers, but bad food, accidentally poisoned through spoilage during the blockades of the Seven Years' War.

The confession of the slave Médor, which started the panic, is the sole interrogation text that was not elicited by torture. In May 1757 Médor confessed that he had poisoned his master Delavaud, Delavaud's family, and slaves on the plantation.[23] He noted that some of the twenty slaves he poisoned were dead, as were livestock and poultry. It is not clear if this confession was a spontaneous act on his part or whether his master had accused him of killing plantation residents. But we do know he had not been recently tortured. Médor's interrogators, four neighboring planters, described his body as scarred by whipping, but these wounds were healed over, not fresh. Médor's testimony planted the idea of a slave poisoning conspiracy that ultimately led to the execution of Macandal and dozens of other poisoners.

He started by confessing that his poisons had caused the recent deaths and by naming some other local poisoners. But he told them that "if he identified all the slave poisoners and malefactors he would never finish, since they were found on all the plantations." Then he described an alarming "secret" among Saint-Domingue's enslaved and free people of color:

He only knows one way to stop the course of these poisons and spells.... To not promise freedom to nègres, négresses and mulattos and that it is only to get this sooner that they poisoned several whites and that this is what made them undertake all their artifices [*stratagèmes*] in order to be able to dress like the whites by the bad dealings that they have when they are free and that there is also a secret among them which is designed to finish the colony, of which the whites know nothing and the free blacks are the principal cause, taking these actions to increase their numbers in order to be able to stand up to the whites if necessary.[24]

Médor's words made it clear how he thought that those who shared this secret would "finish the colony": by increasing the free population of color to the point where its members could stand up to the whites. French colonial authorities quickly became convinced that slaves had a powerful toxin, but it seems more likely that Médor's poison was a magic substance administered to heal or strengthen a slave's relationship with his or her master.[25]

Médor's confession and the death of slaves on the plantation so alarmed Delavaud and his neighbors that they planned to send him to the local court in nearby Fort Dauphin. But on 29 May 1757, Médor's master found him dead in the room where he was being held, with a knife in his ribs. He described this event as suicide, arguing that Médor was upset about being sent to Fort Dauphin, where an exemplary public execution probably awaited him. But his death might have also been murder, as he was shackled and isolated in a room, after naming nearly a dozen fellow poisoners. We do not know why Médor began confessing; his suicide suggests that he initially hoped telling the whites everything that he knew would save him from being tortured to death. Delavaud said Médor wanted to "salve his conscience and prevent the death of innocents," which suggests that he blamed his poisons for the deaths of fellow slaves. We cannot rule out that Médor had finally found a toxin powerful enough to achieve his stated goal of killing his master and that he was discovered after successfully testing it on fellow slaves. It seems more likely, however, that when twenty slaves on Médor's plantation sickened and some began to die, he interpreted these deaths and illnesses as the result of his magic powders. Convinced that his poisons were dangerous, Médor described to his interrogators his two decades of buying and administering magic substances, and by doing this exposed the secret community of poisoners.

As Diana Paton has pointed out, French Caribbean planters were far more likely than their counterparts in Jamaica to associate slave religion with poison.[26] To an extent, French Caribbean concerns with poisoning reflected anxieties held also in metropolitan France. The highly publicized Affair of the Poisons (1676–82) at the court of Louis XIV was not only about toxic materials but also about witchcraft and sacrilege. It ended with the arrest and execution of dozens of persons, including forty-two priests.[27] The legal result was a 1682 French royal edict that differentiated poisoning from sacrilege, and levied specific punishments for each crime. French colonial planters did not immediately adopt this metropolitan law. In May 1720 the Martinique Council sentenced three slave poisoners to be burned at the stake, and in August it passed its own decree establishing this penalty for slave poisoners.[28] Meanwhile, on 4 June 1723, Saint-Domingue executed Colas Jambes Coupées, a maroon convicted of various crimes, including poisoning several slaves.[29] It was only in 1738, however, that Saint-Domingue's high courts registered the 1682 poisoning statute. Around the same time, in 1739, the colony expanded its *maréchaussée* or slave police.[30] In 1748 judges' concern with poisoning laws reached a new level, near the end of the War of Austrian Succession. In that year Saint-Domingue adopted a 1746 law from Martinique that modified the 1682 law for colonial use, prohibiting slaves from trying to cure any illness except snakebite, and outlawing any laboratory that was not operated by a registered doctor or apothecary.[31]

This legislative history shows that Martiniquan planters worried more about slave poisoning in the first half of the century than did their counterparts in Saint-Domingue. The intendant of the larger colony remembered only three accusations of poisoning in the seven years before Macandal's arrest, less than Martinique had experienced in the same time.[32] The events associated with Macandal, however, fundamentally changed attitudes in Saint-Domingue. In a 1762 memorandum the royal attorney of Fort Dauphin reported, "It was only in 1757 that the courts of the North Province of Saint-Domingue were convinced of the hold the slaves had on the whites."[33] Sebastien Jacques Courtin, the royal judge who interrogated Macandal, echoed his colleague's conclusion. But he went further, stating in the opening of his memorandum about Macandal, that African-style religious practices were a central part of the problem. It is worth citing at length:

> For a long time, we have attributed all the long illnesses that have killed so many whites and nègres either to Saint-Domingue's climate

or to scurvy. In the same way for a long time we have scorned all the superstitions of the nègres, their talismans, their idols and the rest that the whites regarded as a consequence of their unthinking devotion to paganism. In the past year discoveries have been made which prove conclusively that slow-acting poisons are the true cause of nearly all the languishing and almost incurable illnesses that attack whites and nègres. And, at the beginning of this year, the trial of François Macandal and his accomplices has proven that all nègres, in their superstitious practices, successively pass to all crimes, a progression that is so well explained in the edict of July 1682 and consequently, instead of ignoring their so-called superstitions, we must take every measure to stop them.[34]

Colonists stopped ignoring slaves' "superstitions" in 1757 for two reasons. One was the sheer number of the deaths they witnessed, which we suggest below was due to a food crisis but which they connected via Médor's confession to a poisoning conspiracy. The second was that Macandal exposed them to a different kind of African religion, one that used items that masters as well as slaves believed had spiritual power. Courtin's notes on Macandal's interrogation illustrate how different the maroon's talismans were from Médor's powders and potions. Tellingly, Courtin called Macandal's objects *gris*, a West African word for small bags containing charms, including pieces of paper with magic symbols.[35] His use of that term suggests he had never before encountered a charm like a macandal: a finger-size bundle of human bones, nails, roots, communion bread, and consecrated incense, all of it wrapped in cloth and twine, and soaked in holy water. These macandals sometimes included a small lead crucifix, like the ones Jesuit missionaries passed out to Saint-Domingue's slaves. Courtin described how he "saw, horrified, a crucifix entombed in the coating and the bonds of this loathsome composition, in which only the end of the arms and the head, with the crossbar of the cross, were visible." This reaction calls to mind Moreau de Saint-Méry's more metaphorical description thirty years later of the "monstrous assemblage" produced when Congos and other West Central Africans in Saint-Domingue mixed their ideas of Catholicism with "idolatry."[36] The wrapping of Christian elements with cords evokes the seventeenth- and eighteenth-century Congolese brass crucifixes that are known as *nkangi kiditu*. Hein Vanhee suggests that this name means "the bound Christ." Surviving examples of these crucifixes do not show cords, but the act of binding and tying was spiritually significant in West Central Africa. It was an important part of many Kongo *minkisi* (singular *nkisi*), ritual objects used for healing, protection, and divination.[37] Even today Haitian Vodou altars contain bundles wrapped in cord, some of them with crucifixes, known as *paquets kongo*.[38]

Courtin and his contemporaries in the 1750s apparently did not know that Portuguese missionaries had introduced Catholicism into the Congo River basin centuries earlier.[39] When Courtin forced Macandal to reveal the sacred words he spoke while making his bundles, the judge heard the words "alla, alla," and described the incantation as being in the "idiom of the Turks," a claim that some historians have used to argue that Macandal was from Islamic West Africa. But Macandal claimed the phrase invoked the blessing of God and Jesus Christ, which makes sense if he was from the Congo. Two of his followers gave what may have been a more literal translation: "God who knows what I do; God give me what I request."[40] Most of the rest of the incantation was in the creole language developed in the colony by slaves as a fusion of French and African languages, but Courtin reported a mysterious word "Ouaie" the meaning of which, witnesses said, was only known to the sorcerers. David Geggus points out that *waáyi* means "slavery" in Kongo.[41]

One reason colonists in 1757 began to feel vulnerable for the first time to slaves' magic—or poison—was that enslaved people from the Congo region, with their "monstrous" mix of Christian and African symbols, were on the point of becoming the new majority in Saint-Domingue. In the 1740s and 1750s the Congo region began to replace the West African slave ports in the Bight of Benin as the prime zone for captives arriving in Saint-Domingue. The Benin area produced 45 percent of the slaves embarked for Saint-Domingue in the years 1713 to 1742, while only 23 percent came from West Central Africa, or the Congo. In the period from 1743 to 1757, however, Benin's share of captives declined to 33 percent while those from West Central Africa rose to 43 percent. For the rest of the century, captives from this latter region dominated Saint-Domingue's slave imports.[42] Macandal therefore represented a new cultural movement within the black population, at least as far as whites were concerned.

In fact, Macandal's spiritual secrets were so different from those of other Africans like Médor[43] that Courtin could not prove that his captive was a poisoner in the toxicological sense. He had no evidence that Macandal distributed powders, just charms. When other slaves testified about these macandals, they claimed the talismans could not cause death or even illness, only good or bad luck. Courtin instead accused Macandal of mocking true religion. He relied on the arguments French authorities had wielded in 1682 to condemn the European folk magic traditions associated with poisoning in order to explain why Macandal's use of sacred Christian objects threatened the colony.[44] He argued, "A slave wearing a *macandal* . . . has only to take a single step to become [a poisoner]. Accustomed to desecration and impiety, and to cold-bloodedly doing evil to others by occult methods, if this practice does not succeed to his liking, he has no qualms about turning to poison, and this progression is almost necessary."[45] Unable to prove that Macandal possessed deadly toxins, Courtin instead condemned him for poisoning the minds of his followers, preparing them for future crimes.

We believe that Macandal was a charismatic leader in a Congo-influenced tradition, part of a fundamental shift in the black culture of Saint-Domingue. He was a *nganga* in the Congo tradition or a healer, diviner, or judge who made minkisi objects out of spiritually charged materials and helped nonspecialists use them. Colonists were correct to recognize Macandal as an important figure, not because he was a poisoner but because his activities marked the moment when captives from West Central Africa began to dominate the ethnic composition of the colony's enslaved population and when their cultural practices started to become more important in the colony. Macandal aroused fear in his captors because his spiritually charged bundles contained "their" symbols, incorporating Christian iconography infused with African meanings.

Whites were determined to see what Macandal was doing within the prism of poisoning—an activity they were convinced slaves practiced all the time against whites. In 1757, the prosecutor in Fort Dauphin, near Médor's plantation, reflected on the history of slave poisoning in Saint-Domingue: "We must consider as an established principle what people have long said, that in these lands the blacks poison each other, since in different eras there have been immense losses on many former plantations whose owners have been ruined."[46] But other explanations for the "poisoning" episode of 1757–58 are more convincing than those advanced by planters. What masters perceived as "poisoning" may have been slaves using spiritual techniques to survive famine conditions. Many of the deaths attributed to poison were probably food related. Macandal may also have been more a religious mendicant than a rebellious poisoner. The deaths attributed to him can be more easily explained through problems brought on by wartime blockade than as slaves' deliberate design to bring the plantation system to its knees.

Even in times of peace and good weather, Saint-Domingue's workers lived near the edge of starvation. In 1733 Pierre-François Charlevoix wrote that "many masters do not feed their slaves and settle for giving them some respite [from work] to look for or earn their living; but as much as we look into this, we have not been able to discover what they live on in these cases." Colonial culture taught that this cost cutting did not affect workers' productivity. Charlevoix noted, "They say that what is barely enough for a white man's meal would feed a slave for three days; . . . but however little they eat or sleep, they are just as strong and hardworking."[47] While heavily worked slaves fed themselves by scavenging for food in their free time, a drought or bad storm easily produced famine. In 1737 after hurricanes devastated the southern peninsula colonial authorities permitted foreign merchants to deliver food on an emergency basis; in 1737 Versailles allowed French merchants to ship salted beef and other comestibles from Ireland to Saint-Domingue.[48] There is no evidence explicitly connecting this famine period to slave poisoning, but it was in 1738 that Saint-Domingue formally adopted Louis XIV's 1682 poisoning law.

Similarly, the coincidence of the Macandal episode and the Seven Years'

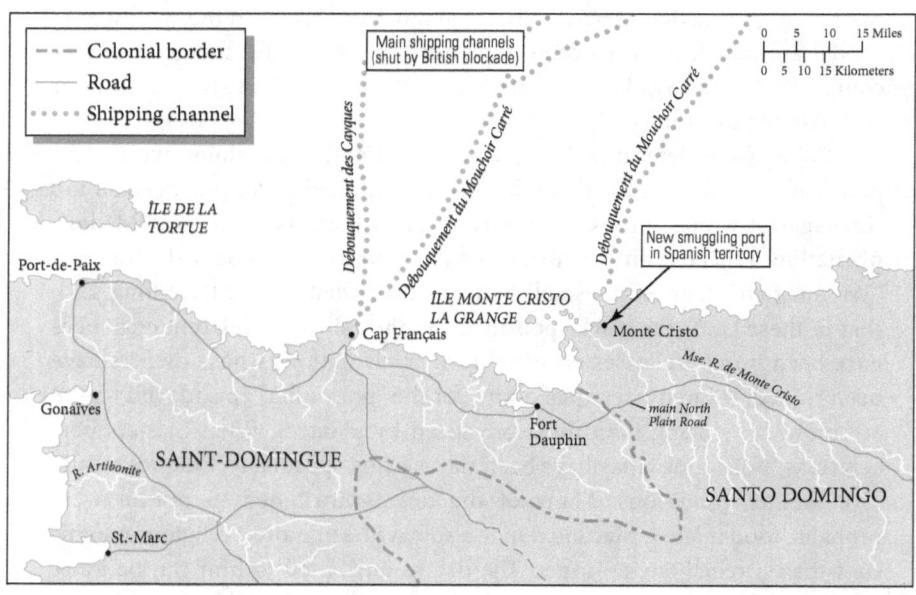

Map 2. The Blockade of Cap Français, 1755–58

War has largely escaped historians' notice. But war and famine went together in the Greater Antilles. Saint-Domingue was badly affected by blockade during the Seven Years' War. By 1757 a British naval blockade had virtually eliminated trade between Saint-Domingue and France.[49] On 5 January 1758, shortly before Macandal's arrest, the intendant wrote, "All our landings [*atterrages*] are covered by cruisers. They form a chain that covers the entire island, and we have never before been held as closely as we are now."[50] (see Map 2) On 15 July 1758, Lambert, the colony's acting intendant, wrote Versailles that the British had turned away eight to eleven Dutch ships carrying food to Saint-Domingue and forced them to sell their cargoes in Jamaica.[51] He pointed out that a few small-scale smugglers had been able to penetrate the blockade. Describing events that occurred months earlier, he explained to Versailles that the colony was "on the verge of a complete shortage, although a decision to "allow small neutral boats from neighboring islands to carry foodstuffs to our ports" had brought some relief.[52]

Smuggling was a critical element of the French West Indian economy, particularly in wartime. During the War of Austrian Succession (1740–48),

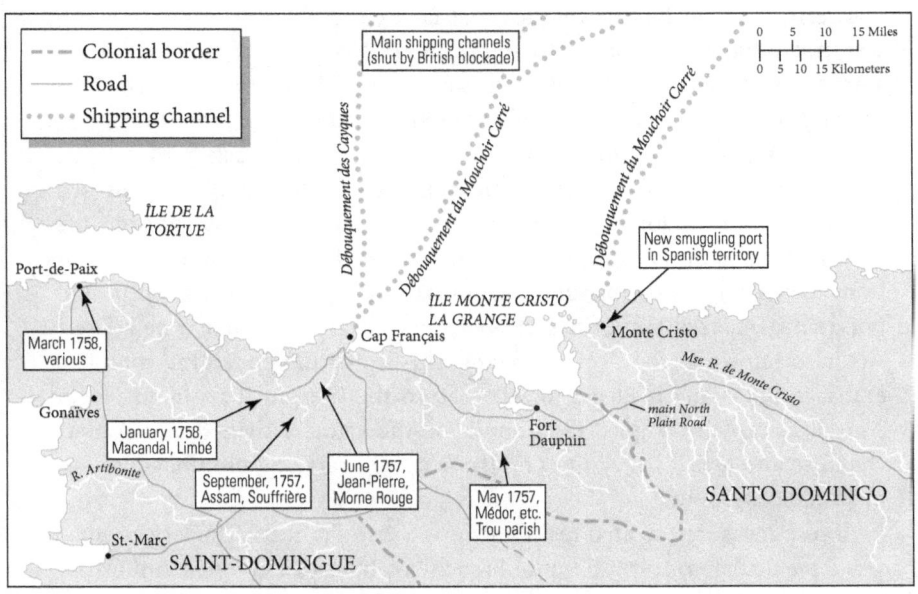

Map 3. Progression of Poisoning Arrests, Saint-Domingue, 1757–58

merchants in New York, Boston, Philadelphia, and Newport developed a number of techniques that allowed them to trade food and other supplies for French sugar, molasses, and indigo. These included transferring goods to French commercial agents in Dutch Saint Eustatius or in Saint Thomas and Saint Croix in the Danish Virgin Islands. In 1755, as war became inevitable, Dutch merchants in New York City shipped thousands of casks of flour to these colonies. By December 1755, French traders in Saint Eustatius were already reporting delays in these shipments.[53]

Such inconveniences grew far more serious in 1756 and 1757 after war was formally declared. The situation in Cap Français became especially dire, for it was more dependent on transatlantic shipping than were Saint-Domingue's three other major ports. In 1767, after the war, Cap Français received 195 ships from Europe and twenty-six from the Americas, while ports along the southern peninsula received forty-nine ships from Europe and twenty-three from the Americas.[54] When the British navy cut off European shipping to Saint-Domingue, the colony's North Province had to turn to smugglers to replace its usual French suppliers.

The British supplemented their naval blockade, however, with commercial restrictions in North America that far exceeded those of the previous war. In December 1756, imperial authorities in New York began to enforce a wartime embargo against trade with the French islands requiring all ships bound for foreign ports to post bonds of £1,000 to £2,000. From the beginning of 1757, New York merchants reported that it was much harder to get their cargoes to the Dutch and Danish islands. In March 1757, about two months before Médor began spontaneously confessing about poison, the British stopped all North American ships from sailing to foreign destinations. On 11 July, as the prison in Cap Français began to fill with suspected poisoners, imperial authorities in New York began to enforce a new Flour Act, also known as the Provisions Act. This legislation ended the complete embargo but made it illegal to sell food to the French. It required North American captains to post high bonds on the value of their declared merchandise and levied heavy fines on those caught carrying prohibited goods, most critically, flour.[55]

These measures created real hardship in Cap Français. The lack of flour was a particular concern, because French law obligated colonial authorities to supply soldiers with imported wheat flour.[56] Like soldiers, Frenchmen new to the Antilles disdained yams, plantains, cassava bread, and millet. Bread was a staple of white colonists' diet, and the city of Cap Français had twenty-five bakeries, which needed seventy barrels of flour per day, around twelve thousand pounds.[57] Some colonists believed that local provision crops would make them sick, and European food was also an important status marker. In 1764 Saint-Domingue's Councils described how "there were a great number among us who during the entire war subsisted on damaging provisions. Skin color alone distinguished them from their slaves."[58]

The colony's enslaved people did not have the luxury of choosing their food. After living in Saint-Domingue from 1769 to 1773, the Dominican priest Jean-Barthélemy Nicolson described slaves' nutritional state in dire terms. He noted, "Nothing is more terrible than their situation. . . . Their food is the same as that given to the most unclean animal, and yet they almost never have as much as they want. I am not exaggerating here; I know there are masters who take a little better care of their slaves than of their livestock, [providing them] with hardtack when ground provisions lack, . . . but these are the best masters, whose numbers are very small."[59] With whites reduced to eating slave provisions during the war because of the blockade, slaves used to eating the lowest quality food might well have scooped up the staples that whites discarded.

In fact, because of the shortage of imported provisions, even royal officials sold spoiled food on the open market. In November 1758, four months after Lambert described how important smuggling had become for Cap Français, a French naval squadron arrived in the port after running the blockade. It was carrying wine, hardtack, and dried peas, but during the voyage a leak had ruined these sorely needed provisions. They were too spoiled to give to soldiers and were mouldering in royal warehouses. Lambert admitted that he was planning to sell this bad food in the markets of Cap Français, hoping the funds would allow him to buy better supplies from smugglers.[60]

Contemporaries recognized that poisoning deaths occurred exclusively in the North Province. They failed, however, to make any connection between these deaths and the hastily improvised smuggling networks straining to supply this region with flour and other goods. The royal attorney of Fort Dauphin, thirty miles east of Cap Français on the island's north coast, used regional differences to explain the alleged poisoning cases, but he also made a moral argument. He believed that planters in the West and South Provinces were more likely than his neighbors in the North to have "criminal and immoral intrigues" with enslaved women, who therefore protected them from poison.[61] The more plausible conclusion, however, is that they were saved by their taste for local food, and by their access to high quality smuggled flour.

Royal officials in charge of the grain trade had long recognized the dangers of spoiled flour. So too did French consumers, who often rioted when royal grain depots delivered moldy grain. Over the course of the eighteenth century, as the inter-European food trade increased, grain spoilage became a prominent problem. It was especially likely to occur when ships were used to transport food; on longer voyages and in smaller vessels, flour was exposed to more of the heat and moisture that nourished mold spores. France had the most problems with grain arriving from northern European ports like Amsterdam and Danzig. Northern grain fared especially poorly at sea because, unlike harvests shipped from Spain or Italy, it was not well dried in the fields.[62] To provide bread for overseas troops, in the 1730s the French naval ministry began to invest in expensive kilns used to dry Caribbean-bound grain. They had flour milled in a special way in Poitiers, La Rochelle, and Bordeaux to prepare it for transatlantic transport. They hired gangs of workers to turn it regularly with shovels in order to dry it and packed it in special watertight casks. The British naval blockades that began in 1756 would have stopped most shipments of this specially prepared French flour.[63]

Although contemporaries did not understand it in these terms, they were taking measures to prevent microscopic fungi from blooming in the flour and producing the deadly alkaloid compounds known as mycotoxins. It was not until 1783 that the French medical press reviewed a German physician's study of a 1770 epidemic of ergotism, which he determined to have been caused by moldy rye flour.[64] Over two centuries later scientists are still discovering new varieties of the microscopic fungi that grow in damp flour. Some of these produce alkaloid compounds known as mycotoxins, which cause a wide range of symptoms in humans and animals. Depending on the fungal species, the type of grain, and the quantity ingested, myotoxicity can kill either quickly or slowly, much as Macandal's poison was reputed to do. Slow, painful deaths occur from the poison, as victims' immune systems are weakened, or as their bone marrow and internal organs become damaged.[65]

In the 1970s, as scientists revealed more of the characteristics of these natural poisons, some historians suggested that the Salem, Massachusetts, witchcraft cases of 1692 were caused by ergot, a hallucinogenic compound produced in moldy rye grain.[66] Historians now think that there are excellent reasons why ergotism is not a good match with events in Salem.[67] And because it is so difficult to identify poisons through documentary records,[68] the most reliable examples of mycotoxins are all from recent history. In the twentieth century, their effects have been particularly deadly when wars have delayed harvesting or disrupted food transport. During World War II, the trichothecenes in moldy millet, wheat, and barley may have killed 10 percent of the population of Siberia. In 1993 in southern Tajikistan, bread made from spoiled flour killed two dozen people and badly sickened thousands, amid a regional war.[69] The ideal propagation temperature for these toxic species in Europe and Asia has been identified at around 8 to 14 degrees Celsius (46 to 57 Fahrenheit), too cold for Caribbean ports. Reports from Africa in the 1980s, however, identified mycotoxins growing in wet oats and barley that produced significant amounts of poison at 25 degrees Celsius (77 Fahrenheit).[70]

It is clear that in 1756, 1757, and 1758, the Atlantic blockade and new restrictions on North American trade dramatically reduced the quantity and quality of flour available in Saint-Domingue's North Province. High demand in the French colony led Danish or Dutch smugglers in Saint Thomas or on other small islands to scrape together whatever flour they had and to send it to Saint-Domingue. Grain from Northern Europe was often poorly dried and arrived already moldy in French European ports. Once grain reached the Caribbean, smugglers had to use small vessels, especially vulnerable to grain spoilage, because the British were blocking all large ships from reaching Saint-Domingue. These craft flocked to the coastal hamlet of Monte Cristo on the north coast of Spanish Santo Domingo (see Map 2).[71] From 1756 to 1763, this fishing village just across the border from Saint-Domingue became the source of much of the imported food available in the North Province.

Not accidentally, the arrests and interrogations of slave poisoners followed a definite geographical progression from the Spanish border toward the west (see Map 3). In May 1757 planters living in the districts around Fort Dauphin, like Médor's owner Delavaud, would have gotten their flour from the nearby Monte Cristo, and this smuggled food would have reached them long before it got to Cap Français.[72] It is likely that poisonous flour killed the enslaved people on the estate where Médor lived. The month after his remorseful confession, planters began to arrest slaves in the Plaine du Nord parish, where the planter Delaye testified that his Congo slave Jean-Pierre had killed twenty-six slaves on his plantation.[73]

There were more accusations of poisoning in the Fort Dauphin area in

September 1757, but panic about unexplained deaths spread farther west.[74] In the Souffrière section of Acul parish, in September 1757, Assam, an African woman of the Poulard nation and a former domestic slave, testified that in her attempt to cure sick slaves on the Vallet plantation she had purchased and administered substances that in fact killed them. In October 1757 authorities arrested slaves in Petit Anse, just east of Cap Français, including Hauron, a Yoruba man identified as having supplied poisons to other slaves.[75] In December 1757, authorities captured Macandal in Limbé just west of Acul and Cap Français. Finally, in late March 1758, two months after Macandal's execution, suspected poisoning began in the Port de Paix district, over sixty miles west of Cap Français. After two weeks, Port-de-Paix had only thirteen slaves in prison, but another twenty-two had been denounced, and officials believed that their district might have as many poisoners in Port-de-Paix as in Cap Français, where two hundred alleged followers of Macandal were crowded into the prison at the same time.[76]

The trail of arrests from the Santo Domingo border west to Port-de-Paix reflects the movement of smuggled flour, along the roads and trails that linked Monte Cristo to Cap Français. Most versions of the Macandal legend describe the master poisoner as directing a legion of accomplices who could deliver his concoctions to any plantation. In 1787, the *Mercure de France* identified peddlers as the heart of this network. In fact, in 1757 and 1758 free blacks were particularly numerous as operators of small coastal craft that carried goods from Monte Cristo into French waters.[77] Because of the famine, slaves were more active than ever in buying and selling food informally. The practice was prevalent enough that in 1762, the police judge of Cap Français prohibited merchants or ship captains from selling flour to free people of color. People of color could buy bread but only if they had a note from a white employer. A month later the law was reiterated, now with a stiff fine for masters and the threat of corporal punishment for slaves.[78]

One likely problem with free and enslaved black flour peddlers was that they did not know if they were buying healthy or unhealthy flour. An incident in Port-au-Prince in January 1766 highlighted the different perceptions of whites and blacks in regard to flour. Officials and citizens investigating the cause of some sixty recent suspicious deaths visited warehouses at the port to inspect incoming flour. Moreau de Saint-Méry said that they found sixty out of 5,744 barrels that were damaged, which they immediately threw into the sea. The only merchant to come forward and inform officials that his flour was spoiled was Lambert, a free black, who owned seven barrels.[79] This unusual openness suggests that Lambert may have purchased the flour not knowing it was potentially poisonous, while white merchants tried to conceal their barrels, fearing that officials would destroy them. Smugglers selling spoiled flour in wartime Santo Domingo were under no such scrutiny and were probably even less willing to discard damaged barrels. Moreover, enslaved or free black merchants like Lambert might have retrieved discarded provisions, unaware of their danger, especially retrieving wheat flour because it was both expensive and "an unmistakable marker of [elevated] status."[80]

As we have seen, the mysterious slave deaths did not end with Macandal's execution in January 1758. They continued to occur for almost two more years, finally ending in December 1759, when colonial administrators wrote that the poisoning "conspiracy" no longer threatened Saint-Domingue.[81] Not coincidentally, by that time trade had resumed between the North Province and its preferred source of smuggled food: North America. North American flour was far less likely to be spoiled than the flour supplied by improvised trade routes for a number of reasons. It was more likely than European flour to be milled from field-dried grain, it came in larger ships, and it spent less time at sea. It arrived in large quantities in 1759 and 1760, driving down the price of imported provisions generally, which eliminated people's willingness to buy and consume spoiled food.

By 1759, for example, New York merchants began to export their flour from nearby ports like New Haven, New London, or Perth Amboy, where customs enforcement was less rigorous. Mainland captains learned to sail to Monte Cristo, where they could observe the letter of the Flour Acts by selling to Spanish middlemen who would transport their wares to Saint-Domingue. By 1759 the commander of the Jamaican naval squadron estimated that two hundred smugglers passed through Monte Cristo every year, many of them from North America. Flour from newly constructed mills in the Delaware Valley was now streaming into Saint-Domingue, to great salutary effect. On 2 February 1759 a South Carolinian merchant wrote, "We are well informed that the French, who were during the late embargo almost starving throughout Hispaniola, are now plentifully supplied." By February 1760, the British vice-admiral Thomas Cotes wrote that "flour and all sorts of provisions is 50 percent cheaper at Hispaniola than at Jamaica and money plenty, all occasioned by the trade to Monte Cristi and flags of truce."[82]

The Macandal affair, whatever its underlying causes, triggered important changes in Saint-Domingue because whites believed the poisoning was deliberate. In May 1763, after the end of the Seven Years' War, Saint-Domingue's intendant reminded Versailles of the recent poisoning episode, noting that "if its effects are less frequent today, they are even more dangerous, in that they are more hidden." In September 1787, the *Mercure de France* noted that "for about 25 years the island of Saint-Domingue has shuddered at even the name Macandal."[83] The poisoning scare led to new laws, described below, that restricted the religious and medical practices of slaves and free people of color as well as their access to European medicines and poisons like arsenic.

One of the most significant results of the Macandal affair was that judges and administrators ceded nearly complete control of slave interrogations and punishments to individual planters. Such cession of power was highly significant, as it encouraged planters to assume that it was them, not colonial officials, who had the power to punish slaves as they wished. When, as in the 1780s, imperial officials wished to reassert some of the controls they had given away during the Seven Years' War, they met strong resistance. During the height of the poisoning scare in 1757–58, royal officials punished convicted poisoners in public, hoping to send a message to other poisoners. In December 1757, for example, the intendant Laporte-Lalanne informed Versailles, "whatever the case, [public] examples are the first remedies to use, and we are not sparing them."[84] Six months later, his successor, Lambert, explained that burning alive the principal leaders of the poisoning "will not fail to make a great impression on the others, and there is room to hope that we will be able to stop the progress of this awful crime."[85] The royal attorney of Fort Dauphin argued, however, that these public immolations had little effect. He advocated another terrible public display. Authorities would mutilate poisoners' noses and ears "according to the degree of their crimes" and sentence them to a life of labor on the local chain gang, the men separate from women. These grotesque beings would repair roads and flood works in full view of the local population, their paradoxical social isolation reminding everyone of their crimes against humanity.[86]

As the number of deaths began to taper off, officials backed away from this emphasis on public punishment. By the 1760s, they had informally ceded planters the right to investigate and punish slave poisoners. One explanation for this new policy came from the new intendant, Jean-Etienne Bernard de Clugny, the first parlementary judge to be appointed to a high administrative position in Saint-Domingue. Clugny was well known for advocating the rights of the colonial judiciary.[87] Yet even he noted, in 1763, that "poisoners take precautions that often prevent one from acquiring sufficient proof to condemn them in a court of law. But masters sometimes, with good reason, have powerful suspicions about who is guilty, and, in this case [of a failed legal prosecution] they find themselves hindered."[88]

Saint-Domingue's authorities did relatively little legally to counter the possibility of another poisoning "plot." In March 1758, two months after Macandal's botched execution, the Cap Français Council outlawed the wearing of macandal charms and at the same time made it illegal for slaves to sell or make remedies.[89] In April, they reissued a compilation of existing slave laws, reminding all masters that slaves were not permitted to assemble, play drums, or hold religious ceremonies, even in churches. The compilation forbade slaves from peddling goods from estate to estate and required all free blacks and mulattos to register their freedom papers with the courts.[90] In July, the court fined a plantation manager for allowing a slave calenda, or dance, on an estate that had been named in Médor's testimony.[91] In April 1761, the Cap Français Council had to renew the prohibition about slaves assembling in churches at night. The law mentioned how during the recent Macandal affair "a gross superstition led slaves often to mix the holy things of our religion with the profane objects of an idolatrous cult."[92] Yet for a plantation colony that had lost six thousand workers, supposedly to a conspiracy mounted by a sacrilegious poisoner, imperial administrators and judges did relatively little to protect themselves against the possible danger of slave religion in the 1760s or 1770s.[93]

Instead, in this colony where planters felt the state had too much power over them, officials found it easier to trust planters to punish their own slaves, as had been the custom and was now increasingly the norm. One consequence of this policy, however, was the state's inability to prevent masters' torturing their workers. On 28 May 1771, a dozen slaves went to the judge of Cap Français to complain that their master, the youngest son of the wealthy Dessources planter family, was torturing them. They testified that when "a slave who was being tortured with fire accused the entire workforce of being macandal or poisoners, [their master] had Janneton, the commander, burned [to death], then Azard, and that he had similarly had the feet and legs of the slaves Fanchette and Jeannette burned and that he had then had them buried alive. And that he had burned the feet and legs of a pregnant slave and after that put her in a dungeon [cachot]."[94]

It is very difficult to judge the effect of the Macandal affair on enslaved people. It does seem clear, however, that colonists' hysteria "proved" that the powders and spells of free and enslaved spiritual leaders had caused thousands of mysterious deaths. Because so many of the victims were slaves, not everyone in bondage welcomed this poisoning power. Slave informants helped authorities capture Macandal and other "sorcerers." Moreau de Saint-Méry reported that "Macandal" was one of the worst insults a slave could make against another slave.[95] No new Macandal ever emerged in Saint-Domingue after 1758, but the fusion of European and African folk beliefs and religions that he represented continued, as we explore in Chapter 10. Not only did tens of thousands of fresh African captives constantly reestablish African spiritual practices in the colony, but the expulsion of the Jesuit order in 1763, in part because of its sympathy for accused poisoners in the Macandal affair, removed the colony's most committed missionaries and the most fervent advocates of slave Christianization.[96]

It is as impossible to prove that Macandal's "secret" was spiritual, not toxicological, as it is to establish definitively why thousands of blacks and whites died in 1757 and 1758. Yet it is not implausible that the repression and hysteria of the late 1750s offered enslaved people a glimpse of how spirituality could unite them against the white population, as happened in 1791. As Pierre Pluchon points out, "the poisoning scare amounted to an agreement by all the racial groups that these mysterious sorcerers/poisoners were effective."[97] And after 1758, everyone in Saint-Domingue knew the rebel mythology of Macandal.[98]

Tacky's Revolt was of a different order of seriousness from the Macandal poisoning scare. Although far fewer people died in Jamaica than in Saint-Domingue, there was no controversy about the fact that it threatened the very survival of white rule on the island. It started with acts of great ferocity on 7 April 1760 and was not fully suppressed by chastened whites until nearly a year later. It was, Michael Craton asserts, the most severe challenge to Britain by nonwhite colonial subjects until India's Sepoy Rebellion of 1857. It ranks alongside the 1641 rebellion in Ireland as the most severe wartime challenge before the twentieth century to British imperial rule by internal as opposed to external enemies. Its seriousness can be measured in two ways. First, it was very destructive of people and property. Perhaps sixty whites, sixty free blacks, and four hundred enslaved persons were killed in action or committed suicide, and a further one hundred slaves were executed. At least five hundred slaves were exiled to the Bay of Honduras, a common dumping ground for black criminals. The revolt caused as much as £100,000 or 2.1

million livres in property damage, including substantial sums provided to slave owners to compensate them for slaves who had been killed in battle or executed as rebels. Second, it shocked whites out of any complacency they might have had about the real intentions of their slaves. As Edward Long explained, in a particularly compelling account of the rebellion, rebel slaves aimed to overthrow white authority. They wanted, he argued, "no other than the entire extirpation of the white inhabitants; the enslaving of all such Negroes who might refuse to join them; and the partition of the island into small principalities in the African mode, to be distributed among their leaders and head men." Long added that the "Coromantin" or Akan slaves whose "turbulent, savage, and martial temper" Long considered to be "well known" conspired in secrecy throughout the island to break free from slavery, having seen "the happy circumstance of the Maroons; who, they observed, had acquired very comfortable settlements, and a life of freedom and ease."[99]

Our only firsthand account of the Tacky Revolt is in the diary of Thomas Thistlewood, who lived in Westmoreland Parish, near one of the centers of the rebellion. He describes, in vivid prose, on 30 May 1760 the "Strange Various reports with Tumults & Confusion" that gripped whites as rebels came close to overpowering them on isolated plantations. His account bristles with the tension that he felt as he realized that his life was in imminent danger. On 26 May 1760, for example, Thistlewood described how four neighbors, forced by rebel slaves to flee in states of undress, rode bareback to his estate to tell him that one overseer had been murdered, another "sadly chopp'd" and that if Thistlewood did not flee he "should probably be murdered in a short time." Thistlewood escaped but returned once the immediate danger had ended, in order to defend his property. He may have recalled on his flight to Savanna la Mar some earlier premonitions of how his slaves wanted to do whites harm. His underling John Groves had told him a day or so previously that some slaves had warned Groves that he would shortly "be dead in Egypt," the estate where he worked. Nevertheless, Thistlewood could defend himself and his possessions only by enlisting the help of his slaves. Colonists were forced to rely on slaves whom they armed and ordered to stand guard over plantation property.

The problem masters faced, of course, was that they did not trust their slaves. Moreover, the rapidity by which secret plans had turned into concrete action made a frightening situation desperate.[100] Thistlewood was convinced that several of his slaves "were in the Plot." After the rebellion had been quashed, he recalled that "at the beginning of the Rebellion, a Shaved head amongst the Negroes was the Signal of War. The very day our Jackie, Job,

Achilles, Quasheba, Rosanna &tc. had their heads remarkable shaved." He knew also that his slaves had links with the rebel slaves who initiated the Westmoreland uprising. Quasheba's brother "fell in the Rebellion: he had feasted away Some time before."[101] There were direct links between slaves on his estate and rebels living at the estate of Captain Arthur Forrest, an absentee proprietor and naval captain in the Jamaica squadron. On 22 May, "2 of Capt. Forest's Negroes at our Negroe houses" consulted with Jackie (whom Thistlewood suspected later on of being involved in the rebellion). Three days later, at 9 P.M., Thistlewood "heard a blast of a horn at our Negro houses." He later remembered that Lewie had been at Captain Forrest's plantation a day or so before the uprising and that "Coffee & Job [were] also very outrageous."[102]

When on 29 May he "arm'd our Negroes and kept a Strict guard and a Sharp lookout; all afternoon and Night; being in dreadful apprehensions," he had every reason to worry that some of these armed slaves might join rebel insurgents. After all, he concluded, "when the report was in of the Old Hope Negroes being rose, perceived a strange alteration in ours. They are certainly very ready if they durst, and am pretty certain they were in the plot." Nevertheless, only one slave on his Egypt Estate—"a Negroe fellow named Achilles of the Papah Country"—ran away, and he may have done so not to join the rebels but to escape punishment, after having told a slave driver that "if he did not take Care he would Cutt his head off in the Bush." Thistlewood, however, had his doubts about several other slaves' loyalty. For unexplained reasons, which may have been the revolt but which also might have been mere opportunism, he managed to include two slaves—Abraham and Quaw, both persistent runaways—in a shipment of slaves transported to the Mosquito Shore after the revolt had been quelled.[103]

Owners feared that slaves might be more sympathetic to the ambitions of the rebels than scared about the dire consequences that befell slaves who rebelled unsuccessfully. A dominant theme that emerges in the writings on the rebellion is white peoples' horror at how "loyal" slaves were willing to turn on their masters. Edward Long, for example, related how a planter armed twenty Akan slaves "of whose faithful attachment to him, he had the utmost confidence." But once they had their guns, the Akan slaves told their master (who they promised not to harm) that "they must go and join their countrymen, and then saluting him with their hats, they every one marched off." The message was that no slave was faithful to even those masters who treated them well. Slave owners' fears about the intentions of their slaves remained, even after whites had subdued, tortured, and killed hundreds of rebels.

Thistlewood was convinced that it was only good fortune that had prevented slaves from achieving their aim—to set fires "in many places at once" so that "all the Whites who Come to help Extinguish them, were to be Murdered in the Confusion." He thought that another revolt would soon follow. Dining with a fellow overseer in October 1760, he was told by his companion of an old proverb "which frights many people: One thousand seven hundred and sixty three, Jamaica no more an Island shall be (not for the whites)."[104]

The newly arrived Bryan Edwards—nephew of the merchant prince Zachary Bayly, on whose St. Mary estate the revolt had begun, and the successor to Long in the late eighteenth century as Jamaica's principal historian—was shaken badly by the revolt. It made him question deeply held beliefs about the civilized British and the barbaric African. Edwards, just seventeen in 1760, was so moved by the horror of rebel slaves being burned alive and put in gibbets until they starved to death that he wrote an early poem in the voice of a condemned slave. Alico was undaunted in the face of torture and painful death, a martyr to liberty:

Firm and unmov'd am I
In freedom's cause I bar'd my breast,
In freedom's cause I die.

Edwards later repented for his empathic identification with Alico. But he remained impressed, even after many years' residence in Jamaica, with the stoicism by which African rebels faced death. He noted thirty years after the event how "the wretch that was burnt was made to sit upon the ground, and his body being chained to an iron stake, the fire was applied to his feet. He uttered not a groan, and saw his legs reduced to ashes with the utmost firmness and composure; after which one of his arms by some means getting loose, he snatched a brand from the fire that was consuming him, and flung it at the face of his executioner."[105]

What we know about the rebellion emphasizes that the rebels stunned whites with their sudden ferocity and by the unexpectedness of their assaults. The careful coordination of the first strikes, the large numbers of slaves involved in rebellion, and the speed with which the revolt spread across the island suggests a remarkable degree of organization and secrecy. On 7 April 1760, ninety slaves from five estates that belonged to planter grandees, Ballard Beckford and Zachary Bayly, rose up, burning, looting, and killing several white plantation functionaries. Their leaders were Tacky and Jamaica, two of Beckford's Coromantee slaves, who, according to Long, had "long been concerting a Rebellion with three other Chieftains of their Country, who were each of them to have an Estate for his good Services."[106]

Within a few days, the number of rebel slaves had swelled to several hundred. Governor Henry Moore quickly mobilized all the men under his command—British regulars from the Forty-Ninth and Seventy-Fourth Regiments, reluctant local militia units, Maroons, and free black rangers.[107] After several skirmishes, a decisive battle took place. Soldiers "lined the Outside of the Wood where the Rebels were posted." In addition, a "party of stout Negroes," or free black rangers, along with a body of Maroons, cornered the rebels and killed Tacky and Jamaica. The ears of the rebels were cut off by Maroons as proof of their victory (and to guarantee payment from the government). The dead Tacky was decapitated and his head sent to be displayed at Spanish Town. By 21 May 1760, Moore reported to William Pitt that "most of the Rebells in Saint Mary's have either been kill'd or taken and Parties from the [Maroon] Negroe Towns are now after the few that remain of them who have secreted themselves in the Woods."[108]

Moore was overconfident. Four days after Tacky was killed, enslaved persons on Captain Arthur Forrest's plantation rose up, "sadly Chopp'd" a visiting ship captain and caused panic in the far west of the island. It put a lie to Moore's statement to Pitt that all that was happening in Westmoreland was that after "a Disturbance on the Estates of the late Mr. Richard Beckford . . . the fears of the People were improv'd into an Insurrection of all the Negroes in that part" and that in fact "everything is quiet, and appears to be so in the other parts of the Island." Moore quickly sent the royal navy and regular troops to assist local militia, as well as detachments from "the Negro Towns," as the Maroon settlements were called. The Westmoreland revolt was reckoned "more formidable than the first" and caused "great Terror and Consternation," as it seemed to confirm that "Disobedience & Revolt being intended to be Universal." For several days, western Jamaica was imperiled. The day of greatest crisis was 29 May 1760, when a body of local militia was routed. Thistlewood noted gloomily, "Our Negroes have good intelligence [of the rout], being most elevated, and ready to rise, now we are in the most imminent danger." It was only after 2 June, when the rebels were defeated by the same combination of troops as in St. Mary, with the Maroons to the fore, performing "with great bravery," that the immediate crisis passed.[109]

Nevertheless, another uprising soon occurred in Manchioneal, in the parish of St. Thomas-in-the-East, requiring Moore to send a detachment of Kingston militia and forty British marines to quell it. The extent of the rebellions made whites believe that slaves in every parish were preparing to attack them. Plots were discovered in the central parishes of Clarendon, St. Dorothy, and St. John. Moore reported to the Jamaica Council that slaves in the northwest parishes of Hanover and St. James intended to join Westmoreland slaves in rebellion. Fortunately, he declared, "their plot was discovered the day before it was to carried into execution and the Principal Persons concern'd in it were taken up and Executed."[110] It was not until early July that the danger finally passed, after which executions could start in earnest. It took another year before all the rebels hiding in the woods were gathered up.

White Jamaicans may have survived this "formidable" challenge to their authority, but the revolt was an immense shock to their complacency. They could no longer believe, as James Knight wrote in 1748, that they dwelled "in greater Security . . . than People do in England."[111] They immediately started searching for explanations, although such a search was hampered by the rebels' secrecy and by the army's decision to kill or transport prisoners immediately, without any interrogation. Why did slaves rebel when they did and how were they able to keep their conspiracy so secret? Subsequent historians have been as interested as were Henry Moore, Edward Long, and Thomas Thistlewood in understanding the causes of one of the two largest revolts (the 1763–64 revolt in Berbice was as large and was more devastating) in the mid-eighteenth-century Caribbean.[112] Long's explanation—that it was a general conspiracy orchestrated by Coromantees, as the British called speakers of Akan—has proved to be most attractive. Thistlewood's theory—that it was a general uprising of Africans, rather than one ethnicity—has also had adherents. In either case, Tacky's Revolt stands apart from later revolts in being both African led and also African in its ideology. Vincent Brown also draws attention to the role of obeah in shaping the conflict, seeing Tacky's Revolt as a competition between whites and blacks over the superiority of various forms of sacred authority. Scholars interested in Maroon history are concerned to explain Maroon participation with whites in putting down a black revolt.[113]

Yet what is seldom mentioned in these analyses is that the rebel leaders—Tacky and Jamaica in St. Mary; Pompey in St. Thomas in the East; Cubbah in Kingston; and Wager, Aguy, and Simon in Westmoreland—declared war against Jamaican planters during the Seven Years' War.[114] Tacky's Revolt is virtually never mentioned in general accounts of the Seven Years' War, but it was one of the more deadly and violent events that occurred during that conflict, at least in the Americas.[115] In the Seven Years' War in the Americas, only the 1758 Battle of Ticonderoga, where nearly two thousand soldiers were killed or wounded, was more violent. There were no other encounters where casualties exceeded the six hundred deaths and five hundred transportations in Jamaica in 1760–61.[116] In comparison fewer than 120 British and French soldiers died at the decisive Battle of Quebec in 1759. Some circumstantial evidence suggests that rebels took advantage of weakened defenses as a result of the Seven Years' War in planning their attacks. It may not have been a coincidence, for example, that the rebellion in Westmoreland began on the estate of Captain Arthur Forrest, a naval captain who had distinguished himself in 1757 and 1758 in actions around Cap Français and who had served extensively in the Caribbean during the Seven Years' War. He had acquired, according to Long, several French slaves who had been taken prisoner at Guadeloupe and sent them to his Westmoreland plantation. Long alleged that "these men were the more dangerous as they had been in arms at Guadeloupe, and seen something of military operations, in which they acquired so much skill." Long related that these Guadeloupe slaves helped greatly in the rebels' defenses, building "a strong breast-work across a road, flanked by a rocky hill; within this work they erected their huts, and sat down in a sort of encampment."[117]

These Guadeloupe slaves would have come into contact with the most important rebel leader, Wager, or Apongo. Thistlewood thought him "chief of the rebels." He had a high opinion of him as a man who had formerly been "a prince in guinea, tributary to the king of Dahomey." He told a curious story about Wager visiting with his employer's father—John Cope Sr. Cope had once been governor of Cape Coast Castle in West Africa. He had treated with Wager when Wager was a man of substance, "attended by a guard of 100 Men, well arm'd." In Jamaica, the two men became reacquainted. The Guinea prince, now enslaved, "used when a Slave Sometimes to go to Strathbogie to See Mr. Cope, who had a Table Set out, a Cloth laid, &tc. for him, and would have purchas'd him and Sent him home had Capt. Forest Come to the Island." Wager clearly had extensive military experience in Africa. However, he may also have seen action in Atlantic conflicts. Thistlewood noted that he got his unusual name, Wager, from the ship that Forrest had sailed in actions against the French and Spanish in the War of the Austrian Succession. The evidence here, admittedly, is slight. But it does seem important that several slaves on the estate where the rebellion started in western Jamaica, including "the chief of the insurgents," had been exposed to the larger geopolitical contexts of imperial warfare.[118]

The circumstances of the war were a significant factor in the successful repression of the rebellion. Governor Moore could respond rapidly to the uprising because he had more troops than normal in peacetime. Admiral Thomas Cotes reported to the Admiralty on 19 April 1760 that the rebels "would certainly have done more mischief" if Moore had not quickly proclaimed martial law (easier to do in wartime than during peace) and if he had not sent "two strong Parties of Regular Troops to support the Militia and free Negroes."[119] Thistlewood's observations during the height of the Westmoreland rebellion that there were "Frequent Alarms fired" and that "Vast Numbers of dispatches [were] passing" suggest that Jamaica responded quickly to the imposition of martial law. Westmoreland resembled a military theater with "Vast Numbers of people, belonging to the Troops, Militia" tramping through the countryside.

One of the main effects of the slave rebellions of 1760 was to make a link in white minds between the external threat of enemy attack and slave rebellion. The Jamaican Assembly made the link evident late in 1760 when in an address to the King it diagnosed Jamaica's problems as being mainly due to a shortage of British troops, "as evinced by the late rebellion." The Assembly laid out the terms of an argument that was to be repeated, with little variation, throughout the 1760s. They placed much of the blame for the revolt on large planters, especially nonresidents. Absent planters attracted their particular ire by not abiding by the provisions of the Deficiency Act, first passed in 1703 and amended several times thereafter, which fined planters not employing sufficient numbers of whites in proportion to the slaves that they owned. People not living on the island contributed to slave rebellion, it argued, by not exerting "influence over their slaves." The second contributing factor was a shortage of regular troops, the insufficiency of which "hath, by fatal experience, in the late dangerous rebellions, been fully evinced."[120]

It was obvious to some observers that the French or perhaps the Spanish would take advantage of internal travails in Jamaica and invade the island. Thistlewood wrote in July 1760: "It seems in Trinidada, they are overjoy'd to hear the News [of Tacky] and Say now is the time to take Jamaica." Governor Moore was blunt in his calculations of the impact of Tacky on Jamaica's foreign enemies. He told the Jamaican Assembly that "the warlike preparations carrying on by our neighbours ... point out the dangers to which we may be subjected to without proper precaution." It may be that the Jamaican response in "our present difficulties" was likely to turn inward rather than to worry about the French or Spanish, but Moore warned that dealing with Tacky "should never prevent a constant attention being given, to every measure, which could render ineffectual the Attempts of a powerful enemy."[121]

Getting more troops onto the island became a constant goal of the Assembly and governor throughout the 1760s. The problem was especially acute during 1762. Governor William Lyttleton pleaded with the Board of Trade to send more troops to Jamaica, claiming that the island "is at present in a state of extreme weakness." It had only one thousand soldiers healthy enough to fight; its forts were in a poor state; and notwithstanding the just recently quelled "insurrection of the Negroes," the Militia was "ill train'd, and will with difficulty be brought into tolerable order." The naval forces were not much better equipped: Forrest had only seven ships of the line and nine frigates, scarcely enough to hold off a French or Spanish expeditionary force. These linkages between internal and external challenges and the conduct of delinquent absentee planters were repeated throughout the 1760s in messages from the Assembly demanding more troops and more ships of war for protection. A petition received by the Privy Council from the Jamaican Assembly in 1764 was typical. The Assembly insisted on two thousand troops, "a number they apprehended absolutely necessary and scarce sufficient to secure and Preserve the Peace and tranquility of the Island from the secret Machinations and open Insurrection of their Internal Enemys."[122]

Colonial authorities responded ferociously to the revolt. After Tacky and Jamaica were killed in battle, their heads were severed and stuck on poles near Spanish Town. Whites decided that they needed more awesome displays than usual, punishing rebels in ways that made clear their willingness to inflict cruelty on slaves who dared to engage in rebellion. They made some slaves endure horrific ceremonial executions in which they were burnt alive by the application of a slow fire or were placed in gibbets hung up in public squares and left to starve to death. A remarkable feature of these executions is how they were made part of normal life. Thomas Thistlewood described how Apongo, or Wager, was condemned to "hang in Chains 3 days then be took down and burnt." Thistlewood respected Wager, as we have seen. But he had no doubt about his leadership of the rebellion. He repeated on 12 July 1760 a rumor: "It is said, Wager and his wife had a Quarrill in the Negroe ground, the Sunday they began, and that She threatened to discover the Plot, was the Occasion of beginning that Evening, other ways not to have been till the

Shipping had Sail'd." But the death of such a prominent rebel happened with minimal ceremony. Thistlewood saw Apongo "gibbeted alive" in Savanna La Mar and wandered over to ask him whether he knew "any of our Negroes." Apongo replied that "he knew Lewie and wished him good bye." The way that Thistlewood records this episode suggests that Apongo was languishing in the gibbet unattended and unwatched while most people went about their business.[123]

The extent of the violence meted out to slaves impressed and horrified white Jamaicans. Observers were amazed at the fortitude with which rebels met their fate. After watching one slave being burnt alive, Thistlewood, like Bryan Edwards, marveled at how "altho' [being] burnt by degrees with a Slow fire, made out a distance from him," the rebel "Never flinch'd, moved a Foot, nor groan'd or cried out." Even Edward Long, who believed that enslaved rebels deserved every punishment thrown at them, admitted that they died bravely. Describing how the rebel slaves Fortune and Kingston suffered during seven and nine days of being starved to death in a gibbet, he noted that although they deserved "this cruel punishment inflicted upon them in terrorem to others" because of the "murders and outrages they had committed," the two men were "little affected" by their torments, "behaving all the time with a degree of hardened insolence, and brutal insensibility."[124] But the men being tortured to death were not brutes. Some even managed to keep their spirits up and engage in banter while in extremis. Thistlewood saw Davie, "Jibbetted Alive" on 17 July 1760. When Davie saw a monkey upset a bucket of water, he "laughed heartily," declaring "that Monkey damn'd Rogue true, &tc. &tc." There was an edge, however, to Davie's comments. Seeing two of his white tormentors come to blows, he yelled "Tha's good, me love for See So." Davie was broken but he was not beaten, retaining his opposition to whites until the bitter end.[125]

The revolts started by Tacky and his lieutenants confirmed for white Jamaicans that maintaining the strictest discipline over enslaved people was absolutely necessary. We get a sense of this from Thistlewood's wistful comment in 20 December 1760 about an unfounded rumor: "It is Said 6000 Negroes rebell'd in Hispaniola, and that the French there have made a Law for them to be hang'd up immediately, when off their owner's Land without a Ticket." No such revolt in Hispaniola had occurred: possibly Thistlewood heard an echo of the Macandal poisoning scare. Thistlewood would have liked the punishments inflicted on rebels in Jamaica to have been even more far reaching and violent than they in fact were. He disapproved of the decision made by Brigadier General Norwood Witter, a local planter and head of the militia in the Cornwall region of western Jamaica, to pardon rebels who surrendered to him. Witter promised rebels who surrendered that he would commute their death sentences to a verdict of transportation. Thistlewood favored instead the policy of Colonel Spragg, who "off'd with the heads of all the Rebells who fall into his hands immediately: he Says he is Sent to destroy the Rebells, & destroy him he will to the utmost of his power, except Women and Children." He also approved of the legislature's unsuccessful efforts to revoke Witter's promises.[126]

Beyond punishing the rebels, the Jamaican Assembly cast around for people and policies that could be blamed for allowing slaves to think they could revolt. Assigning blame, it seems, was more important than understanding what slaves believed had caused the rebellion. It is striking the degree to which whites responded to Tacky's Revolt by evoking various ideas that white planters and imperial officials already wanted to push; they showed little interest in working out why slaves had rebelled as a means of stopping future rebellions. In this way, the differences between the French response to Macandal and Jamaica's response to Tacky were remarkable and telling.

The Assembly concluded that planters had relaxed their guard, allowing slaves access to guns, letting them wander unsupervised around the island, and failing to prevent them from gathering in groups. New laws after 1760 strengthened planter responsibility for slave discipline. The Assembly also discovered that whites were delinquent in their militia duties, and it passed laws reinforcing these obligations. The Assembly petitioned the Crown for additional regiments of regular troops.[127] Most important, the shock of Tacky's Revolt encouraged white Jamaicans to pay new attention to the spiritual ideologies that they believed motivated Africans to rise against them.

Edward Long believed that Tacky's conspiracy was an Akan conspiracy, aided and abetted by religious specialists—obeah men—who were "chief in counseling and instigating the credulous herd." He went into lurid detail about "execrable wretches" who convinced blacks that they "would be invulnerable to the white man; and, although they might in appearance be slain, the obeah-man could, at his pleasure, restore the body to life." For Long, Tacky's Revolt illustrated the danger that these "priests, or conjurors," posed to white safety. Before 1760, colonists in Jamaica treated slave spiritual ceremonies as quaint primitive customs similar to the folk beliefs they knew from Europe. Thomas Thistlewood wrote of "obia" in his diaries in 1753, and this term was known among whites, at least in Barbados, from the second or third decades of the eighteenth century.[128] For Thistlewood, seven years before Tacky, "obia" was a mostly harmless activity. The actions of the obeah man Guy reminded him of English con men, such as Black Lambert, a conjuror from Ackworth, Yorkshire, whom Thistlewood remembered being driven out of town. Moreover, obeah was not just a practice. As with "macandals" in Saint-Domingue, "obeah" also referred to charms enslaved people wore to ensure good luck. These talismans, or obeah, were made up of various materials such as feathers, broken glass, animal teeth, and grave dirt.[129]

Tacky's Revolt transformed the way white colonists saw obeah. It convinced the Assembly that black shamanism posed a grave danger to the colony. In December 1760, the Jamaica Assembly approved the first law in the British West Indies describing obeah as a crime. It was part of a series of legal measures designed to prevent slaves from congregating, and it remained essentially unchanged until the Consolidated Slave Act of 1809, which adopted the same approach with even greater specificity.[130] The preamble to the 1760 law is quite explicit about the connection between slave spiritual practices and the recently put-down revolt. It claimed that obeah men and obeah women were present on "many estates and plantations," and that through their "influence over the minds of their fellow slaves . . . many and great dangers have arisen destructive of the peace and welfare of this island." The act was remarkably specific, far more than equivalent statutes in Saint-Domingue after Macandal. Obeah was practiced by people "pretending to have communication with the devil and other evil spirits." The law provided such a long and detailed list of substances associated with "any Negro or other slave, who shall pretend to any supernatural power" that it suggests some legislators were not quite sure that such practices were merely "pretending." Obeah was to be confirmed if slaves "be detected in making use of any blood, feathers, parrots beaks, dog's teeth, alligator's teeth, broken bottles, grave dirt, rum, egg shells or any other material relative to the practice of obeah or witchcraft, in order to delude and impose on the minds of others." The penalties to be enacted against such practitioners of the supernatural were severe—death or transportation.[131]

This legislation was framed in this way because those convicted of practicing obeah after 1760 were defined as ideological and political enemies of the state. Obeah men claimed that their powers were superior to Christian religion. Colonial authorities tried to force them to realize that the imperial state was yet more powerful, subjecting convicted obeah men to death by slow fire or even to punishment by rudimentary forms of "electrocution."[132] In short, whites decided to treat obeah as treason. As Diana Paton notes, "Punishments involving destruction of the convicts' bodies communicated specific meanings for the English planters who designed them. In English penal tradition, attacks on convicts' bodily integrity had come to signify by the eighteenth century that the convict was a traitor, a rebel against legitimate authority."[133]

Jamaica's legislation inspired similar laws throughout the British Caribbean. What is more significant, however, is that Jamaican colonists redefined obeah as a political, not religious, crime. By the mid-eighteenth century, metropolitan Britons had started to refuse to prosecute people for witchcraft. They characterized witchcraft practices as symptoms of mass hysteria about imaginary problems. After 1760, by contrast, Jamaican colonists came to see obeah not as an African religious practice or even as a "superstition." Instead the law defined it as a crime perpetuated by men who took advantage of the superstitious nature of Africans to foment dissension. As Paton explains, "Obeah itself is a construct produced through colonial (and postcolonial) law-making and law-enforcement over more than two centuries." Paton echoes here Kenneth Bilby and Jerome Handler, who have argued that whites used "obeah" as a catchall term for supernatural-related practices that whites believed were intended to harm other slaves and more specifically themselves. Handler and Bilby speculate that written laws and regulations intended to suppress obeah may have done more to make the term official and to help diffuse the term regionally than did black usage, explaining why statutes against obeah in the English-speaking Caribbean survived emancipation, independence, and, in some former colonies, cultural nationalism.[134]

In the 1760s, therefore, Jamaican colonists followed their counterparts in Saint-Domingue in concluding that African spiritual practices that they had long tolerated were in fact very dangerous. As in Saint-Domingue, the problem was not the beliefs or practices themselves but the power they gave charismatic leaders. As a witness in a later inquiry into conditions of slavery in Jamaica in 1791 argued: "The Negroes in general, whether Africans or creoles, revere, consult, and abhor them; to these Oracles they resort, and with the most implicit Faith, upon all Occasions, whether for the cure of Disorders, the obtaining of Revenge for Injuries or Insults, the conciliating of Favour, the Discovery and Punishment of the Thief or the Adulterer, and the Prediction of Future Events."[135] It was not an accident that colonial planters in

Saint-Domingue and Jamaica started to take an interest in African-style spirituality only when their power was challenged. Saint-Domingue's whites found it easier to ascribe "poisonings" and vague "secrets" to the power of a man who had charisma and made supernatural claims than to try to understand the roots of slave spirituality. In Jamaica, Tacky's Rebellion could be at least partly explained by the actions of obeah men who spiritually empowered enslaved people to fight fearlessly against white enemies.

In Saint-Domingue, the government opted to give planters carte blanche against slaves' transgressions, including the power to determine which spiritual practices were dangerous. In an absolutist system where everything passed before agents of the Crown, royal officials nevertheless agreed that such problems were best handled privately. Although the law prohibited slave "dances" and slave doctors, some planters tolerated such practices while others punished them severely. Masters argued that it was impossible to prove poison in a court of law and that only they could investigate and punish such crimes. French authorities did not even have an overarching term to describe slave spirituality; the word "Vaudoux" rarely appeared in print until the 1780s, and even then it had a very specific meaning.[136]

In Jamaica, on the other hand, colonists saw African-style spiritual practices as a public threat, to be punished publicly and harshly. One reason for this difference is that Jamaica experienced a terrifying rebellion based on a well-concealed conspiracy. Although Saint-Domingue had lost more lives in their so-called poisoning episode, some residents there, including the colony's top administrators, denied that there had been a conspiracy. Second, as Protestants, Jamaican colonists had an easier time separating African and English belief systems than did Catholics in Saint-Domingue. Saint-Domingue's planters abhorred how the Jesuit order tried to convert enslaved people, and this was one of the charges that allowed them to expel the Jesuits in 1763. Yet some French colonists, like the controversial missionary order, took a more syncretic approach. As we see in Chapter 9, some even endorsed the ability of slave healers to detect poisoners.

One great irony of the Seven Years' War in the Caribbean is that the major challenge French and British planters faced during this conflict came from within their societies rather than from foreign invasion. Yet this new alarm about African-style spirituality in well-established plantation colonies confirms the global nature of the war. The enslaved majorities in those societies, replenished year after year from a wide and changing variety of African locations, were generating their own American cultures that were deeply alien to the Europeans who believed that they controlled the West Indies. We may never know exactly what Macandal or Tacky hoped to achieve in their dealings with colonial society. But their actions led to remarkable and parallel reconfigurations of social and cultural values in Jamaica and Saint-Domingue, especially relating to race. It is to these reconfigurations that we now turn.

CHAPTER 6

Racial Reconfigurations Before the American Revolution

The Seven Years' War is rarely depicted as having had a lasting impact on the Caribbean. Yet the final years of this conflict and the immediate postwar period produced momentous political shifts in Saint-Domingue and Jamaica. One major change was in race relations. In Jamaica in 1761, and then less than a decade later in Saint-Domingue, colonial legislators and administrators created laws that made it difficult or impossible for wealthy people of color to pass as white. Before this, both colonies had formal and informal mechanisms that allowed wealthy Europeanized families of partial African descent to enter the colonial elite. The new laws set aside social class, assigning free people identities solely on the basis of appearance or genealogy. This postwar shift marked the rise of a pernicious "white purity" form of racism that shaped Atlantic societies into the twenty-first century. These two societies hardened their racial classifications not so much in reaction to developing ideas of scientific racism but in response to the particular political problems each colony faced. The problems were different in Jamaica and Saint-Domingue, but the results were similarly transformative.

This chapter traces how these changes occurred in Jamaica and prepares the way for describing a similar change in Saint-Domingue, in Chapter 7. By the start of the American Revolution, new racial orders were in place in Jamaica and Saint-Domingue. They were among the first European colonies where the ruling minority defined its position in racial terms, and that deliberately excluded wealthy families of mixed ancestry from positions of authority. In this and the following chapter, we argue that this development was closely related to colonists' new imperial assertiveness after the Seven Years' War. In both societies the new political climate arose from a mixture of confidence and anxiety. For the planters and merchants of Jamaica, the anxiety concerned their susceptibility to attack from slave rebels, and their confidence arose from Britain's victory in the Seven Years' War and from the prosperity of the plantation economies. In Saint-Domingue, conversely, imperial administrators were the anxious ones. The British had nearly destroyed France's overseas empire, and Saint-Domingue, some believed, remained French only because it had not been attacked. Versailles wanted to strengthen colonial loyalty to the empire and was resolved to liberalize absolutist political and commercial regimes. French administrators, observing what had happened in Jamaica in 1760, were keenly aware of the danger of slave revolts. Saint-Domingue's colonists, on the other hand, had few such fears. Even as they maintained their own right to torture and execute their most dangerous workers, they argued that they were ready to live under civilian laws rather than military rule. With Saint-Domingue's prosperity advancing every day once the war ended, colonists felt they could assert themselves against government officials and enslaved people without much fear of either group pushing back.

Tacky's Revolt coincided with a profound reshaping of opinion in Britain around the time of the Seven Years' War about the relative rights of Britons, white imperial subjects, and nonwhites. This reshaping occurred during a period when new conquests, notably in India, brought vast territories and millions of nonwhites into the British Empire. The result was that more people traveled to more places, traded more goods, and published more works on their experiences than ever before. It meant that Britons became more accomplished in developing hierarchies of peoples, in ways similar to and influenced by the biological taxonomies created by Carl Linnaeus in the 1730s.[1] At the same time as Britons were experiencing more fully what it meant to be an imperial nation, they began to develop new understandings of their duty to strangers in foreign lands. One reason why many Britons came to see American colonists less as compatriots was revulsion against American cruelty to Native Americans and Africans. In the 1760s, as thoughtful Europeans began to see empathetic linkages between themselves and all peoples of the world, white colonists in North America and in the West Indies hardened their attitudes toward nonwhites. They began to talk about Native Americans and enslaved Africans as if they were different species to Europeans, as irredeemable and animalistic savages. Nevertheless, Europeans who witnessed the wanton killing by colonists of Native Americans in backcountry Pennsylvania, or the state-sponsored terror that followed Tacky's Revolt found it increasingly hard to tell who were savages in America—whites or nonwhites.[2] Humanitarianism was one reason why Britons gradually came to think of Americans and West Indians as "them" rather than "us." Moreover, the cruelty of British American colonists toward Indians and Africans formed part of a hotly debated argument between Frenchmen and Britons in the immediate aftermath of the Seven Years' War about whether the English national character was naturally barbaric. Were the English, in short, the savages of Europe, as a popular French text by Robert Martin Lesuire had it? Britons were especially sensitive to accusations that their behavior meant that they had replaced the Spanish as the tyrants of the New World.[3]

Jamaica's repression of Tacky's Revolt confirmed British suspicions about the inherent barbarism of life in the New World. Several British and North American papers described in detail the gruesome nature of executions and the amazing endurance and humor of slaves tortured to death. Typical was a report in the *London Evening Post* where the correspondent noted how gibbeted slaves were not allowed "any Kind of Liquid or Sustenance" and how they "commonly live from four to eight days." What was especially amazing, the newspaper thought, was how they could live so long without water in a hot country "where Thirst comes so Importunate." When they reflected on the appalling punishments given to rebels, however, Britons began to contrast the resolution and bravery of the slaves to the brutality of the whites. In September 1760, the *London Evening Post* reported, "A Gentleman lately come from Jamaica . . . says it was amazing to observe the Firmness and Resolution with which some of them had suffered, and that he was afraid the great Number of executions would have a contrary effect to that intended, as some of those of a resolute Turn prided themselves in observing the Courage of the Sufferers; and those that were timid, were driven to a Kind of Despair."[4]

The more Britons thought about it, the more disturbing Jamaican actions seemed. Treating slaves so badly seemed to guarantee further revolts and the potential loss of Jamaica to France or Spain. That Tacky occurred during wartime was not lost on observers. In March 1762, when an invasion of Jamaica seemed imminent, the *London Chronicle* lamented the "Importunate and defenceless State of Jamaica." Jamaica was threatened most, the correspondent believed, from internal rebellion. Tacky's Revolt had shown Jamaica's "merciless masters" where the real danger to their colony lay. They "know too well," the writer opined, "the cruelties they have practiced upon these miserable wretches, to expect any other return than betraying them to an enemy, or cutting their throats." He went on: "The men who would rob us of Jamaica are already there, and want only arms and ammunition with a few good officers and men to head and conduct them."[5]

Britons were well aware of both how frequent slave revolts were in the New World and also how viciously they were put down. Between 1737 and 1773, fifty-two articles about forty-three slave revolts appeared in the *Gentleman's Magazine*, the most widely read periodical of its time. It suggested to well-read men, like Samuel Johnson, that America was irredeemably violent. Famously, Johnson quipped about American protestations against British tyranny in the 1760s that it was surprising that Britons heard "the largest yelps for liberty among the drivers of negroes." It followed on from his remark in 1758 that "slavery is now nowhere more patiently endured, than in countries once inhabited by the zealots of liberty." Johnson took the common assertion made by American colonists and made the link that historians now more commonly make, which is to claim that slaveholders were hypocrites—people who were especially sensitive to their claims to liberty being contested because they were themselves very keen on taking liberty away from Africans "patiently enduring" a particularly brutal form of enslavement.[6]

Johnson's dislike of West Indian and American slave owners fitted in with his fervent endorsement of his friend's Richard Savage's 1737 poem "Of Public Spirit in Regard to Public Works." Savage's poem sympathized with "Afric's sable Children," inflicted with "nameless Tortures" and ended the poem with a blood-curdling (to white West Indians, at least) prediction that "yoke may Yoke, and Blood may repay." Johnson's recommendation of such sentiments was of the same order as his remarkable identification (during wartime, no less) with Native Americans in a work in the *Idler* in 1759 in which, writing in the voice of an ancient Indian, he declared that as far as Indians were concerned it was good news that "the sons of rapacity [the French and the English] have now drawn their swords upon each other." Indians should remember, he provocatively declared, "that the death of every European delivers the country from a tyrant and a robber."[7]

Indeed, metropolitan responses to reports from Jamaica in 1760–61 show that Europeans were more concerned about white brutality than about the inequality of slavery. The early sentimentalist novelists described differences between whites and blacks as produced by environmental factors rather than as arising from inherent biological diversity. They believed in ameliorating slavery and disagreed with colonials that only violence could keep slaves in order, or that white supremacy had to be defended at all costs. Commentators assumed that Africans would respond well to kind treatment. As they came to understand the nature of Jamaican slavery, novelists argued that masters' relentless violence meted out to slaves made slave rebellion inevitable and suggested that planters' cruelty was unconscionable. Sir George Ellison, the reluctant slave-owning sentimental hero of Sarah Scott's 1766 novel, declared, "The thing which had chiefly hurt him during his abode in Jamaica was the cruelty exercised on one part of mankind; as if the difference of complexion excluded them from the human race, or indeed as if their not being human could be an excuse for making them wretched."[8] In addition, news of the post-Tacky executions helped writers incorporate rebels into sentimentalist texts that featured Christian symbols of martyrdom. For example, in Thomas Day and John Bicknell's popular poem *Dying Negro* (1773), an enslaved man commits suicide, proclaiming the superiority of Christian death to the prerogatives of property ownership. As Vincent Brown notes, this work "helped to establish a recurring pattern in abolitionist literature, glorifying the African who chooses Christian redemption in death over servitude in life."[9]

Tacky's Revolt supported ideas that white men in Jamaica were unparalleled tyrants. Charles Leslie wrote of Jamaican slave owners in 1740 that "no Country excels them in their barbarous treatment of Slaves, or in the cruel Methods they put them to death."[10] Thomas Thistlewood's diaries provide ample evidence that Leslie was correct. Thistlewood constantly whipped and sexually molested his slaves. His diaries are the most thorough surviving documentation of the relentless violence permeating slave society in the eighteenth-century Atlantic World. Numbers alone convict him. At Egypt sugar estate, where he was overseer between 1751 and 1767, he gave out more than one whipping a week. In 1756, he gave a slave population of sixty slaves fifty-seven whippings, gagged four slaves without whipping them, and put eleven slaves in stocks overnight. Adult slaves, especially men slaves, who received more whippings than women, could expect to be flogged at least once a year.[11]

Given how Thistlewood treated his charges, it is little surprise that some of them were prepared to join the Westmoreland rebels against him, and his fellow colonists, in 1760. We have written elsewhere of the degradation that Thistlewood subjected his slaves.[12] Thistlewood, however, was representative, not aberrational. He was a vicious but rational master. Though he acted savagely he rarely did so in hot blood. He was not like his subordinate, John Groves. At the height of danger in Tacky's Revolt, Groves lost his composure, shooting at rebel and faithful slaves indiscriminately "in a frenzy." Groves left Thistlewood's employ in 1761 "because he might not flog the Negroes as he pleased." Thistlewood had chastised him for being "like a madman amongst the Negroes flogging dago, primus etc without much occasion." His next situation ended for a similar reason. Thistlewood noted that Groves was dismissed from William Beckford's Roaring River sugar estate "for beating the Negroes impudently in the Boiling House." Thistlewood's sadism, moreover, was not as pronounced as some of his contemporaries. In 1765 Harry Weech related to Thistlewood, seemingly proudly, that he had "cut off the lips, upper lip almost close to her Nose, off his Mulatto sweetheart in Jealousy, because he said no Negroe should ever kiss those lips he had." To the sadism could be added here a (literally) disfiguring racism.[13]

At the same time, commentators noted that white Jamaican men bridled when other whites treated them as subordinates. Bryan Edwards noted that "a marked and predominant character to all the white residents" was "an independent spirit and a display of conscious equality throughout all ranks and conditions" so that "the poorest white person ... approaches his employer with an extended hand."[14] The fundamental equality of all white men in Jamaica, whatever their wealth and social standing, was rooted in material conditions. High mortality made white men a scarce commodity. Because planters always needed white managers to oversee their slave forces, white men easily found employment. Rich white men were so anxious to secure the services of white male workers that poor white men had significant economic bargaining power, especially if they were tough enough to be able to take on the daunting task of controlling, punishing, and terrorizing black men as plantation overseers.

That economic bargaining power was transformed into social, if not political, power. Poor white men insisted that the legendary hospitality of planters be extended to all whites regardless of circumstance. White men were always entitled to free lodgings, free meals, and often admittance to planters' social entertainments if they turned up unannounced at a planter's house. It was important being English, or British, on an island where only 5 percent of the population in rural areas was white. In 1793 Bryan Edwards acknowledged as much when he attributed Jamaicans' sense of conscious equality to "the pre-eminence and distinction which are necessarily attached even to the complexion of a white Man, in a country where the complexion, generally speaking, distinguishes freedom from slavery."[15] This racial sense of whiteness, based on an even deeper sense of belonging to an English heritage, was a logical outcome of the changes in racial classification embarked on by white Jamaicans at the start of the 1760s. A white male colonist in Jamaica was thus almost by definition a master of slaves. Mastery, the ability to control other men, was a key component of masculinity. In Jamaica, most colonists had that power, whether owners of slaves or merely men in charge of other men's slaves. Being a "master" in eighteenth-century society allowed ordinary colonials to appropriate the language and behavior of patriarchs, even when they themselves were ostensibly dependents. Their ubiquitous control over slaves contributed to "impatience with subordination" and meant that "men accustomed to be looked upon as a superior race of beings to slaves submit with reluctance, if they submit at all, to be treated as if they enjoyed no will of their own."[16]

A pugnacious attitude to authority was also manifest in how the Jamaican Assembly reacted to perceived or real slights cast against them by their traditional nemeses, colonial governors. By the early 1760s, the combativeness of the Assembly in advancing its rights was so notorious that Governor William Lyttleton claimed that their "Aspiring endeavour to acquire in their Assemblies & within the Sphere of their activity the same Powers and Privileges as are enjoy'd by a British House of Commons was universal[ly] . . . Espous'd by the Inhabitants." These combative attitudes, he argued "[were] too deeply rooted & founded upon Doctrines & Practises of too high & complicated a nature to be removed or much lessen'd by any means in my power."[17]

Lyttleton came to experience the full assertiveness of the Jamaican Assembly three years after he had taken office in Jamaica in 1761. On 8 December 1764, the provost marshal seized coach horses belonging to John Olyphant, a planter and assemblyman, for nonpayment of debts. Olyphant complained that this was a breach of his privileges, as the rules of the Assembly prevented members from being arrested or having goods seized while the Assembly was in session. Merchants often complained that planters used this law to escape debts.[18] The Assembly believed that the provost marshal's action was a direct challenge to its rights. It made clear that it alone would define its own privileges. The impounding of Olyphant's horses became a constitutional crisis when the Assembly ordered the provost marshals to explain themselves to a House committee. The provost marshals applied for and received a writ of habeas corpus from Governor Lyttleton. The Assembly accused the governor of not respecting the rights of the assemblymen as "free born Britons" to make laws for themselves while Lyttleton accused the Assembly of not obeying the royal instructions that regulated his conduct as governor.[19] The Assembly challenged Lyttleton by issuing ten strongly worded resolutions, asserting their rights in bombastic fashion. They dared Lyttleton to challenge those rights, which he did, dissolving the Assembly on 24 December 1764, and calling a new election.

The Jamaican electorate supported their intransigent representatives by re-electing them to office, and battle was joined for the next two years. The governor's two leading antagonists in Jamaica were Charles Price Jr., elected speaker on 13 August 1765, and his father, Charles Price Sr., who came out of retirement for the conflict. Price Sr., an old foe of the royal prerogative, relished the fight, as did other major political figures. Indeed, Nicolas Bourke's polemics against Lyttleton were probably the most compelling defense of settler rights made in any British American colony before the American Revolution.[20] Just as important were the skillful machinations of Price and his supporters in Jamaica. The powerful leader of London's radical merchant class, William Beckford, led opposition in Britain. The Assembly then made the strategically brilliant move of putting forward a motion asking the House of Commons to support them against a governor they argued to be tyrannical.[21]

Jamaica's request arrived just as a new administration, the Rockingham-led ministry, arrived in office, replacing the Grenville administration that had been forced into humiliating retreat over colonial reactions to the Stamp Act. The Rockingham administration was in no mood for another constitutional conflict, especially with Jamaica's rambunctious Assembly. On 22 May 1766, the Secretary of State for the Southern Region, Henry Seymour Conway, answered Lyttleton's pleas for support, urging him to seek some "mode of Conciliation." Isolated and beaten, Lyttleton advanced his planned retirement, taking immediate sick leave and returning to London on 2 June 1766. The victory of the Assembly was almost complete. As was the Assembly's wont, it was quick to rub its triumph in the nose of its opponents. It authorized, at considerable cost, jubilant events in Kingston and St. Jago de la Vega, where the Jamaicans celebrated, as historian Bryan Edwards later commented, in a manner "we may imagine were displayed by the English Barons on receiving the *magna charta*."[22]

Colonists had good reason to celebrate. They had shown, as Jack Greene notes, that "local consent was crucial to the implementation of metropolitan initiatives."[23] North Americans were not able to establish such certainty in their constitutional arrangements: the result was the American Revolution.

But a combination of heady economic clout, influential allies in London, and a long tradition of defending its legislative powers allowed the Jamaican Assembly to force Britain to respect its power and privileges.[24]

The question of slavery's inherent violence also began to weigh on French thinking about Saint-Domingue after the Seven Years' War. Significantly, these concerns did not emanate from the colony. Colonists continued to talk about the danger of another Macandal, but their solution was to remain vigilant about rooting out poisoners on their plantations, often using torture to do so.[25] They saw the Macandal episode as an illustration of the power a charismatic personality could exert over other slaves, not as an expression of the fundamental dangers of a slave system. It was for this reason that French colonists fixated on poison, rather than on slave revolt, as the major threat that enslaved people posed to them.

Instead, postwar concerns about slave revolt came from France, not Saint-Domingue. After over a century of slavery in the French Caribbean, metropolitan intellectuals began to understand that their kingdom's colonies were not that different from those of the British in Jamaica. This countered the portrait of French Caribbean slavery disseminated in the 1720s by Jean-Baptiste Labat, as discussed in Chapter 2. French planters, claimed Labat, did not face revolts as in the British islands because they had fewer slaves, but because "the religion in which we raise them inspires more human feelings in them, and because we treat them more gently and more charitably than the English do." Labat noted, however, that as more slaves arrived in the French islands, they too were facing problems of revolt. He intimated that slaveholders were courting disaster by their incessant desire to acquire more and more Africans.[26] By the 1740s, twenty years after the publication of Labat's *New Voyages*, the enslaved population of the French colonies was growing rapidly. There were no new French-language missionary accounts from the Caribbean to reassure metropolitan readers that this flood of African captives did not portend disaster.[27] Rather, there was an influx of translations into French of texts describing slavery in the British Caribbean, including Aphra Behn's novel *Oronoko* (1688, translated 1745); Hans Sloane's *Natural History of Jamaica* (1707 and 1725, translated 1740); and Charles Leslie's *History of Jamaica* (1740, translated 1751).[28] These accounts were influential. For example, Montesquieu, who knew Sloane and owned a copy of his translated book on Jamaica, described the dangers masters faced when they made slavery too unbearable for their workers in his 1748 *Spirit of the Laws*.[29]

Around this time Antoine-François Prévost's journal *Le Pour et le Contre* described the rise of Jamaica's Maroon population and the inability of the British to defeat this internal enemy. Prévost provoked in metropolitan intellectuals "the persistence of a troubling idea: the frequency of the troubles, the impossibility of ending a rebellion ... all indicating an disease with no cure and creating fear of contagion."[30] Prevost's 1735 French translation of the apocryphal speech of Moses Bom Saam, said to be a Jamaican Maroon chief, just months after it appeared in *Gentleman's Magazine*, shows that French intellectuals were exposed to many of the same literary images of slavery as their British counterparts. News of Tacky's Revolt compounded their concerns. In his 1768 history of England, Jean-Baptiste Targe noted that as the British were celebrating their conquest of a "sterile" Canada in 1760, they almost lost Jamaica, their richest colony. Drawing on British newspaper accounts, Targe claimed the British planters were saved by "the free blacks who are commonly called wild negroes and who live in peace in their establishments under the protection of the government." He went to describe the punishment of the rebels during which "their cruel masters, according to our English sources, enjoyed the barbaric pleasure of seeing the slow death of beings like themselves, whose only crime was to have sought to recover the liberty taken from them under no other law than that of violence."[31]

It was no accident, therefore, that the first French antislavery story set in the Caribbean—Saint-Lambert's 1769 "Ziméo"—was about a Jamaican slave revolt.[32] It is significant that even in a period when contact between Jamaican and Saint-Domingue was relatively slight, except for smuggling, French perceptions of the similarities between Jamaica and Saint-Domingue grew.

As these images of Jamaican slave unrest replaced Labat's rosy picture of contented and well-run French plantations, French metropolitan officials increased their efforts to convince Saint-Domingue planters to treat enslaved workers with greater humanity. Versailles's official instructions to Saint-Domingue's new intendant Clugny in 1760 urged him to fight marronage by requiring masters to provide slaves with enough food and clothing. In 1762 Choiseul, the minister of the navy, urged the colonial Councils to "produce guidelines for a commission to investigate the welfare and treatment of slaves." This initiative followed on the heels of similar anti-torture initiatives in 1712, 1727, and 1747.[33]

Nevertheless, imperial administrators after the Seven Years' War worried more about colonial loyalty than they did about possible slave revolts. The idea that slavery was an obstacle to patriotism helped inspire France's attempt to launch a new colony on the coast of Guyana in 1763. Rather than emulate Saint-Domingue, the organizers of the failed Kourou River colony designed a colony intended for European farmers and their families, who Versailles imagined would defend its territory from British attack, rather than capitulate so as to protect their investments.[34] The death of close to ten thousand settlers at Kourou, caused by disease but also by poor planning, as noted in Chapter 4, illustrated how a sense of urgency that overrode more practical considerations shaped Versailles's approach to colonial reform after the Seven Years' War.

Working out how to prevent slave revolt was a more urgent concern in Jamaica than in Saint-Domingue. After the near-disaster of Tacky's Revolt in 1760, white Jamaicans came to believe that their only protection against a massive black population was to enshrine in law white supremacy. They understood that the sheer scale of Jamaican slavery, alongside continuing high white mortality, meant the island would never become a settler society like South Carolina or Virginia.[35] This realization meant that slave discipline would have to be ferocious to keep Jamaican colonists safe. What was extraordinary in Jamaica is that colonists did not pretend that their rule needed to be justified. They also cared little about what blacks thought about their treatment. White Jamaicans acknowledged with equanimity that they were tyrants, which meant that no one could stop their depredations. As the historian Bryan Edwards put it, not entirely disapprovingly, the occasional planter kindness "affords but a feeble restraint against the corrupt passions and infirmities of our nature, the hardness of avarice, the pride of power, the sallies of anger, and the thirst for revenge."[36] Tacky's Revolt did not make Jamaican masters more violent; that was scarcely possible. Instead, it fostered a belief in white caste solidarity that was so strong that it could overcome every other social or economic consideration. After Tacky, therefore, "whiteness" became more important. Specifically, a law passed by the Assembly in 1761 made Jamaica's traditions of "passing" from colored to white—formally, through legislation, and informally, through social acceptance and inconsistent record keeping—much more difficult.[37]

After 1760 Jamaicans began to sharply differentiate between those residents who had rights (whites only) and those who were excluded from having any rights (all slaves and increasingly most free people of color). White Jamaicans were committed to the primacy of the law and insisted on the rights and privileges that were their birthright as freeborn subjects of the

Crown. As the Jamaican Nicholas Bourke wrote in 1765, "If our lives, liberties and properties are not our inheritance, secured to us by the same laws, determined by the same jurisdictions, and hence in and defended by the same constitution, as the wisdom of our ancestors found it necessary to establish, for the preservation of these blessings in the mother country," then the condition of British colonists resembled that of "our neighbours, the American Spaniards" in terms of the "oppression, injustice or hardship" they "do at this time endure." If colonists were not British, then they were "not freemen but slaves; not the subjects but the outcasts of Britain."[38]

What most concerned Jamaican whites was the thought that white settlement on the island had failed. They had hoped before Tacky that Jamaica would become a prosperous island, in which enslaved blacks would produce valuable plantation goods, as their white owners created a civilized, improved settler society resembling England. To that end, planters embarked on a series of expensive schemes designed to attract settlers to the island, especially white tradesmen with wives and children. All these projects failed. Governor Charles Knowles, a vociferous critic, estimated that in the late 1740s and early 1750s the colony had wasted as much as £30,000 or 633,000 livres trying to bring seven hundred families to Jamaica. By the late 1750s, Jamaica had given up on settlement schemes, such as a 1750 act giving migrating white tradesmen a bounty. The Assembly accepted that these schemes were ruinously expensive and ineffective.[39] As slave imports rose and as white settlement schemes failed, Jamaican colonists feared that they were being swamped—culturally as well as physically—by African ideas, values, and influences. They did not like the direction in which they saw matters drifting. The only resistance possible to such Africanization was to seal off as much as they could all avenues whereby people of color, especially women, could masquerade as whites.[40]

In 1761, the growing community of free people of color became the scapegoat for Jamaica's failure to become a society of European settlers. Such ideas had been bruited before the Seven Years' War and before free people of color had started to become a growing population. In 1746 the anonymous author of the *Essay Concerning Slavery* explained why Jamaica's rising free colored population was an obstacle to increasing the number of whites, especially white tradesmen. Free people of color, like white Jamaicans, he argued, were devoted to slavery and were convinced that the only way work could be done was through the labor of enslaved people. It was a fact, the author declared, "that whatever you allow to be done by slaves, you will never thereafter get a

White Man to do." This fact was true for free coloreds too, he pointed out. They were "ruined" by their devotion to acquiring and exploiting slaves, becoming a "lazy debauched Race, a very Nuisance to the Community." If, however, "they were obliged" to work themselves rather than to have slaves work for them, Jamaica's free coloreds could become an "adornment" to society. The author of the *Essay* advocated making free coloreds into an intermediate class between whites and slaves. The way to do this, the author argued, was to eliminate skilled slave craftsmen. The *Essay* argued for the immediate manumission of all "negro tradesmen," who would then serve seven years as indentured servants. After such a period, the colony should reward these newly freed blacks with fifteen or twenty acres, taken from the abundant land that large planters had engrossed. This policy, the author thought, would have multiple benefits. First, it would promote the migration of white "artificiers" to Jamaica. White tradesmen did not come to the island because when "a new artificier arrives" he was forced to be "a Drudge to some of the old Standers and grandee tradesmen," all of whom had slave artisans. By eliminating skilled slaves, Jamaica could become a society in which "an *Englishman*, just arrived, would be as able to hire as well as the best old Stander, and he would be likely to be thought, or really to be, as good a Genius as any one of them."[41]

Second, by ensuring that no slaves could practice skilled trades, the author contended, free coloreds would stop aspiring to own slave craftsmen and would live on earnings from their own labor. Working for their own account would transform free coloreds into a useful yeomen class, working on the colony's estates. Like traditional English yeomen, they would be encouraged to serve in the militia, leading "to a tolerable Company of free negroes fit for arms, which would be of great service." The problem the author envisaged was "fixing" the "free negroe to his estate." The solution, the author believed, was a law: "no free Negro should be allowed to be in towns that was not an Artificier." The author thought that authorities would not need to enforce the law heavily for it was "likely negroes would stick to their estate, which he would probably be fond of too, as having been bred to it."[42]

An idea implicit in the *Essay* is that the way to prevent Jamaica from being Africanized was to give special privileges to Europeans, notably to poor white migrants, that elevated them above the brown and black populations. White emigration thus would increase, the author of the *Essay* thought, if Jamaica became a caste society, in which social mobility was confined to the white population. These policy prescriptions involved a good deal of wishful thinking about the limited future of slavery. The author declared that "of this I am persuaded, whatever Nation in America declares first for liberty and absolutely abolishes Slavery must be Masters of the whole Continent."[43] This statement suggests that the author of the *Essay* was Governor Edward Trelawney. Four years after the *Essay* was published, Trelawney informed the Board of Trade, in a message he carefully concealed from the Jamaican Assembly, that plantation slavery should not be established on the Mosquito Shore, in Central America. The English, he averred, were "the worst managers of slaves under the Sun." Their policies might work, barely, on an island. But on a continent, where slaves could easily flee to more lenient Spanish masters, trying to establish Jamaican-style estates would be a disaster: "once the negroes have run away, our fruitful plantations will become wilderness again, [and] meanwhile the movement to the Shore will alarm Spaniards." The best idea, Trelawney thought, was to keep the Mosquito Shore as a refuge for a small colony of poor people, forced out of Jamaica by large planters' land monopolization and extreme devotion to slavery.[44]

Trelawney's vision of a Jamaica without slavery was morally brave, intellectually consistent, and hopelessly impractical. Britons moved to Jamaica to plant, to make money, and to escape the social and political constraints of the motherland. As the subsequent history of Jamaica showed, whites resisted any attempt to limit their rights to own slaves. They would have condemned Trelawney's vision, if they had known about it, as profoundly reactionary. He envisioned, for example, fixed and separate social classes or estates. Under Trelawney's scheme, poor planters in the Mosquito Shore, for example, would stay poor, for without slaves they could never move into the ranks of larger planters. Moreover, Trelawney argued for the formation of a free black peasantry, a fanciful notion that was a perpetual favorite of imperialists who were troubled by the moral effects of slavery on colonists.[45] Conservatives in Britain and in Jamaica understood that slave owning and the wealth it produced threatened British social hierarchies. William Beckford of Hertford Pen, for example, believed that "negroes could be transformed into a peasantry" if there was an "alteration in the manners and pursuits of Europeans." Slavery, he thought, was both cause and effect of the dread "leveling principles that obtains among the white people of Jamaica." In Europe "the chain of subordination that descends from link to link ... preserves the strength of the whole," while in Jamaica, the tyranny that accompanied slavery "entrenches upon the duties of society and annihilates the bonds of power and the good effects of subordination."[46]

In December 1761, the Jamaican Assembly passed a law, or act, that altered the place of the colony's free people of color in two critical ways. The law changed the long-standing method of "passing," moving the color line one generation further than had been previously accepted. It was now "persons of the fourth degree (from a negro), when granted the rights and privileges of whites" who could be legally exempted from regulations applying to free people of color, rather than the third degree as established in 1733. More critically, the 1761 law stated that "no land, cattle, slaves, money or real estate be granted for any Negro whatsoever or to any mulatto or other persons not . . . born in lawful wedlock." As the act explained, these bequests "tend greatly to destroy the distinction requisite and absolutely necessary to be kept in the island between white persons and Negroes, their offspring and issue."[47]

This act was the most important legislation that the Jamaica Assembly passed in the aftermath of Tacky's Revolt, more significant than the anti-obeah measures it passed twelve months before. The colony's agent, Lovell Stanhope, later replaced by Stephen Fuller, explained its actions to the Board of Trade, which was concerned that this act took away men's legal right to devise property as they pleased. The Assembly was worried that unpatriotic and morally scandalous rich whites were confusing racial hierarchies by bequeathing large amounts of property to their colored illegitimate children. The Assembly not only limited free colored inheritances to £2,000 but also denied free coloreds the right to defend their property in civil suits. Nothing in the text of the act connected these restrictions to the fear of slave rebellion. But it was no coincidence that it was passed during the same October 1761 session when Lieutenant Governor Henry Moore announced that the slave revolts had been completely suppressed and that the Assembly charged a committee to research how many wealthy mulattoes there were in Jamaica and investigate the consequences of limiting white bequests to nonwhites.[48]

Stanhope defended the restriction of bequests by claiming that it would discourage white men from forming extramarital liaisons with black women. The law would encourage whites to marry other whites and create legal heirs. The main thrust of Stanhope's argument was that stricter racial distinctions were necessary to maintain white control of Jamaica. He was convinced that "laws of colour" were absolute. Just as slaves wanted to form their own black kingdom under Akan rebel leaders, so too free coloreds, "the Descendants of slaves," would seek power. As "power ever follows property," he argued, the frequency by which large estates were devised to "Negroes and Mulattoes" meant that eventually free coloreds would control most Jamaican property. When that happened, Stanhope insisted "it will be impossible to keep them out of Power, the laws ag'st them must be repealed, the Power will then be in their hands and the Island become a Colony of Negroes and Mulattoes." If that occurred, the British may as well abandon Jamaica, as under black rule "all distinction of Colour will at least be levelled and the Inhabitants of Jamaica, like their Neighbours the American Spaniards will, too probably, become one day a People without Spirit, Religion or Morality" as the Spanish Americans are "the most degenerate and dastardly People upon Earth."[49]

In fact, the 1761 policies limiting the wealth of free people of color worked, at least in the short term. Eighteenth-century Jamaica never developed a free colored planter class like Saint-Domingue; after 1761 its free population of color was composed of artisans, plantation employees, small farmers, and poverty-stricken men and women with neither land nor a trade. Daniel Livesey points out that those free coloreds who had already received special privileges were able to escape most of the 1761 limitations. While the Assembly gave voting rights to only a few individuals, it was more liberal in granting privileges like the ability to testify in court or allowing eligibility for a variety of jobs normally reserved for whites. Moreover, Livesey notes, the great majority of free people of color who received special privileges from the Assembly got them after the act was ratified by the Board of Trade in 1768.[50] In other words, the 1761 act drew a strict line that prevented newly freed families of color from attaining the wealth that might bring them elite status. But the Assembly continued to promote free coloreds who were already wealthy or who had strong family connections to elite white Jamaicans, allowing them limited and specific avenues of social promotion into elite status, while controlling which individuals and families were eligible for these privileges. The same limited opportunities were available for free people of color outside the legal and political sphere. In education, for example, the directors of Wolmer's, a charitable school founded in 1735, resolved in 1777 that "none but children of white Parents be admitted." The institution, however, continued to accept paying nonwhite pupils until 1798.[51] After 1761, therefore, the line between "white" and "nonwhite" in Jamaica was not impermeable, but it was carefully policed. Free people of color had to carry papers of registration and pay extra taxes, and they found it harder to get poor relief or other forms of charity. They presumably found it harder to get employment on estates, for after 1761 they no longer met the requirements of the "deficiency" laws requiring plantations to have a minimum number of white employees. Jamaican free families of color might still hope that one day their descendants might be regarded as white, but that status became far more difficult to achieve after Tacky's Revolt.

The relationship between these changes in how Jamaicans treated free people of color and new racial ideas circulating in Europe is not entirely clear. Peter Fryer argues that it was West Indians writing about the controversial *Somerset* case of 1772 who introduced a more biological form of racism to Great Britain. It is possible that there was a peculiarly colonial construction of racial difference that prefigured later, more explicitly racist, theories of biological determinism. Continental intellectuals had been dissecting the bodies of Africans since the beginning of the eighteenth century, and in 1768 the Dutch writer Cornelius DePauw published his *Recherches philosophiques sur les Américains*, taking the environmental arguments of the Count de Buffon and transforming them into a theory of the inherent physical and cultural "degeneracy" of Native Americans and Africans. Voltaire, among others, had insisted since 1734 that Africans had a different origin than Europeans.[52]

Jamaican historian Edward Long, who came close to this "polygenesis" stance at several points in his *History of Jamaica* of 1774, was a pioneer in asserting that Africans had ineradicable physical differences to Europeans that rendered them suitable for perpetual servitude. Long, however, was in the minority. Moreover, he was so attached to humoral interpretations of race and ethnicity that he cannot be seen as a fully-fledged biological racist.[53] Most people in Jamaica as in Britain held to long-standing views of race that emphasized that complexion was changeable by climate and environment and denied any suggestion that Africans were of a different race, or of a different kind, to Europeans.[54] It appears, rather, that the move to more rigid racial classification in both Jamaica and also later in Saint-Domingue was rooted in the exigencies of practical politics. If planters' safety could be best maintained by colonists sticking together at any cost, then "whiteness" had to be a clear legal category. The law had to clearly demarcate whites from nonwhites. Racial mixing and the social and economic prominence of some biracial families might lead to the gradual Africanization of Jamaican colonial society.[55]

Whites in Jamaica had only themselves to blame for the eventual unwillingness of free people of color to "naturally attach themselves to the white race as the most honourable relations, and so become a barrier against the designs of the Blacks," as James Ramsay put it.[56] White colonists treated every attempt by free people of color to advance their own position as an attack on racial barriers and answered such efforts by tightening the restrictions that made free people of color's lives so precarious. Jamaica's reform of 1761, restricting the ability of free people of color to move into the elite ranks of colonial society, was also another step toward white egalitarianism. In other words, by making "whiteness" a category that was legally defined and heavily policed, the Assembly promoted the benefits of being white. Clerks became more rigorous in describing free people of color by their racial status in birth, marriage, and death records, noting that they were not white rather than leaving their racial designation vague, as had often been the case in records from the 1750s. When Mary Augier died in Kingston in 1760, for example, she was described as a widow rather than as a free mulatto. After 1761, the Augiers, a prominent and comfortably wealthy free colored family who regularly sought white partners, could not pass as whites, even though by the late eighteenth century many in the family were quadroons rather than mulattoes, using the nomenclature of the time. Both Jane Augier and her quadron son William, for example, were denoted in the Kingston parish register when they died in 1766 by color rather than status.[57] By this date, Jamaica had effectively turned itself into a color-conscious society where caste overruled class.[58]

A similar process of racial segregation did not occur in postwar Saint-Domingue until after 1769. When it did happen, however, the change was more radical than Jamaica's 1761 reform, in part because Saint-Domingue's free colored population was proportionally larger and possessed more property than its Jamaican counterpart.[59] Imperial politics, not a slave revolt, motivated the advent of Saint-Domingue's new stricter color line, as discussed in Chapter 7.

In the mid-1760s, however, patriotism rather than racial purity was the paramount concern of French colonial authorities. The Seven Years' War surrenders of Guadeloupe (1759) and Martinique (1762) were compounded by the attempts of planters in Saint-Domingue's southern peninsula to negotiate surrender terms with the British in Jamaica. One Jamaican privateer even built a country house on Ile à Vaches (Cow Island), within sight of the major southern port of Les Cayes.[60] In the face of the pressing issue of imperial loyalty, Versailles accepted, for the moment, Saint-Domingue's claim that planters, not royal officials, were best suited to monitor, judge, and punish their own slaves for Macandal-style offences. Moreover, since 1721 Saint-Domingue had had a professional rural constabulary, the *maréchaussée*, which took over the task of hunting escaped slaves from the militia. Because white men came to Saint-Domingue to make a fortune, not to track other men's slaves, in 1733 and 1739 officials reformed the institution so that it was composed mostly of free men of color. These constables received a salary, performed weekly searches, wore distinctive bandoleers, and collected a bounty when they apprehended an escapee.[61]

The problem of securing Antillean loyalty to France was not a new issue. Well before the Seven Years' War, Dominguan intellectuals had rationalized planters' refusal to sacrifice themselves, their enslaved workers, and their estates in order to defend the colony. Emilien Petit's 1750 *Le Patriotisme américain* argued that planters' French loyalties would triumph over self-interest only when France removed the colony's "tyrannical" system of military government. He advocated replacing the "arbitrary" personal power of parish military commanders with a more liberal and consistent "rule of law." Freed from the harassment and unpredictability of these local potentates, planters would become attached to the colony and rational self-interest would strengthen their attachment to France. In his book, Petit described how racial segregation might place the interests of whites ahead of free coloreds and relieve French colonists of the militia duty they hated. Like the anonymously authored *Essay concerning Slavery* of about the same time, Petit advocated banning free coloreds from the cities, so as to encourage white craftsmen to come to Saint-Domingue. Confined to rural zones, free coloreds would become a productive peasant class that could shoulder Saint-Domingue's militia duties. But his main point was constitutional, not racial. Metropolitan-style absolutism could not function in the self-interested world of sugar planters; the colony needed a new governing system.[62]

Petit was part of a new generation of elite colonists, a generation born in Saint-Domingue and educated in French legal circles. This generation came into its own after the Seven Years' War. Another elite colonist, trained in France as a barrister, was Jean-Pierre Desmé-Dubuisson, whose father, Claude, had come to Saint-Domingue in the 1720s. The older man established himself as a merchant in Cap Français and married a planter's daughter, eventually inheriting a valuable sugar estate. In the 1740s, during the War of Austrian Succession, Claude Desmé-Dubuisson founded a unit of musketeers and paid to have four thousand pikes made for the white and black soldiers defending Cap Français. He served as parish sexton and commissioner and as treasurer of the city's charity hospital, and he contributed money for Cap Français's first theater.[63] He gave to his three sons, born in

Saint-Domingue, "all possible education to make them capable of serving the colony and attaching them to it." In 1752, his two eldest sons finished their legal training in Paris. He informed the naval minister that "I want to bring them with me next November to Saint-Domingue to make them heads of family and counselors in the Superior Council of Cap Français if you deem it appropriate."[64] One of those sons was Jean-Pierre Desmé-Dubuisson. He returned to Saint-Domingue in 1753 and became a magistrate on the Cap Français Council almost immediately. In 1760, he was named royal attorney at the unprecedented age of twenty-seven.[65]

That same year, as the Seven Years' War dragged on, Saint-Domingue welcomed a new intendant, or head of civil government, to the colony. Like Dubuisson and Petit, Jean-Étienne Bernard Clugny de Nuits was a creole, born in Guadeloupe. The owner of a colonial plantation, he trained in the law and was a member of the Parlement of Burgundy. The choice of Clugny as intendant signaled that the Naval Ministry had heeded colonists' complaints about military administrators, for he was Saint-Domingue's first intendant who was not a naval officer.[66] Clugny was true to his civilian roots. Despite being an intendant during a war, he quickly set about creating and reinforcing civilian institutions. He found a meeting place for the newly formed Cap Français Chamber of Commerce and Agriculture, and he suggested the creation of a commercial exchange.[67] In 1762, even as Martinique surrendered to a British attack, Saint-Domingue's shift away from military government was confirmed when the naval secretary Choiseul announced he was creating a Legislative Commission, with a brief to restructure colonial government. He appointed Emilien Petit to represent Saint-Domingue's Councils on this body. At about the same time, he ordered Saint-Domingue's Council judges to review all colonial laws and to describe the reforms they believed would be best for the colony "with regard to the administration of justice, the maintenance of order, the increase of commerce and the population of the colony." In October 1762 colonists learned that the Crown had prohibited military officials from interfering in legal affairs, resolving a long-standing complaint about the "arbitrary" power of military governors.[68]

In other words, in the early 1760s, as France's other Antillean colonies were throwing themselves into British hands, Versailles was addressing Saint-Domingue's long-standing constitutional grievances, primarily to secure the loyalty of the kingdom's most valuable West Indian possession. Imperial officials, rather than challenging planters over the treatment of their slaves, as they did in the 1780s, worked to end what colonists described as their own "slavery" to the unchecked power of colonial military officials. Nevertheless, the war intruded on these ambitious schemes to strengthen colonial self-government. Despite official movements toward reducing the power of Saint-Domingue's military government, the threat of a British invasion reinforced the status quo. The colony's top military commander, the Vicomte de Belzunce, brought the civilian militia and the constabulary under military command and created a five-hundred-man experimental free colored army unit known as the *Chasseurs volontaires*.[69] In December 1762, anticipating a British attack on Cap Français, Versailles promoted Belzunce to governor-general of Saint-Domingue, setting aside, as described in Chapter 4, Gabriel de Bory, the previous governor, who shared Intendant Clugny's interest in reform.[70] Belzunce was an authoritarian. Indeed, in February 1763, he began a series of new defensive projects, requiring planters—including Council judges—to send him their slaves to cut timber and aid in construction. Neither the end of the war nor the high death rate of the eighteen hundred European troops stationed at the camp was enough to stop this unpopular project. Construction came to a halt only in August, after disease killed the governor himself.[71] An anonymous writer wrote decades later, "The power of the king's officers was nearly intolerable during the last two wars. They showed that their talent was not so much to repel the enemy as to vex the citizens. Every day they held reviews, drills. . . . At the end of a review, one [officer] was heard to say, all puffed with pride, how glorious it was to make men who possessed millions stand for two hours in the sun."[72]

Such complaints were exactly why the Naval Ministry decided, finally, to bring civilian government to Saint-Domingue. Thus, on 24 March 1763, just weeks after concluding the Treaty of Paris, Choiseul signed new legislation that narrowly defined the areas of authority of Saint-Domingue's military and civilian governments, overseen by the governor-general and the intendant, respectively. Of the 119 articles in the law, the most important was article 4: "His majesty wishing in the future that his military service be done in Saint-Domingue by regular troops, there will be neither general nor particular militias in this island."[73] This article signified that henceforth civilian rule would be paramount in the colony. The new law spelled out in detail the powers of military and civil administrators and devoted one article to limiting the power of local military officers: "the local commanders . . . will have no authority over residents except regarding matters that may interest local security; they will in no way interfere in anything that might involve the colony's police or the civil administration."[74] Most of the text was devoted to describing the duties of a host of new bureaucrats who would serve under the intendant, replacing the wide powers previously given to military officers.

On 15 June 1763, the Cap Français Council registered the reform. Two days later it approved another law, this one written locally and presented by Intendant Clugny. This new statute set in place a locally elected official, the *syndic*, who would take over the contentious responsibilities previously managed by the militia commander. This official would arrange military lodging; requisition slaves, carts, animals, and provisions; plan road work; compile census reports; and oversee butchers and taverns. The syndic would serve for three years and report to the local representative of the intendant. The law also contained the seed of what would become a major issue for wealthy free men of color, who had previously commanded their own militia units. The militia was abolished, but the syndic law provided for three "mulatto or negro" men to be assigned as messengers to the offices of the syndic and the local military commander. The law was unclear about whether these would be free or enslaved men of color, but it ordered the syndic to draw up a list of men eligible for this work in each parish. Even as the 1763 law liberated all free men from the aggravations and humiliations of militia service, this messenger service showed that men of color would continue to serve the state in ways that would be both aggravating and humiliating.[75]

Versailles wanted colonists to know that these long-desired changes were expensive. The day after Choiseul signed the government reform, he finalized a second law spelling out the salaries of royal colonial troops. In August 1763, he sent the colony a request for 4 million livres in taxes a year, which would offset some of the costs of defending a colony that had no local militia.[76] Yet even with this new colonial revenue, Choiseul soon regretted his reform. From the mid-1760s a new and more spirited atmosphere of contention in Saint-Domingue's elite ranks was evident, especially as more colonial judges than ever before had been trained in metropolitan legal courts.

The result of this immersion in French legal traditions was that by the 1760s colonial Council judges came to see themselves as equivalents of French parlementary judges, the magistrates of the kingdom's thirteen provincial law courts. Moreover, they saw the Councils as having a responsibility parallel with that of the parlements to defend Saint-Domingue's traditions and privileges against the royal bureaucracy. In France, after 1763, a handful of parlements fought Louis XV's attempt to increase provincial taxes and strengthen ministerial power. In 1766 Brittany's royal governor, for example, closed that province's parlement and sent its leading orator to the Bastille.

The naval minister Choiseul, who served as Louis XV's chief minister until 1770, was deeply identified with royal power in this struggle. By adopting such pretenses in the colonies, the Council judges of Saint-Domingue were sending what royal officials thought was a worrying sign of their willingness to assert themselves politically.[77]

Moreau de Saint-Méry, another creole barrister like Desmé-Dubuisson, described how Saint-Domingue's judges, who had originally been chosen from planters and militia officers, continued to wear swords and elaborate costumes of their own individual design right down through the Seven Years' War, in order to insist on their own status vis-à-vis the elaborate dress of the naval officers who ran the colony. Significantly, in 1766 the judges adopted a black robe like French parlementary judges, though they continued to wear a sword under those robes.[78] Like Jamaican colonists in their struggle with Governor Lyttleton, the Dominguan judges were asserting that their institutions were similar in form to those of the homeland. In 1763, Saint-Domingue's judges, informed by this oppositional culture and influenced by their aspiration to be considered one of France's "sovereign" courts, refused to register Choiseul's reform of the constitution as law in the colony, even though that reform would have considerably increased the power of their institution. The reason they gave was that the document came to them directly from the Naval Ministry and not from the King's Council of State. Choiseul was infuriated at what he saw as both nitpicking objections and also unwarranted assertions of colonial autonomy.[79] On 6 September 1763, he rebuked the judges for their attempts to amend the reform law, fulminating that "[the Councils] have gone so far as to assume legislative powers . . . which [the king] can only regard as a formal attack on his authority."[80]

As the stalemate dragged on, in February 1764 the colony's outgoing governor and intendant convened an unusual plenary meeting of the Councils of Cap Français and Port-au-Prince. They asked the judges to register the four-million-livres tax, reminding them of the recent reforms that France had been pleased to grant them. After the governor's speech, Jean-Pierre Desmé-Dubuisson addressed his assembled colleagues. Raised by his father to be a creole patriot and trained in Paris to defend the prerogatives of the kingdom's parlements, Dubuisson spoke to a number of other judges from similar backgrounds. Philippe Galbaud du Fort, for example, was the son of an official in the Brittany Parlement's Chamber of Accounts and had served in that body himself from 1745 to 1762. In 1762 Galbaud came to Saint-Domingue to manage the family's sugar and coffee estates and was named a member of the

Port-au-Prince Council.[81] Desmé-Dubuissson's speech conveyed the postwar self-confidence characteristic of Saint-Domingue planters enjoying, as in Jamaica, an unprecedented economic boom. He reminded his colleagues that Saint-Domingue's founding buccaneers had not been conquered but had voluntarily allied themselves with France, advising the judges not to surrender "the advantageous privilege that their ancestors earned by giving themselves to France, the right to impose their own taxes; a precious right."

It was a touchy point for both Jamaican and Saint-Domingue legislators. They insisted that they were not conquered places but had been settled freely by Europeans who had never given up their rights when migrating across the ocean. Desmé-Dubuissson hinted that the two Councils were not just high courts, but in fact sovereign bodies with the ability to delay and criticize royal laws, just as the French parlements claimed in their battle against Louis XV's centralizing measures. He continued: "how consoling and encouraging for the colonists to see . . . that His Majesty appears to join and link the tax increase demanded of you to the suppression of the militia and the exemption from special labor obligations!" He urged the judges to ratify the new taxes, not because the Crown ordered it, but in a quid pro quo for the suppression of the militia system and associated abuses—"insurmountable obstacles to the growth of agriculture and population."[82] His arguments were startlingly reminiscent of the arguments made by Jamaican patriot Nicholas Bourke in 1765, remonstrating against Governor Lyttleton's attempt to limit the prerogatives of the Jamaican Assembly. Like Jamaican planters flexing their muscles against Governor William Lyttleton and proclaiming boldly all the rights and privileges of Englishmen, Desmé-Dubuisson used the quasi-mythical story of the buccaneers to suggest that Saint-Domingue's relationship with France had always been based on consent, thereby directly challenging French imperial authority.

This constitutional conflict charged the political atmosphere that greeted Charles-Henri Théodat, Count d'Estaing, Saint-Domingue's new governor-general, when he arrived in Cap Français about a week later. D'Estaing was a high-born army officer who had spent the Seven Years' War in the Indian Ocean, where he became notorious for zealous and unorthodox behavior. Contemporaries disagreed whether it was these patriotic exploits, or the fact that his illegitimate half-sister had given birth to two children by Louis XV, that had earned him the unusual dual rank of lieutenant general in the navy and in the army by 1763.[83] D'Estaing was Saint-Domingue's first high-born noble governor, and he played this role well, arriving with twelve domestic servants, armorial china, a large library, a fine wine cellar, and an elaborate wardrobe. In France d'Estaing's debts were such that creditors had claimed much of his landed property. But Versailles did not want its top administrators to be awed by planters' wealth. After 1759 the Naval Ministry paid Saint-Domingue's governor 150,000 livres a year, or £6,500, more than five times d'Estaing's income in France. At age thirty-five, the new governor had a libertine reputation that suggests he had much in common with Saint-Domingue's planter class. He had once, for example, written a twenty-five-hundred-line poem called "Pleasure." In Saint-Domingue d'Estaing and his handpicked intendant, René Magon, bought a country house on a hill overlooking Cap Français for their assignations. For their official duties they remodeled the city's former Jesuit residence into a governor's mansion at considerable expense.[84] Yet d'Estaing's political aim was anything but hedonistic: he intended to assert imperial and gubernatorial power at any opportunity, just as Lyttleton was trying to do at the same time in Jamaica.

On 7 May 1764, d'Estaing appeared formally before the Council of Cap Français.[85] Tensions were so high that the judges requested that he speak from a written text, so they could enter it into their registers for future debates. For his part, the new governor asked the judges to give him, in writing, their thoughts about the 4-million-livres tax. Representing his office as an extension of the monarchy, d'Estaing explicitly challenged Desmé-Dubuissson, denying his claims that the Council had the power to critique and delay royal laws. This assertion by the royal attorney amounted to "conditional obedience," d'Estaing argued, an attempt to put the colonists on the same level as the monarch. On 21 May 1764, after the Council still had not moved on d'Estaing's request, he ordered Desmé-Dubuisson to board a ship for France to explain himself to the Naval Ministry.[86] The outspoken creole attorney general duly went into "exile" in France.

From this moment forward, the battle between the Councils and the governor was joined. The two Councils were legally obliged to register a royal letter appointing d'Estaing and describing his powers, but they delayed this for nearly a year.[87] They did agree to the 4-million-livres tax, on Desmé-Dubuisson's rationale that it was the price for abolishing the militia.[88] They did not know yet, however, that Choiseul had secretly ordered d'Estaing to reinstate the militia system. There was fierce colonial opposition, therefore, in January 1765 when d'Estaing offered a new militia ordinance, complete with the return of local parish commanders. But as late as March 1765 colonial judges had still not formally registered d'Estaing's letter of instruction, which allowed them to claim that he did not have the authority to make these changes.[89] D'Estaing had such poor relations with Saint-Domingue's colonists that newspapers in Europe in March 1765 claimed that Saint-Domingue stood on the verge of a revolt against the Crown.[90] Choiseul recalled him at the end of the year. Although the circumstances were different, this was a similar scenario to what was happening almost simultaneously in Jamaica: powerful colonial potentates using their resources against metropolitan attempts to rein them in. The fate of the exiled Jean-Pierre Desmé-Dubuisson illustrates that royal officials were unsure about how far they should push planters. Arriving in Paris in 1764, Dubuisson quickly rejoined the Parlement of Paris, where he had been trained and been accredited. In December of that year, Louis XV named him an honorary member of the Council of Cap Français, noting that he had "voluntarily resigned" from the office of attorney general in that body. His honorary status gave him full voting rights in the Council.[91] By 1769 Dubuisson was back in Cap Français, sitting in the Council, and he was still there in January 1772, though he died in France at the end of the year.[92]

By the middle of the 1760s a series of laws and political compromises had created a new racial order in Jamaica.[93] In Saint-Domingue a bitter struggle between Versailles and colonial elites over postwar constitutional reforms threatened to exacerbate the problem of planters being more loyal to their investments than to the empire. White colonists in Jamaica shaped new racial laws and customs that made "whiteness" a precondition of membership in the polity. In Jamaica, the architects of the new racial order tried to unify the white community against any future slave attack. After 1769, as we argue in the next chapter, an increasingly rigid system of racial demarcations also fostered white support for the French regime in Saint-Domingue. The precondition for this action was a growth in colonial political assertiveness as colonial jurists tried to claim that they, like French *parlementaires*, could negotiate with the Crown over what laws they would follow. Politics came first; then came racial profiling and increasing caste consciousness. Moreover, it was people of mixed race—exactly the sort of people who might rise from one racial category to another if racial taxonomies were not fixed and regularly policed—who were the main subject of attention. As might be expected given the reactive nature of these racial policies, they were often haphazard, inchoate, and even contradictory. What is important is that race was replacing all other ways of determining status and standing within Jamaica and

Saint-Domingue. It happened more quickly in Jamaica, because Tacky's Revolt galvanized planters into action and because colonial elites in Jamaica, until the late 1760s at any rate, had a much greater ability to shape legislative policy than did their counterparts in Saint-Domingue. Money might be the measure of all things in the Greater Antilles, but that measure worked only in certain circumscribed circumstances. The result was two societies that by the eve of the American Revolution were remarkably caste conscious, racially obsessed, and racially exclusive.

CHAPTER 7

The Golden Age of the Plantocracy

The years from 1763 to 1776 were extraordinarily prosperous for white colonists in both Saint-Domingue and Jamaica. That prosperity was based on the increasing efficiency and scale of production on their estates, and on their growing numbers of enslaved workers. In both empires, colonial leaders recognized the importance of keeping planters happy, productive, and loyal. In the mid-1760s colonists and their local representatives in both societies had successfully defied governors' attempts to control them.

But this changed in the late 1760s. In 1769 Saint-Domingue's planters and colonial judges failed to defeat an unpopular imperial reform, even though some parishes went to the point of outright revolt. In the same period, Jamaica's colonial elites accepted, more or less quietly, the imperial reforms that convinced mainland colonies to unite in rebellion. They did so less because they accepted constraints on their freedom of action than because their reflex position during the 1760s and 1770s was toward loyalism. Saint-Domingue's planters and colonial judges pushed even harder against imperial reforms than did Jamaican planters, to the point of violent protest. But they too eventually accepted imperial reforms, adopting new laws limiting the social mobility of free people of color. Like Jamaica after Tacky's Revolt, Saint-Domingue solidified the caste lines between white and mixed-race families in order to resolve tensions about colonial identity. By the 1770s, both colonies had developed firm ideologies of whiteness, emphasizing that only those of "pure" European ancestry were full citizens. Thus, in both Jamaica and Saint-Domingue the movement toward caste consciousness can be explained only through close attention to particular political events at particular times.

Saint-Domingue's prosperity after the Seven Years' War brought a new level of interest from France. The Crown recruited governors among high-ranking nobles in the French navy and paid them handsomely. The colony's profitability also attracted investments from leading figures at court, a trend that accelerated after the American Revolutionary War. In 1768, for example, the financier Jean-Joseph de Laborde, who had made his fortune lending money to Louis XV during the Seven Years' War, purchased a plantation and lands in Saint-Domingue's southern peninsula. His estate eventually included three adjacent sugar plantations with a total of fourteen hundred slaves, making him one of the very largest slave owners in the Greater Antilles.[1] In 1766 and 1768, contemplating the postwar growth of Saint-Domingue, and the high price of lumber there, the new naval secretary, the Duc de Praslin, a cousin of the outgoing secretary Choiseul, granted family members exclusive rights to the undeveloped coastal islands of Tortuga and La Gonâve, which contained valuable timber. In 1770, Praslin stepped down from the ministry and received a royal grant for another unsettled island, Ile à Vaches. He used profits from the lumber there to help supply his nearby sugar estate.[2] Although high-ranking colonial administrators were theoretically forbidden to purchase colonial properties during their tenures, in 1769 the duc de Rohan-Montbazon, governor of Saint-Domingue, wrote Praslin that he had just purchased a plantation outside Port-au-Prince: "I cannot imagine that this acquisition might displease His Majesty. Your example and that of several persons at court who own property in the colony led me to believe that I was equally allowed to acquire one."[3]

At the center of these growing connections between Saint-Domingue and Versailles in the 1760s was Jean-Baptiste Dubuc, a Martiniquan planter and barrister. Chosen by Choiseul in 1764 to be his chief secretary for colonial affairs (no colonist attained such a close connection with the imperial government in Britain before the American Revolution), Dubuc was a creole whose family had established itself as sugar planters in Martinique before 1672. His grandfather led a colonial revolt in 1717, fighting against the restoration of the French commercial monopoly after the War of Spanish Succession. The next generation of Dubucs kept out of political affairs. Instead they built prosperous sugar plantations on Martinique's Atlantic coast, a region notorious for selling its sugar to foreign traders.[4] In 1761, in the middle of the Seven Years' War, Jean-Baptiste Dubuc, who had studied law in Paris and had qualified for a seat at the Paris Parlement, left home to represent Martinique in France's Bureau of Commerce. Within a year of his departure, the British attacked Martinique, provoking the crisis of loyalty that Versailles had hoped such new schemes of representation would avoid. Dubuc's younger brother, as well as the island's governor, who was another kinsman, was deeply implicated in Martinique's rapid surrender.⁵

This apparent lack of patriotic feeling did not affect Dubuc's career in Paris. It may have made his advice more influential with Choiseul, who hoped to understand the phenomenon of planters throwing in their lot with France's enemy. The two men met in 1764, and after a one-hour discussion, Choiseul hired Dubuc to be his top aide for colonial affairs.⁶ In this position, Dubuc helped shape high-level decisions at the Naval Ministry, most of all a 1767 liberalization of the French commercial monopoly, a reform that strengthened Saint-Domingue's commercial ties to British North America. Moreover, Dubuc's wit and social skills distinguished him in Parisian intellectual circles. He frequented the salon of Mme. de Necker, where Denis Diderot and the Count de Buffon were constant guests. Necker's papers mention Dubuc thirty times, describing him as a brilliant conversationalist and deep thinker. Diderot concurred, calling him "a man of infinite *esprit*." It was probably Dubuc who made colonial records available to Diderot, and it was possibly Dubuc who influenced Raynal's eventual advocacy of free trade for France's colonies.⁷

While the Seven Years' War raised profound questions about French colonial loyalties, it also highlighted a new aspect of Saint-Domingue's commercial potential. Before 1756, many French planters had resigned themselves to jettisoning their molasses, which they produced in much greater quantity than Jamaica, which exported only brown sugar. The North American smuggling trade in the latter years of the Seven Years' War encouraged merchants from the mainland to buy Dominguan sugar at Monte Cristo and to pay good money for molasses. After the war, the possibility of a profitable North American molasses trade tantalized Saint-Domingue's elites. France's colonial trade monopoly, however, prohibited such commerce.⁸

The postwar presence of Dubuc and other colonials at the Naval Ministry, and the promise of Saint-Domingue, prompted Choiseul to reform the French commercial monopoly. In 1763 the Crown liberalized the grain trade within France, and in 1764 it allowed grain to be exported from twenty-seven different French ports.⁹ In 1767, Choiseul put forward a new Caribbean commercial policy shaped by Dubuc, overriding the protests of France's port cities. This reform was not the complete trade liberalization that colonists wanted, but it did establish two Caribbean ports, including one in Saint-Domingue, where sugar syrups could be traded for a list of North American products needed to build the plantation system: livestock, raw and finished wood, and provisions.[10] These growing ties with North America were an important part of Saint-Domingue's postwar prosperity. By the 1780s, some plantations estimated that the sale of molasses and tafia, a crude form of rum, made up between 10 and 33 percent of their profits.[11] These new sources of income help explain the ongoing creation of new sugar plantations, the number rising from 599 in 1754 to 636 in 1771.

The rise of a new crop, coffee, was even more important to Saint-Domingue's postwar prosperity. Coffee was introduced to Saint-Domingue in the 1730s, and by 1771 the island had 3,503 coffee estates. In 1767, the colony exported twice as much coffee as it had before the Seven Years' War: 15.6 million pounds versus 7 million. In many parishes this new crop replaced indigo. In 1754 there had been 3,379 indigo plantations, but in 1771, there were only 1,027. Both crops could be grown by planters who had only thirty or forty slaves, rather than the hundred or more workers recommended for a sugar estate. Moreover, coffee planting supplemented rather than competed with sugar production because coffee thrived in the hills, where slopes were too steep for sugar and where land prices were lower.[12] All of this meant that by 1770, even with the loss of Canada, France's colonies had nearly doubled the prewar value of their exports to the metropole. These were now up from 77.6 million livres per annum or £3.38 million in 1755 to 147 million livres or £6,400,000, and the great majority came from Saint-Domingue.[13]

For Saint-Domingue's colonists, the postwar years, therefore, were as transformative as the war years were frustrating. With the abolition of the militia system, the welcoming of colonial spokesmen at Versailles, the partial liberalization of the molasses trade, and the emergence of coffee, the colony was reborn after the painful experiences of the Seven Years' War. Investment flooded into the colony, including money from highly placed court insiders like La Borde. The leading men of the colony—planters, merchants, and judges—were prosperous, confident, and eager to take advantage of these propitious conditions. There was, however, a lingering fear in the minds of imperial officials. Given the experience of the French West Indies in the Seven Years' War, and colonists' hostility to Governor d'Estaing, Versailles doubted that Saint-Domingue's planters were as committed to the empire as they were to their commercial ambitions.

For white Jamaicans, the years between the Seven Years' War and the American Revolution were also "a brief golden age for the plantocracy."[14] Matters were so good that in 1772 the Jamaican Assembly made the rare gesture of making a governor the subject of particular commendation. Governor William Trelawny, a relation of the very successful Edward Trelawny, governor between 1738 and 1752, was praised in fulsome terms: "We cannot but acknowledge with the greatest Justice that the present Harmony and general Tranquillity of this Country are the natural effects of Your Excellency's steady and impartial Administration." As a measure of respect, after his death a month later, the new western parish cut off in 1770 from St. James and St. Ann was named after him. For Edward Long, the planter-historian, such praise (and the gift of £1,000 sterling, or 23,100 livres, for "giving his remains an honourable interment") was signal evidence that "there are no people in this world who exceed the gentlemen of this island in a noble and disinterested munificence." Alluding to William Henry Lyttleton, Trelawny's controversial predecessor, Long admitted that planters might be accused of "wilfully seeking occasions to quarrel with their governors." But this was true only for governors they did not respect. When, as with Trelawny, planters thought a governor was willing to accept their constitutional positions, they displayed "liberality and a just deference to those governors who have deserved well by the mildness and equity of their administration."[15]

Long's eulogy to Trelawny was pointed. He had been an active member of the Jamaica Assembly during the so-called Privilege Controversy and was a determined opponent of Lyttleton. Indeed, it was governors such as Lyttleton that Long had in mind when at the start of his monumental 1774 history of Jamaica he denounced "provincial governors, and men in power" as "a portrait of artifice, duplicity, haughtiness, violence, rapine, avarice, meanness, rancour, and dishonesty, ranged in succession; with a very small portion of honour, justice, and magnanimity, here and there intermixed, to lessen the disgust, which, otherwise, the eye must feel in the contemplation of so horrid a group." Trelawny was one of just three West Indian governors Long thought deserved praise. Lyttleton, on the other hand, belonged to the general "tribe of hirelings, tools and sycophants," determined to attack the liberty of freeborn Britons living in Jamaica.[16] Not coincidentally, it had been Trelawny who had suggested to the British government in May 1770 that the last claim of the Assembly still contested by the governor be abandoned; he recommended that they not be required to replenish money Lyttleton had drawn from the colonial treasury to pay British troops. The victory of the Jamaican planter class over the British governor was now complete.

The planters were at the height of their power on the eve of the American Revolution. Long's remarkable history is one long paean to their virtues.[17]

For Long, there were no firmer defenders of liberty in British America than Jamaican planters. He even had an answer to critics such as Samuel Johnson who derided Jamaicans for their attachment to liberty while denying it "to so many thousand Negroes, whom they hold in bondage." Like the Romans and Greeks, who also depended on slavery for their prosperity, Jamaicans were more conscious than Britons of the importance of liberty because they saw servitude all around them. He hoped that "they will never surrender their birth-right, but continue to maintain the sacred charter [of liberty], with equal fortitude, to the end."[18]

Long elaborated on planter virtues and planter deficiencies at length. The Jamaican planter, he declared, was a paragon of an English gentleman. He was "tall and well-shaped" with "penetrating sight." In addition, he was "sensible, of quick apprehension, brave, good-natured, affable, generous, temperate and sober; unsuspicious, lovers of freedom, fond of social enjoyments, tender fathers, humane and indulgent masters, firm and sincere friends" and marked most of all by an almost compulsive hospitality. Indeed, he concluded "there are no people in this world that exceed the gentlemen of this island in noble and disinterested munificence." His beau ideal was the Speaker of the Assembly, Sir Charles Price, made a baronet on a recommendation by Trelawny and acclaimed as "The Patriot" by his compatriots. Price died in the same year as Trelawny. Long eulogized him as "endued with uncommon natural talents" with "a truly patriotic attachment to his country." By patriotism, Long meant that Price, "though ever ready to assist and facilitate administration, while conducted for the great principle of public good," was "always the steady, persevering, and intrepid opponent to illegal and pernicious measures of governors." He was the owner of an elegant and "well finished" country house in central Jamaica, which Long thought "an agreeable retreat." As a sign of his Roman rectitude and generosity, he manumitted every year a slave he deemed especially faithful or worthy. His son, also later Speaker of the House of Assembly, constructed for him an elaborate tomb in the parish church of St. Catherine with his father's virtues spelled out in the language of Livy. Long noted that with Price's death "Jamaica [had] lost one of its best friends."[19]

It was not just great planters, however, who prospered in the years immediately before the American Revolution. Thomas Thistlewood had arrived in Jamaica as a close to penniless immigrant in 1750. After spending fifteen years as an overseer on a sugar estate in western Jamaica, he established himself as a pen keeper with a small property and a first-rate garden, where he entertained local dignitaries and their womenfolk to sumptuous meals—roast goose with paw-paw sauce was a favorite dish, while his guests also enjoyed an abundance of fresh fruits, exotic vegetables (such as asparagus), stewed fish, and crabs. He made a comfortable income of between £275 and £300 per annum with debts amounting to less than 1 percent of his overall wealth. His principal source of wealth—his slaves—amounted to twenty-nine by the American Revolution. He was a justice of the peace and a lieutenant in the militia of Savanna-la-Mar and had established a local reputation as a man of science and an accomplished horticulturalist. Some idea of his local standing came on 29 March 1772, when his employer, John Cope, entertained Governor Trelawny. Thistlewood furnished the table, including vegetables and fruit from his garden. Trelawny and his wife were so impressed, according to Thistlewood's friend Harry Weech, that they "several times expressed a great desire" to see Thistlewood's garden. They were stopped only by the warnings of the wealthy planter Richard Haughton, who told them that the road was "so very rocky and bad." For Thistlewood, Jamaica in the 1770s was very much the land of opportunity.[20]

Planters got much of what they wanted from Jamaican governors in part because they had influential allies in Britain. Although there was no exact British equivalent to Dubuc, there were perhaps as many as seventy West Indians in the House of Commons, with Jamaicans prominent among them.[21] This meant that Jamaicans, mainly owing to the different nature of the British and French political systems, were far more able than their counterparts in Saint-Domingue to be able to make their voices heard within metropolitan politics. Their influence was somewhat obscured by a contemporary discourse that labeled Jamaican planters living in Britain as morally delinquent "absentees." Contemporaries, and later historians, have fulminated against the supposedly deleterious effect such "absentees" had on Jamaican politics and society.[22] We prefer the more neutral term of "transatlantic broker" and maintain that this group probably did the colony more good than harm. They worked to maintain Jamaica's privileged position within Britain's mercantilist framework and to ensure the smooth transit of Jamaican produce into British ports. They enjoyed singular success from the 1730s to the 1770s over legislation involving the West Indies, far more so than their North American counterparts. The islands retained their preferential tariff on sugar and rum despite its costs to consumers because they recruited skillful agents, notably Rose and Stephen Fuller, who constantly solicited the government on West

Indian concerns, cultivated friendships with prominent politicians, and submitted evidence to parliamentary committees.[23]

Like Saint-Domingue, Jamaica's wealth was rooted in sugar and slavery. Though it exported little to no indigo, cotton, or coffee, important crops in Saint-Domingue, Jamaica exported other products. And Kingston's trade with Spanish America had no parallel in its French neighbour.[24] Slaveholding was a major component of colonists' personal wealth. Slave prices increased dramatically over time, even if slave mortality rates remained stubbornly high, making a slave purchase potentially risky. High slave prices reflected foreign demand and planters' evolving management techniques, which allowed them to get more value out of slaves' labor. Land also increased in value, though not at the same rate as did the price of slaves.[25] Yet sugar planting was a risky business. Even an accomplished planter could meet with unexplained bad luck that might destroy all his hard work. Thistlewood noted in 1784 how Sir James Richardson, whom Thistlewood thought "a sensible man," had lost 141 of 190 slaves purchased in the previous fourteen years, "such bad luck has he."[26] Profits varied not only from estate to estate but from year to year. Mumbee Goulbourn, an absentee planter during the age of abolition, noted to his attorney that "whoever has a property in the West Indies must make up his mind to the sudden transition from good to indifferent news."[27]

But over time the number of years in which there were productivity, production, and profitability gains outnumbered losses by some margin in the sugar industry. Between 1748 and 1776, the total production of muscovado sugar more than doubled, from 450,000 hundredweight at midcentury to a peak of just over 1 million hundredweight of sugar in 1773, with very similar production levels in 1774 and 1775. Richard Sheridan estimates that the gross profits accruing to the empire from Jamaican plantation production and from trade was of the order of £1,547,000 per annum, or 35.7 million livres, between 1771 and 1775. Rising sugar production came from both an increase in the number of sugar estates (from 566 in 1763 to 775 in 1772) and a rise in the size of individual estates, with larger slave forces than in the past. Moreover, new management techniques improved productivity per slave from just over five hundred pounds of sugar produced per slave in 1755 to nearly seven hundred pounds per slave in 1775. This meant that Jamaican plantations dramatically increased their average output (from 168,000 to 265,000 pounds per estate between 1755 and 1775). This surge in productivity meant that despite slightly declining sugar and rum prices in London in the first half of the 1770s these were profitable years for Jamaican sugar planters. J. R. Ward estimates that between 1763 and 1775 average rates of profit on Jamaican sugar estates were 9 percent per annum, down from the 15 percent per annum between 1756 and 1762, but still much higher than possible in other economic activities.[28]

Analyses of profitability on individual estates bear out the general prosperity of Jamaica in the years immediately preceding the American Revolution. On the small Spring Estate in St. Andrew parish, near Kingston, net profit between 1772 and 1775 was £1,784 per annum, or 41,000 livres, over double what it had been in the early 1750s. On the much larger Penrhyn Estates in central Jamaica, owned by absentee Richard Pennant and managed by attorney (and wealthy planter in his own right) John Shickle, profits were more considerable. From averaging net profits upward of £6,400 for nearly twenty years before Pennant inherited fully in 1771, the next few years were much more profitable. In 1771, net profits on the Pennant properties jumped to £9,886 and in 1772 reached £15,740, or 363,000 livres, a peak that was not to be reached again until the late 1780s. Between 1771 and 1776, the Penrhyn properties averaged net profits of £12,843. It is difficult to establish what the average rate of profit was on these estates, as we do not have figures on the overall value of the estate, but in 1775 Pennant made a net profit of £13,679 from the labor of 1,016 slaves, meaning that the average slave produced a net return in that year of £13.46, or 311 livres, or nearly a third of the capital value invested in each slave.[29]

A full appreciation of what was involved in managing slaves and forcing them to produce at maximum levels so that estate profitability could be enhanced is available through the extensive documentation of the management of Golden Grove Estate in St. Thomas-in-the-East, a property owned by Chaloner Arcedeckne and managed on his behalf by Simon Taylor, later to become the richest resident planter in early nineteenth-century Jamaica. Barry Higman has brilliantly explored this rich evidence for the years 1765 to 1776. In these years Golden Grove's production of sugar doubled, even though Taylor only slightly increased the acreage under sugarcane. Taylor was an extraordinarily capable and hard-headed manager, determined to extract the last ounce of labor from the four hundred enslaved people on his estate. He maintained his sugar mills scrupulously and developed innovative ways of getting water to these mills; he established rigorous accounting practices and used them to monitor the effectiveness of his management arrangements.

Taylor, with his four hundred slaves and numerous hired workers, produced remarkable crops. In 1765 the estate produced 368 hogsheads or 602,784 pounds of sugar and 103 puncheons or 11,451 gallons of rum. After disappointing crops in 1768 and 1769, Golden Grove produced 630 hogsheads of sugar in 1770, which Taylor considered "more than any estate has ever yet made in this island." He produced a similar sized crop in 1774 and then in 1775 had a crop that amounted to 740 hogsheads or 1,212,120 pounds of sugar, which Taylor celebrated as 'the most extraordinary crop that ever was made on any estate in Jamaica." The year 1775 marked a high point for both planter and property. It gave Taylor high confidence in the future, even though he described the looming conflict between Britain and the thirteen colonies as "truly alarming." But he did not think the conflict would affect Jamaica very much, as the island could "do tolerable well without America." He even felt that the oncoming war might lance a boil and that "good will arise from evil." Jamaica, in his opinion, could cope without its dose of provisions from the North America trade: "I have constantly resided in Jamaica near sixteen years and when there has been no hurricane know we can supply ourselves with provisions if we will but plant them.... [We have] land enough for provisions lumber etc. ... [and] the money that used to be paid to the Americans will rest among us and not be carried to Hispaniola to purchase sugar molasses and coffee there to smuggle into America to the ruin of our colonies."[30]

In Saint-Domingue, as in Jamaica, the prosperity of the early 1770s was accompanied by a sense that planters were in the ascendancy. Moreover, unlike Jamaica, Saint-Domingue did not need royal troops to repress slave rebellions. Postwar prosperity, their victory over d'Estaing, the new influence of creoles in the Naval Ministry, and the limited opening of foreign trade in 1767 further bolstered their self-confidence. As noted in Chapter 6, many members of Saint-Domingue's two Superior Councils saw their courts as part of a kingdom-wide struggle between provincial governors in France and France's thirteen regional parlements. This constitutional conflict coincided with the perennial issue of loyalty—to whom were the residents of Saint-Domingue loyal and for what reason? In the years after the Seven Years' War, Dominguans embraced the arguments put forward by metropolitan parlements in favor of provincial autonomy against royal absolutism. When the Crown decreed new metropolitan taxes in 1764 to stave off a postwar fiscal crisis, regional parlements delayed their official registration of these measures, which would give them force of law in their provinces. In Brittany and

Normandy—regions that produced many planter families—parlements used their constitutional right of remonstrance to issue fierce written critiques of their provincial governors.

In 1766, after the Parlement of Paris sided with its sister courts, Louis XV made a rare personal appearance before the Parisian judges, commanding them to register long-delayed laws and condemning their opposition to his government. In 1771 the Crown dissolved French parlements and sent activist judges into internal exile.[31] Saint-Domingue's judges were connected to this historic conflict in a number of different ways. Many judges had family roots in Brittany and Normandy. Others, like Desmé-Dubuisson, were trained at the Parlement of Paris. In 1766 this link to the centers of French opposition to royal authority was strengthened, when Choiseul sent eight advocates from the Parlement of Paris to Saint-Domingue to replace judges like Desmé-Dubuisson whom d'Estaing had sent back to France. Although some advocates died in the colony, like so many newcomers, at least one of these transplanted parlementaires survived and led the judges of the Port-au-Prince Council in opposition to the new reforms.[32]

This assertive stance, successful against Governor d'Estaing in 1764, caused difficulties for his replacement, the Duc de Rohan-Montbazon, a cousin of Louis XV and brother to the Cardinal de Rohan, the Grand Almoner of France. The political situation in Saint-Domingue was quite different from that in Jamaica, where London appointed the locally connected Trelawny after the controversial Lyttleton: Rohan was much more of an ambitious outsider. Having given up a comfortable life at court to work his way up the naval hierarchy, Saint-Domingue's new governor followed his predecessor's high-handed approach. Colonists soon stopped deferring to Rohan because of his social pedigree, focusing instead on his autocratic style and naked self-interest. One of his opponents dismissed the thirty-five-year-old governor-general as "a young lord fond of parties and pleasures."[33] Indeed, Rohan appeared to enjoy colonial life, despite his political troubles. Within a few years of his July 1766 arrival, he was rumored to be living with a woman of color. As we have seen, in 1769 he purchased a colonial plantation despite a prohibition on such investments. The following year he received an additional 13,501 hectares of colonial land, or 33,601 acres, most of it from the recipient of a royal land grant who passed the property on to Rohan. The governor was determined to make the best of opportunities that fell his way.[34]

Rohan shared planters' interest in profits and pleasure, but his orders were to restore the hated militia system that Choiseul had abolished in 1763.

In May 1767, he laid out the Crown's arguments for reviving the militia before a "national assembly" of parish representatives. They voted overwhelmingly against it. Their intransigence so angered Rohan that he arrested three leading antimilitia speakers and sent them to France, with orders not to let them return. In 1768, to counter the claim that he was reestablishing the militia to build his own personal power, Rohan delivered to the Councils Louis XV's formal orders to revive the militia system. Refusing to accept this document would have been treason. The Cap Français Council complied. It registered the document, thus giving it legal standing in its jurisdiction. The Port-au-Prince Council was less cooperative. On 14 October 1768, despite an antimilitia riot three days earlier, Rohan presented the royal orders to the Council. After debating the document's authenticity, the Council registered the law. The judges, however, embedded the text of their acceptance within a long legalistic protest, in the style of a *remonstrance* from a metropolitan parlement, an obvious overstepping of their constitutional authority.[35]

Léger, the Council's acting attorney general, played a central role in these events. In 1765 the intendant Magon had described Léger, a creole who owned a small plantation in the southern peninsula, as "always disposed to lend his office to the violent demands and pretensions of the Port-au-Prince Council."[36] Like Nicholas Bourke in Jamaica at about the same time, Léger reminded the Crown of colonists' constitutional rights. His strategy included explicitly reminding Versailles that Guadeloupe and Martinique had prospered by surrendering to the British. In 1764, facing off against d'Estaing, Léger had argued that "your other colonies while ... passing under enemy domination by force of arms, had changed the sad state to which war had reduced them ... but ... by the imports of the enemy itself, they surpassed their losses. Their misfortune made them wealthy. Your colony of Saint-Domingue has kept only its honor, all the rest is lost. No such prosperity for us!"[37] Citing the privilege of a member of the Council, Léger had refused d'Estaing's order to come to Cap Français and explain his opposition to the governor's reforms.[38]

Indeed, while Versailles recalled d'Estaing, Léger kept his position. Though he cooperated with Rohan at first, the Port-au-Prince attorney general again soon became deeply identified with the antimilitia movement. On 31 October 1768, Rohan returned to Port-au-Prince from an unsuccessful attempt to muster the new militia in Croix-des-Bouquets, a neighboring parish. He learned that the Council's 14 October 1768 criticisms of the new militia law—ostensibly secret—had been circulating through the city. He summoned Léger to explain the unauthorized publicity of the Council's opposition. Léger claimed innocence, but the administration was convinced that the Council was encouraging resistance.[39] On 24 November 1768 Rohan returned from a visit to the southern peninsula, where, according to his informants, Léger was in close contact with planters. That region's military governor described the attorney general as "the author of all the schisms in the colony."[40] Rohan charged that in this southern region "bad-intentioned persons and disturbers of the public peace . . . had engaged whites, negroes and mulattoes to assemble with arms and by their statements had made it understood that they were authorized by the decrees of the Council of last October 14 and 31; that the free people of color had an additional reason [to protest] because there was an effort to re-enslave them [by means of the revived militia]."[41]

Around this time in Mirebalais, a mountainous parish behind Port-au-Prince with a large free colored population, a man later identified as "L" had circulated a petition opposing the militia. He forced a number of residents to sign it and appeared at the head of a large antimilitia assembly, wearing a mask.[42] As in the southern peninsula, whites, free negroes, and mulattoes were alarmed by allegations that there might be an approaching "enslavement" in Mirebalais, and they "assembled, armed and in a tumult, in large numbers." This gathering elected a leader and "controlled the plain for three days."[43] A similar event occurred around the same time in Cul-de-Sac, a fertile plain near Port-au-Prince. Moreover, in December 1768 an antimilitia riot occurred in Croix-des-Bouquets.[44] By this date Rohan and his aides had begun giving militia commissions to prominent planters. In Cul-de-Sac, when one plantation manager refused the commission, Rohan had him arrested and sent to prison in France.[45] On 19 January 1769 an anonymous poster appeared in Port-au-Prince threatening Rohan and his officials with bodily harm if they did not take measures to "restore public tranquility," meaning to stop restoring the militia.[46] Although Léger and the Council officially condemned this text, royal officials were convinced that they were stirring up all the unrest. Similarly, when Rohan ordered the southern peninsula to assemble its new militia units, prominent colonists refused officers' commissions while poor whites and free men of color refused to attend the musters. Some who did attend were threatened and harassed by their antimilitia neighbors.

In January and February 1769 antimilitia whites and free men of color began assembling in the hills above the city of Les Cayes, collectively harassing promilitia colonists. Rohan declared the region in revolt and sent royal troops, who quickly crushed the movement. In early March 1769 Rohan had Léger and the rest of the Port-au-Prince Council arrested. They were deported to Bordeaux and eventually sent to the Bastille.[47] Seventeen men in the West and South Provinces were punished with varying degrees of severity, including time in the royal galley fleet.[48] In military terms Saint-Domingue's 1769 militia revolt was a trivial event, much like the Boston Massacre of 1770 was trivial in military terms in British North America.[49] But, like the Boston Massacre, the revolt was a thunderbolt in terms of colonial culture, dividing local society along factional lines that were still being observed twenty years later.[50]

Jamaican colonists faced an even greater political problem than did colonists in Saint-Domingue. Unlike the conflict over the militia in Saint-Domingue, the problem—the growing dispute between North American colonists and the British Parliament that led to the American Revolution—was not one that could be resolved locally. The question of why Jamaica did not join in the North American rebellion has vexed historians, and Saint-Domingue's turbulence during these very years adds a new dimension to this problem.[51] It is not enough to point to Jamaica's prosperity vis-à-vis the economic hardship that North Americans experienced in this period. In Virginia, for example, the years following the Stamp Act crisis were economically difficult, as the colony felt the effects of the 1772 British Atlantic credit crisis.[52]

But Saint-Domingue's prosperity did not prevent unrest there in 1769. Moreover, no British colony had been more vociferous than Jamaica for a longer period of time in defending its liberties against perceived or real metropolitan intrusions. Colonial settlers in Jamaica insisted that they were freeborn Britons who were prepared to stand up to London, royal governors, and any local leaders who they felt willing to give away Jamaica's vaunted freedom for personal advantage. Jamaica's Assembly had confronted England in the 1670s over Stuart encroachments on their legislative privileges; had been engaged in prolonged disputes with Britain over whether the Crown had the right to a permanent revenue in the 1720s; had promoted its right to make any laws it pleased despite constitutional rules prohibiting colonies from enacting new legislation in the 1730s; had resolutely refused to insert a suspending law (a law allowing a governor to "suspend" any legislation he disagreed with, thus referring the legislation back to Britain for adjudication) into any legislative act; had fought (and won) bitter battles against colonial governors in the mid-1750s and mid-1760s; and had the dubious distinction in 1757 of being the first colonial assembly to be formally censured by the House of Commons.[53]

Moreover, white Jamaicans shared similar republican ideological predispositions to their North American cousins. The best exposition of these beliefs, as noted in Chapter 6, was put forward in 1765 by expatriate Irishman and wealthy Jamaican planter and assemblyman Nicholas Bourke, in a polemic against Governor Lyttleton. He used the language of slavery. As a large slave owner, Bourke appreciated that this would have particular resonance. If Jamaicans did not resist government encroachment on individual liberties then Jamaican settlers would be reduced from being proud and free Britons into the condition that Spanish Americans (virtual slaves, he felt) endured. If Jamaicans did not enjoy British liberty then Bourke argued that they were "*not* freemen but slaves; not the free subjects, but the outcasts of Britain," because it was "the part of slaves, to submit to Oppression."[54]

Given these kinds of sentiments, it is not surprising that many Jamaicans were broadly sympathetic to the British North American position even if that sympathy did not extend to support for independence. Andrew O'Shaughnessy notes that even though British West Indians were bitterly divided about whether to support Britain or America in the impending conflict, more were opposed to British policy and the drift to war in 1774 and 1775 than were in favor of British actions.[55] Individuals continued to be pro-American throughout the dispute. The wealthy planter Florentius Vassall, scion of seventeenth-century New England merchants as well as a member of a distinguished Jamaican family, confided to Thomas Thistlewood in 1778 that he hoped that "the North Americans might beat the English else they will be enslaved & ruled with a rod of iron, and next us." He thought that the Americans would probably win, "as the Americans will never bear it long, as what army we can keep there will never be able to keep in awe an extent of 2 thousand miles" of American territory."[56]

But opposition to new colonial taxes, like the Stamp Act (1765) and the Townshend Acts (1767) was more pronounced among the West Indian lobby in Britain than in Jamaica itself. On the island, sentiment was muted, and it weakened over time. Jamaicans were indifferent, for example, about the Coercive Acts, which ended self-government in Massachusetts in 1774 and provoked outrage throughout North America. By this date many Jamaicans felt that the Americans had lost the argument by engaging in violent resistance against British authority. Simon Taylor, for example, felt by 1774 that "after what the Americans have done Britain cannot give up the Point [as] it would only be making them more arrogant than they are at present and I look upon them as dogs that will bark but dare not stand when opposed [and] loud in mouth but slow to action."⁵⁷

This hardening of Jamaicans' thinking against North America developed over time. In 1770, Thomas Iredell, a wealthy and irascible Jamaica legislator and planter, was willing to gently lecture his nephew, James, in North Carolina about "meddling with politicks." Iredell Sr. lamented "the spirit of possession that has possessed your people pretty generally" and assured his nephew that "the great at home will soon knock in the head those visionary notions that have led your people to such unwarrantable lengths" because "no Parliament nor Ministry will ever give up the power of taxing." By 1775, his mood had turned to anger, and his tone to his nephew was not gentle but hectoring: "The People of America are certainly Mad," he thought. The dispute between America and Britain was a concocted conflict "under the direction of mercenary Men in Trade" who wanted to break with Britain so they could trade freely with the French, Spanish, and Dutch. "I protest to God," he fulminated, "I would rather live in Turkey than in N. America where a Man is obliged to give up his free Agency [and] must think with the mob or have his Person and Property torn to pieces." Iredell advised his nephew, "Keep yourself perfectly Neuter in those disputes both in words and Actions unless you choose to see yourself adrift with it, maybe a Family at your Heels." The veiled threat was real: as the war began, Thomas Iredell disinherited his revolutionary nephew for "having taken an oath of Allegiance to Congress in violation of your first to this Country."⁵⁸

One of the major differences between white Jamaicans and their counterparts in the French West Indies is that the loyalty of the former was secure while that of the latter was uncertain. Jamaican loyalism was an instinctive response from a population that was largely British in origin and in ideology. It was also based on white Jamaicans' belief that they were sufficiently in control of local politics and able to defend their most vital imperial interests through their representatives in Britain. Unlike Saint-Domingue's planters, Jamaicans did not believe they had any better options: there was no rival empire that would offer them higher sugar prices or a more plentiful supply of enslaved Africans.

Jamaica's elite did not seriously consider rebellion itself or offering more to North America than moral support. They were convinced that the security of the island was too precarious to render them able to help outsiders. A report of a committee of the House of Assembly on the state of the island's defenses on 18 December 1773 spelled out Jamaicans' concerns. They concluded "that this island is in a situation of the utmost necessity and danger, as well as from foreign as from intestine enemies." The state of the fortifications, despite massive sums being poured into their maintenance, was poor, it argued, and the troops provided by the British government for defense were "sickly and reduced" while "the impracticability of compelling free subjects to form a well-disciplined militia" made augmentations to imperial defense from within difficult. The strength of French and Spanish regulars on nearby islands, the Assembly continued, made Jamaica vulnerable to attack, as did "the Great disproportion between slaves and whites" in Jamaica. Because the island was potentially subject to the "insults and invasions from enemies without, and in constant danger of devastation and destruction from within," the committee asked London to station a fleet of men of war near Jamaica to "yield more effectual support in the suppression of internal commotions than can possibly be derived from the present, or even a much greater, number of regular troops."[59]

In contrast to North America, where the deployment of imperial troops was met with great resistance, Jamaica not only encouraged more troops and warships to be sent to Jamaica but was willing to pay handsomely for the privilege of having them there. The cost of the regular army was second only to the expenditure on fortifications in the annual budget. It reached £72,000 Jamaica currency in 1774, or 1.65 million livres, most of which money was spent on inland barracks able to hold 2,572 troops (amounting to over one-sixth of the resident white population).[60] White Jamaican attitudes toward having British troops on their island were diametrically opposed to the attitudes expressed by British North American colonists, for whom the presence of British troops on their soil seemed an almost deliberate provocation.[61] Jamaicans welcomed Britain sending more troops to the island right up until the start of the Revolutionary War. In November 1775, for example, the Jamaica Assembly authorized a measure to raise money to support two British regiments, as well as allocating more money to improve the island's defenses.[62]

One reason why Jamaica supported a standing army in its environs was that it was in a different geopolitical situation to North America. The conclusion of the Seven Years' War had ended the threat of French expansion in much of North America. North Americans, moreover, confidently expected that the French and Spanish would support them in any battle they had with

Britain, as indeed occurred from 1778. Jamaicans, however, knew that in any European war they would become a target, as they had been in the past, from the Spanish and especially from the French, who surrounded them on nearby islands. The Franco-American alliance of 1778 made it clear that Jamaica was an object of desire for both Spain and France. Spain entered wanting a series of prizes, notably Gibraltar, if they won. They also wanted Jamaica, which they had lost over 120 years previously. France was not prepared to allow Spain that favor (although Spain had put this down in negotiations as a desirable rather than a necessary outcome of a successful war). France wanted Jamaica for itself. In a secret pact that was testimony to Benjamin Franklin's powers of double dealing, the French reserved to themselves the right to seize the British West Indies in return for the new nation of the United States taking Canada.[63]

Governor Basil Keith, a military officer attuned to issues of security, spelled out the implications of Jamaica's supposed vulnerability in a report on the state of the island to the Board of Trade in 1774. Keith felt that "the conquest or desolation of [Jamaica] must be a predominant and desirable idea." He was sure that the French would want "to ruin their only considerable rival in the Sugar Trade, by wresting from him a country that produces the better half of that commodity," while the Spanish, "to the common inducement of distressing an enemy, will be stimulated by national pride, to regain a very valuable Island, the loss of which no length of time can reconcile them to, and is as great an eyesore to the creole Spaniards as Gibraltar is to the Europeans."[64]

Rohan successfully reestablished Saint-Domingue's militia in 1769, but his harsh methods raised fears in Versailles. Perhaps exhausted by the political tension, within weeks of the revolt he took a leave of absence and left for France in early 1770, where he married a wealthy widow. He then prepared to return to the Antilles. But despite his pedigree, court connections, and colonial experience, Rohan could not secure a second governor's commission, which suggests that the Naval Ministry regarded him as a failure.[65] In June 1774, as British North America began to rise in revolt, Louis XVI officially pardoned all those involved in the 1769 antimilitia uprising. Rohan complained about this leniency, but it was clear that Versailles remained ambivalent about crushing colonial dissent. It wanted to hold colonists in check but worried about their loyalty. As with Jamaica, Saint-Domingue's wealth gave colonists strong leverage in the metropolis.

The colony's new governor-general, the Count de Nolivos, was more like Jamaica's popular governor Trelawny than he was similar to Rohan. A Saint-Domingue creole, Nolivos came back to his birthplace after four years as governor of Guadeloupe, where Versailles was even more concerned about loyalty than it was in Saint-Domingue. A year after arriving he married a colonial widow that he likely knew as a child.[66] It was a sign of his commitment to the local community. In his first speech, the governor told the Council of Port-au-Prince in 1770 that he wanted to "forget and bury our late troubles in the deepest silence."[67] Such amnesia would not be too difficult, as Rohan had purged the body of rebellious judges, including Léger. The new attorney general, Guillaume Delamardelle was, like the new governor, a creole, and he had a very different perspective on the postwar reforms from the previous attorney general.

Born in Saint-Domingue, Delamardelle had returned to France with his family at the age of nine. Trained in the law at Paris, in 1757 he purchased a Saint-Domingue coffee plantation but remained in France, becoming royal attorney in the city of Tours in the 1760s. Delamardelle was deeply interested in the Naval Ministry's postwar colonial reforms. In 1766, he proposed to Choiseul the creation of a royal colonial school in Tours, on the model of the newly formed Ecole Militaire in Paris. The institution would train five hundred colonial boys to be naval officers, merchants, and colonial jurists. Delamardelle proposed a more practical kind of education for these children than was typical in the metropole, with more attention to mathematics and modern languages and less time devoted to Latin or philosophy. Keenly aware of the problem of colonial wartime loyalty, Delamardelle believed his proposed school could foster a devotion to France among its students. He stressed the importance of raising colonial boys so that they were, above all, "attached to the state in which they were born." The school proposal did not say much about how he would educate his charges about race and slavery except that they would learn "humanity" and especially "firmness" to "contain that species of man, who, vile to our eyes because of the color difference, bears the daily weight of continually providing for our needs."[68]

At about the time he was musing about how to inculcate loyalty among colonial youth, Delamardelle applied unsuccessfully for a legal position in Saint-Domingue. Instead he was hired into the Naval Ministry. At Versailles, Delamardelle, who later described reform as "his favorite passion," worked on a proposal to unite Saint-Domingue's two Councils into a single body, a project that went squarely against the pretentions of colonial judges. In April

1769, the Naval Ministry named him to replace Léger as royal attorney general for Port-au-Prince. Delamardelle led the prosecution of the antimilitia rebels and remained in this position until 1783. As royal attorney general, he was able to indulge his passion for legal reform. Guillamin de Vaivre, Saint-Domingue's intendant from 1774 to 1780, reported that Delamardelle had been the central figure in drafting "nine edicts, orders or regulations approved by Your Majesty and nine other proposed regulations which will be submitted to the administrators."[69]

In February 1770, Delamardelle publicly welcomed Nolivos to Port-au-Prince with a speech before the Council that was reprinted in the colonial newspaper. The new attorney general was free with his advice to the new governor. Significantly, Delamardelle identified among Nolivos's duties that he should "maintain the necessary superiority of the free and unmixed race over that which still carries on its forehead the mark of slavery."[70] The new governor surely understood this reference. In the years after the militia uprising Delamardelle and Nolivos initiated a wave of new racial laws, aimed at separating whites and free people of color. Since the 1680s, Saint-Domingue had had a growing free population that was recognized to be "of color." But that growth had accelerated around midcentury, as free families that had formed in the 1720s began producing children. Some of these families became moderately wealthy thanks to the availability of land in mountainous or remote parishes, and the willingness of wealthy colonial fathers to pass property on to some of their sons and daughters. Since the early 1700s, Guadeloupe and Martinique had limited the amount of property that could be passed from whites to free people of color, just as in Jamaica after 1761. Saint-Domingue had not.[71] A number of these children were educated in France before the end of the Seven Years' War. Up to the 1760s, officials and white society generally often regarded the wealthiest of these families as members of the colonial elite.[72]

Political tensions in the late 1760s moved Saint-Domingue's French colonists, like colonists in Jamaica, to make much firmer legal, social, and economic distinctions between themselves and free people of color. Originally the process had moved slowly. Although the 1685 Code Noir gave masters the right to free slaves as they saw fit, in 1713 the Crown began to try to control this right, subjecting manumissions to formal approval and, by the 1760s, to a tax. As early as the 1720s, free men of mixed European and African ancestry had split into their own militia units, citing the prejudice they faced from white militiamen.[73] But the state did not attempt a rigorous legal and social exclusion of wealthy free coloreds from the colonial elite until after the antimilitia revolt of 1769. What this meant in practice was that from 1770, free people in Saint-Domingue, like their counterparts in the post-Tacky period in Jamaica, started to experience a drastic change in race relations.

One reason why authorities reduced wealthy free people's ability to pass as white was that they recognized how free colored militia service could reduce colonists' militia obligations. During the Seven Years' War, in an attempt to defend the colony as cheaply as possible, royal officials had formed an experimental free colored militia that operated under the command of army officers, rather than under free colored civilian officers, as had been the case up to that point. The success of this experiment, at least in the eyes of royal governors, led to a series of attempts to integrate free colored militiamen into the royal army. The colonial constabulary was already largely composed of free men of color. In 1764 Governor d'Estaing had tried to form a Legion of Saint-Domingue made mostly of free men of color who would be obligated to serve under white officers for three years.[74] This legion would have taken over many of the functions of the traditional militia.

D'Estaing's departure ended this second experiment, but Rohan kept one key element of the legion—the elimination of free colored officers. He made that decision part of his militia reestablishment. After 1769, Saint-Domingue continued to have white and free colored militia units, but free colored units had white officers. Moreover, these units took over many of the most difficult, humiliating, and time-consuming responsibilities of the traditional militia. One such new duty, for example, was the responsibility of twenty-four-hour *picquet* duty, serving as a kind of guard and messenger to run messages from parish militia commanders to other officials.[75]

Free men of color were well aware that these changes lowered their status in the revived militia. These changes were central to how antimilitia leaders had rallied free colored support in 1768 and 1769, by describing their coming "enslavement."[76] The changes were especially galling to free colored property owners, a group that was even more substantial in Saint-Domingue than in Jamaica. In the 1790s, Julien Raimond, the leading free colored activist, estimated that members of this group owned one-half of Saint-Domingue's property and one-third of its slaves.[77] That was surely an exaggeration made for polemical purposes, but Raimond's willingness to make such a statement points to the greater wealth of free coloreds in Saint-Domingue than in any other West Indian colony. The ready availability of land unsuitable for sugar meant that there were many farmers and ranchers, as well as indigo, cotton, and coffee planters, who were able to prosper outside the sugar economy.

As Delamardelle's comment suggested, many of these small-to-middling planters were not "free blacks" but instead were of mixed European and African ancestry. Contemporaries agreed that the concubinage of white men with free and enslaved women of color was nearly universal in Saint-Domingue, though most observers saw these as purely sexual liaisons. Lieutenant Colonel Desdorides, for example, after describing how the sexual skills of women of color "have therefore given them the elite and most colonists as lovers," concluded, "These arrangements are why most colonists do not marry. Instead of the chains of a permanent union, pleasure seeking bachelors prefer more fragile connections that last no longer than a fantasy."[78] The reality of these sexual connections, however, was that many concubinage relationships created lasting family bonds between men, women, and their children. Free women of color, moreover, were not just sexual partners for French immigrants; they were also valuable economic partners.[79] Saint-Domingue's 1685 version of the Code Noir, unlike that promulgated in Louisiana in 1724, never outlawed marriage between whites and the descendants of slaves. Up to the 1780s, such unions, though socially disparaged, made up one-fifth to one-sixth of official church marriages in some parishes.[80] Moreover, in the 1770s the Councils continued to uphold the right of a white colonial testator to leave a large amount of property to his free colored children.[81]

But after Delamardelle's comment about "the necessary superiority of the free and unmixed race," Saint-Domingue's Councils and governors took a series of measures designed to ensure that the subordination of this "mixed" class to whites became a central aspect of colonial life. Just two weeks before Delamardelle's "necessary superiority" speech, the Port-au-Prince Council, probably at his instigation, gave a powerful example of its new determination to cut wealthy biracial families out of the white population. In early February 1770, the Council ordered the enslavement of a planter, Paul Carenan, a respectable married man and father of six. Carenan was a light-skinned man of color, but he could not furnish documentary proof of his freedom. The court overturned its sentence two weeks later on the appeal of Carenan's wife, also a member of a propertied free family of color. Nevertheless, Delamardelle and his colleagues had made their point: all people of color, no matter their wealth or European ancestry, were just a step from slavery.[82]

In following years, the Port-au-Prince and Cap Français Councils, as well as the governor and intendant, promulgated a number of laws establishing the separation between "unmixed" whites and other members of society in new and rigorous ways. The most telling was a 1773 law forbidding any free person of color from taking a "white name." The text explained that "the usurpation of a white name may cast doubt on the status of persons, confuse the ordering of inheritances, and even destroy that insurmountable barrier between whites and people of color that public opinion has established and which the government wisely maintains." The same law required priests, notaries, and scribes to demand that all people of color provide documents proving their freedom before drafting any official deed. As the Carenan case suggested, colonial courts were coming to consider all people of color slaves unless they could prove otherwise.[83] In 1775, a new law reiterated the process of manumission, making it clear that masters needed to pay substantial taxes and receive official approval before granting freedom to any slave.[84] In 1779, a new sumptuary statute stated specifically that it was designed to prevent "above all the assimilation of people of color with white persons, in their manner of dress and the narrowing of distance from one sort [*espèce*] to the other." The law's first article, however, was not about clothing at all. It noted that "we enjoin all people of color, born free or manumitted of either sex, to show the greatest respect not only to their former masters, patrons, sponsors, their widows or infants but to all whites in general, on pain of legal action, if the case demands it, and punishment according to the rigor of the law, even by the loss of liberty if their disrespect merits this."[85]

Delamardelle's interest in education and legal reforms suggests that this new emphasis on whiteness was in part informed by Enlightenment debates over race and biology. But in Saint-Domingue in wake of the 1769 militia revolt those biological concepts fused with the political problem of colonial loyalty. In July 1769, freshly returned to the colony, Delamardelle again fused biology and politics in a speech installing new metropolitan judges in the Council of Port-au-Prince. Describing the antimilitia revolt, he noted, "the crime of a few individuals is not that of an entire people: submissive to its Sovereign, the colony of Saint-Domingue never wavered, and the French blood that runs in its veins will forever be the guarantor of its love and fidelity."[86]

In Jamaica's 1761 racial reform, that island's planter class had sacrificed the ability of fathers to bequeath property to sons and daughters of color in order to prevent the emergence of a free colored planter class. A decade later,

Saint-Domingue's judges and administrators avoided this check on paternal power. They also did not consider restricting marriages between various categories of people, although other French colonies had taken this measure. Saint-Domingue's judges were reluctant to restrict colonists' economic freedom or their economic opportunities, besides making the status of free coloreds more apparent in all legal documentation. Consequently, wealthy free coloreds were more prominent in Saint-Domingue than in Jamaica, even after new discriminatory measures were installed after 1770. To that extent, Saint-Domingue relied on the emerging concept of racial purity and left class and most other economic considerations aside.

The reasons for the comparatively more generous treatment of free coloreds in Saint-Domingue as opposed to Jamaica was to large extent governed by practical exigencies relating to defense and to the delicate issue of French colonial patriotism. French colonial authorities, unlike the British, were trapped by their realization that free colored militiamen could solve three problems for the empire. Their service could help keep the colony more secure at a minimal cost to the state. Second, their militia duty made white service more tolerable. When free colored officers were eliminated, more white men acquired social prestige as militia officers. Even more important, giving men of color the most difficult militia work lightened white units' burden. Third, by splitting the free colonial population into white and nonwhite, authorities avoided the kinds of coalitions between these two groups that made the 1769 militia revolt so threatening to imperial control.

Changing the boundaries between "white" and "nonwhite" was also important in Saint-Domingue because it created solidarity among various immigrant and "unmixed" creole subgroups. From the 1760s through the 1780s, an increasing population of poor whites, fed by immigration from France, threatened to split the white population along class lines. After 1763, about one thousand French immigrants arrived in the colony every year, drawn by the image and reality of colonial wealth. By the 1780s, perhaps one-third of the colony's white population of thirty thousand was composed of poor whites.[87]

This immigration produced a marked change in Saint-Dominguan society. As in Jamaica, "hospitality" was a much cherished value. In the 1730s, Charlevoix had stressed planters' hospitality: "It seems that one breathes this virtue with the air of Saint-Domingue.... [A] man can go all around the colony without spending anything, he is well received everywhere and if he is in need, [his hosts] will add what is necessary to continue his voyage." But in the

1770s Hilliard d'Auberteuil noted that "the planters of Saint-Domingue are generally less eager than they were to be of service, because they have been fooled too often and too cruelly.... Now a man on foot no longer has access to the main house of a rich colonist; he will be given food and a bed in a separate dwelling."[88] The colonist Lory attributed this change not only to bad experiences but also to a growing social snobbery among planters. He commented that "some [planters] today would blush to allow a carpenter, a mason, a roofer at their table; in the past one would not make this distinction. The traveler or man without a job would easily find shelter among nearly all planters, without describing who he was; it was enough that his face displayed the character of the French nation for him to find hospitality and even all the aide that humanity suggests we give our fellow man." Lory believed "there are thousands of examples of vagabonds who have abused planters' cordiality.... Today anyone who is not known or recommended will have difficulty finding shelter with a planter."[89]

Saint-Domingue had long had a problem with lawless whites. The buccaneer population of the late 1680s had diminished after the 1720s, when the growing plantation economy opened new economic opportunities in farming, ranching, and plantation employment. But from midcentury, land and slaves became far more expensive, while plantation work required greater management and agricultural experience than most immigrants possessed. As the colony's fame as a land of opportunity brought more European emigrants than ever before, many of them fell into unemployment. For example, in 1764 Jean-François Marmontel, newly elected to the French Academy, published a story in which the main character left home to make his fortune. Marmontel noted that "it is well known how easily a Frenchman of good character and good appearance can establish himself in the islands."[90] But in 1776 Hilliard d'Auberteuil bemoaned the hundreds of men who arrived every year, "without status, without employment, often without aptitude or before the age when the seeds of talent are developed, to languish or perish in the colony. Nothing is done to pull them from the poverty into which their ill-considered decision has plunged them.... If the colony continues to be infected by this purging of the metropolis, disorders will grow with the population."[91] Indeed, three years later Lieutenant Colonel Desdorides described "an infinite number of sailors, deserted soldiers, unemployed workers, and adventurers of every kind spreading through the colony begging for bread from plantation to plantation. Losing all shame, these people ask for alms even from the slaves."[92]

These unemployed and underemployed whites were precisely the sort of men who had allied with free people of color in the antimilitia turbulence of 1769. How could these "vagabonds" be made to feel a sort of solidarity with the propertied elements of colonial society that would lessen their class resentment? The new laws, based on color, rather than class, created that kind of unity, especially when free men of color assumed the most difficult militia tasks. It is important to note, however, that this color line was built primarily to serve a political purpose. The rulers of Saint-Domingue adopted laws that emphasized race over class because they had several practical problems to solve—working out a suitable system of defense and keeping poorer whites allied to the ruling elite. The new laws were pragmatic political responses to relatively short-term political and demographic exigencies. Their implementation, however, significantly changed the social structures of the colony in ways that free people of color bitterly resented.

One significant difference between Saint-Domingue and Jamaica in respect to the place of free coloreds in society was the wealth and political self-confidence of this group in Saint-Domingue. Relatively few free people of color in Jamaica were wealthy, and the wealthiest families of color were partially co-opted by various privileges, while the new inheritance laws prevented other families from dramatically increasing their wealth.[93] While the 1761 law in Jamaica effectively split the free population along class lines, in the French colony the post-1769 changes unified the free population of color along racial lines, because wealthy and poor families were to receive the same treatment.

The impact of this can be seen in the conflicted career of Julien Raimond. A wealthy indigo planter, with close to two hundred slaves, Raimond was born in 1744 into the large family of a French immigrant-turned-planter and his wife, a free mulatto woman. For most of his early life, colonial officials and neighboring planters and colonists treated Raimond, who was of one-quarter African ancestry, as they had done his biracial mother, that is, like a French colonist. That acceptance changed abruptly after 1769. The new racial laws had no financial effect on Raimond, or on most free people of color. Indeed Raimond grew much wealthier during the American Revolutionary War and the economic bubble that followed it. But Raimond found the new emphasis on his African ancestry and the unpleasant duties assigned to all members of his class "humiliating." He was now obligated to prove his freedom at every turn and to participate in the time-consuming militia duties now reserved for free men of color. He could no longer hold a militia commission. By the early 1780s, careful investments, a judicious marriage, and use of smuggling networks had given him a fortune of over 400,000 livres or £17,000. This wealth, on top of his French education and deep colonial roots, might well have made him a leader in Saint-Domingue's society if contemporaries would have defined him as white.[94]

As Dominique Rogers suggests, Raimond's campaign for free colored civil equality during the French Revolution led him to overstate his case. Free coloreds, for example, never lost the ability to defend their property in civil cases, as he claimed; some free coloreds took their legal appeals all the way to the royal council at Versailles.[95] In this respect, Saint-Domingue's free coloreds always had more legal rights than their counterparts in Jamaica, where the ability to sue in court was severely curtailed after 1761. That restriction was a significant impediment to Jamaican free coloreds' being able to trade and plant effectively. Starting in 1784, Raimond petitioned the Naval Ministry for new laws that would give wealthy, light-skinned, and legitimately born men of color like him the status of "new whites." Here, Jamaican free coloreds had a theoretical advantage over Saint-Domingue free coloreds. Such passing was possible in Jamaica, though it required the Assembly to pass a special bill on behalf of the person who was four degrees removed from the negro "taint." But Saint-Domingue's racial ideology had no such "mulatto escape hatch." Raimond's pleas met with no response, at least until the French Revolution.[96]

All these changes in the status of free coloreds stemmed, in some way or another, from postwar reforms. In the aftermath of the 1769 revolt Delamardelle and Nolivos solved the question of how "American patriotism," as Emilien Petit described it in 1750, could be transformed from a force that united social classes against unpopular French reforms into imperial loyalty, while reducing colonists' military obligations. Race, or more specifically a new definition of "whiteness," was the answer. As in Jamaica, the effect of these restrictions was to cement into law a colonial ideology in which whites were superior to coloreds, no matter how poor the former or how rich the latter. The twin changes in Jamaica and Saint-Domingue moved each society decisively into an era when race was the principal means of social, political, and, in Jamaica's example, economic division among free people.

Both Saint-Domingue and Jamaica faced a choice in how they determined social structure in the prosperous years following the aftermath of the Seven Years' War. Each could have opted for a system that accepted the most prosperous and Europeanized members of free colored population as full, or almost full, colonial citizens. In retrospect, this policy choice would have been sensible. It would have allied to white interests the only group in either society that was flourishing through natural population growth. But this proved ideologically and politically impossible. The consequences of these policies were not felt immediately. In the long term, however, alienating free people of color from white people and making race rather than wealth the determinant of political and social standing eventually played badly for white colonists, both during the Haitian Revolution, where free people of color opposed white rule, and during the advent of abolitionism in Jamaica in the first half of the nineteenth century.

CHAPTER 8

The American Revolution in the Greater Antilles

The coming of the American Revolution in the 1770s was less important for Jamaica and Saint-Domingue than were cultural changes in Europe, leading to the abolitionist movement in Britain and the Revolution in France. But the start of the American Revolution did affect the colonies. Jamaica faced greater difficulty than Saint-Domingue in the 1770s, in part because it had to cope with the division of British America into loyalist and rebel areas. The American Revolution left British West Indians isolated within the empire as loyalist defenders of settler rights and African chattel slavery. In short, the beginning of the American Revolution highlighted the previously unremarked ways that Jamaicans, Britons, and North Americans were different. For Saint-Domingue, on the other hand, the American Revolution illustrated how important the colony was to French prosperity. It strengthened the colonists' argument for being allowed to trade freely with North America.

In retrospect, the most important events affecting Jamaica's future in the lead-up to the American Revolution occurred neither in America nor in Jamaica but in Britain, notably the *Somerset v. Steuart* case in 1772.[1] In this case, James Somerset, with the help of the humanitarian reformer Granville Sharp, brought a suit against his Virginian master, Charles Steuart, to stop him being sold into chattel slavery in Jamaica as a punishment for running away in England. Lord Mansfield ruled in Somerset's favor: "No Master ever was allowed here to take a slave [from England] by force to be sold abroad because he deserted from service, or for whatever reason whatsoever." Mansfield was circumspect in his legal reasoning. As he later insisted, he had not pronounced on the legality of slavery, either in Britain or in the colonies, but had merely ruled that a slave owner could not remove a slave from England without his consent.²

Despite Mansfield hedging around the implications of his decision, leading some abolitionists to be disappointed that he did not go further in his condemnation of slavery and West Indian planters, the *Somerset* case outraged white British West Indians. They felt that it violated their rights over slaves and that it weakened their hold on slaves as property. It led to the publication of works by Edward Long, Samuel Estwick, and Samuel Martin denouncing Mansfield's dicta.³ All three were distinguished West Indian thinkers. Samuel Martin and Edward Long were rich, well connected, and the intellectual leaders of Antigua and Jamaica, respectively. Estwick was subagent in London for the island of Barbados. They responded firmly to Mansfield's ruling because they recognized that it was the first public airing of metropolitan concerns about the morality of slavery. Significantly, the immorality of slavery was a point that they conceded almost without a fight. Long, though he later defended the institution vigorously in his history of Jamaica, even claimed that "I shall not be misunderstood to stand forth a champion for *slavery*."⁴

Long, Martin, and Estwick all informed Britons that *Somerset* was not only legally wrong but was an attempt to pretend that metropolitan Britain was not complicit in the enterprise of colonial slavery. Long insisted that "the whole nation had been in some way or other interested in the advantages drawn from this [slave] trade" and from the goods produced "through the means of Negroe labour." Estwick put the question more graphically: "whatever property America may have in its drugs, it is Great Britain that receives the essential oyl extracted from them."⁵ Martin fulminated that without tropical produce "Britain itself would become a poor, defenseless country and very soon a province of France." If Britain was foolish enough to free African slaves they would open themselves up to French domination and then themselves become "wretched slaves to arbitrary power."⁶

These authors thus framed their dissent from Mansfield in nationalistic and mercantilist terms. Not only was slavery vital to the British American colonies and to the British Empire itself, but maintaining African chattel slavery was also central to preserving British freedom. *Somerset* was a warning shot across the bow for Jamaican slave owners. It showed that there were limits to their powers over their human property. It introduced into Britain notions of "free soil," at least in popular consciousness if not, in fact, in law.⁷

It forced even fervent slavery advocates to admit that slavery was, in Mansfield's words, "morally odious," thus ceding a major public relations advantage to abolitionists.[8] James Somerset's attorneys skillfully wove into their arguments comparisons with African slaves in Virginia, eastern European serfs, Protestant "galley slaves" held by Catholic nations, and Christian slaves held by Muslims. Granville Sharp tirelessly promoted the case as a successful example of a morally aware England rejecting reprehensible West Indian customs. The case inspired black Londoners to publish works that sold widely, benefiting from the increasingly fashionable sympathy with the fate of the African. It made this period the first flowering of black imperial voices.[9] For Jamaicans, the most significant danger of *Somerset* was its implication that slavery and slaveholders were somehow "un-English." *Somerset* was an early public manifestation of a fear that West Indians' attachment to slavery was degrading British life.

In France, there was no case like *Somerset* that received national attention, in part because France already had a free-soil tradition. Parlementary judges refused to enforce laws that allowed planters to bring their slaves to the metropole and to maintain them in bondage. Yet the idea that "there are no slaves in France" was most powerful as a political metaphor for the power of regional elites against the French Crown. It was not, for most, a message about the inhumanity of Caribbean slavery. When the Naval Ministry introduced a law in 1777 protecting planters' property rights over their slaves in France that did not use the word "slave," it was duly approved by the Parlement of Paris.[10]

As this example suggests, French opposition to slavery was primarily an intellectual discourse rather than a public antislavery campaign. Moreover, the birth of a more humanitarian antislavery literature in France owed as much to news of events in Jamaica as it did to events happening in Saint-Domingue. Jean-François de Saint-Lambert's story "Ziméo," republished seventeen times between 1769 and 1797, was "the first colonial fiction to seize the imagination of French readers."[11] Set on a Jamaican plantation during a revolt led by a charismatic West African, it was clearly inspired by accounts of Tacky's Rebellion and Jamaica's Maroon Wars. In the story, Saint-Lambert suggested that it was in planters' self-interest to treat Africans with humanity and even to free them. His title character is an African prince turned Jamaican maroon who leads a revolt against whites' heartlessness. Saint-Lambert's other heroes are two Quakers, one of them the narrator who is visiting the island, and the other a planter who treats his slaves humanely, freeing them after ten years of labor. When the revolt breaks out, the planter's slaves protect his estate while the rest of the district burns. Ziméo learns of the Quakers' benevolence and comes to the plantation to meet these unusual whites. There he discovers that the Quakers have taken in three black castaways who are miraculously Ziméo's wife, her father, and Ziméo's child, conceived on the slave ship. The Quakers help the reunited family escape into the mountains. The Africans swear to be always faithful to their white friends, and the narrator promises to visit them in their new mountain home when peace is restored, to enjoy their "virtue, good sense [grand sens] and friendship." The tale concludes with a brief essay in which the narrator draws out the lesson of the story. He acknowledges that Africans are in general "less advanced" than Europeans, but it is "circumstances and not the nature of their kind [espèce] that determine the superiority of whites over the negroes."[12]

The popularity of Saint-Lambert's story stemmed from the publicity it received in Pierre Samuel Dupont de Nemours's journal *Ephémérides*.[13] By 1767 *Ephémérides* had become the leading journal of the physiocratic movement, which promoted agriculture as the key to population growth and national prosperity.[14] In 1771 Dupont republished the final pages of "Ziméo" as a preface to his own article describing slavery's economic inefficiency. While Saint-Lambert proved slavery was "odious and detestable," Dupont claimed his own calculations showed it to be "an onerous and unnecessary crime." Like Franklin in his *Observations on the Increase of Mankind*, Dupont calculated that the cost of maintaining an enslaved worker in the colonies was 420 livres per year (or £18.23). Meanwhile, twenty-five million workers in Europe lived on roughly 30 livres per year (or £1.30). Slavery was thus not as economically beneficial as its advocates contended, he argued, ignoring the question of whether the income slaves earned for their master outweighed the cost of purchase and maintenance.[15] After 1771, Dupont de Nemours rarely returned to this topic. But his article and Saint-Lambert's story marked the beginning of a French literary attack on Caribbean slavery. By 1774, for example, Voltaire had also come to oppose slavery on economic grounds, despite his opinion that Africans were a different biological species from Europeans.[16] Caribbean colonization was a major theme of the multivolume *Histoire des deux Indes*, and the antislavery sections of this work grew ever more important from the first edition in 1770 to the third edition published in 1780.

In 1776 Dupont's analysis was refuted by a colonial author, Michel-René Hilliard d'Auberteuil, who at the age of fourteen had arrived in

Saint-Domingue, where he worked in the judicial system.[17] He succeeded in getting the Naval Ministry to support his two-volume *Considérations sur l'état présent de . . . Saint-Domingue*. The book's first volume, in particular, responded to Dupont's economic analysis, by celebrating and explaining the transformative power of free trade and investment in Saint-Domingue's wealth-producing capacities. For Hilliard, Saint-Domingue's principal characteristic was not slavery but cycles of capital and reinvestment. Some portion of colonial profits was returned to the metropolis to pay for subsistence needs, he noted, but "the remainder is put back into the system [*employé à des reproductions*], and therefore there is a seed of fortune in the colony, which will grow as long as its lands are fertile."[18]

Saint-Domingue was different from the colonies that had cost France enormous sums of money, Hilliard observed. Profits from this territory, he suggested, sustained one-thirtieth of the residents of France—"the industrious part of the nation." Using censuses from 1764, 1767, and 1775, he calculated that French colonial revenues were eight times the value of investments in the colony. Noting that people were always talking about making their fortune in Saint-Domingue, Hilliard provided a detailed assessment of what this actually meant. He calculated that for each of its forty thousand free and free colored inhabitants, the colony generated the equivalent of 1,333 French livres or £57.70 per year, proclaiming that "there is no country in the universe that offers wealth in such a proportion." Moreover, he analyzed how slave labor created colonial wealth, which he depicted as a function of the supply of enslaved bodies. He estimated that each slave on a well-run sugar estate generated 600 livres apiece annually, or roughly one-third his 2,000 livres or £86.58 purchase price.[19] If, he suggested, Saint-Domingue was allowed to mount its own slave expeditions to Africa in small ships, French colonists, selling their own tobacco and rum to African merchants, could buy Africans for 1,200 livres each. Revealing his belief that a slave's purchase price was not a labor cost, but rather a capital investment, he commented: "if the price of [colonial] products varies, it is necessary that the price of the instruments used to draw them from the bosom of the earth be as low as possible."[20]

Hilliard did insist on the humanity of enslaved Africans and described them as equal in intelligence and virtue to the French. But like Long in his response to *Somerset*, he explicitly declined to examine slavery's moral legitimacy. It existed, he argued, because of the power of the strong over the weak. Providing humane treatment and a precise and unchanging discipline would produce profits, he insisted. Under a good master, who understood the need to make slaves "happy and rich, according to their condition," a slave in Saint-Domingue might have a better life than a French peasant. Slaves, he also added, should be under the authority of civil justice to weaken the tyranny of cruel masters.[21]

If Hilliard's first volume was inscribed in the new context of physiocratic economics, his second volume drew on the tradition of Dominguan critiques of the French Empire. Like earlier colonial writers, he argued that France had not maintained its obligation to protect the colonists from attack. Indeed, France's commercial monopoly prevented planters from making their full contribution to the prosperity of mainland France. Hilliard emphasized that changing Saint-Domingue required focusing on the capabilities of the creole population, by which he meant island-born whites, not free people of color or island-born slaves. Advocating a gradual move toward colonial autonomy, he suggested that the government reserve important administrative positions for creoles. But, he specified, virtuous French immigrants should also be eligible for this "creole" designation. Island-born whites and long-standing colonists would be deeply involved in administrating Saint-Domingue alongside royal officials. In his most ambitious plan of racial reengineering, he envisioned Saint-Domingue developing a three-part caste system based on African ancestry that would extend into the working classes. Hilliard proposed enslaving all blacks including the thousands whose masters had released them from slavery. All biracial people, on the other hand, would be free, including tens of thousands of mixed-race enslaved people. To maintain racial purity, he would prohibit all intermarriage between whites and nonwhites. And masters would only be allowed to free a slave if the bondsman saved the master's life. In such cases, the ex-slave would be required to marry a person of mixed ancestry, to maintain the biracial character of the free population of color.

Hilliard's vision of Saint-Domingue eliminated the inconsistencies that wealth and family connections had created in the racial logic of society in the colony. Whiteness, he thought, would unify Frenchmen and creoles, overriding differences of wealth, social origin, or regional culture. All whites would have the legal right to use physical violence to correct free persons of color. Because all free coloreds shared a mixed European and African ancestry, there would be no "free blacks." And the rising population of wealthy free colored planters would be no better placed than the poorest mulatto laborer. All free coloreds would be subject to the immediate authority of any white person, without recourse to the courts. Enslaved workers would be easily recognizable by the darkness of their skin. Hilliard gave them access to the court system, as their ability to expose cruelty or neglect reduced the moral corruption of the planter class. Free coloreds too would have access to the courts. But whites of any condition could physically punish slaves or free coloreds before the law's intervention; the courts would only intervene after the fact. Moreover, whites were fully in control of social processes, including how people would be assigned to each of three estates.[22]

Hilliard's *Considérations* celebrated the capitalistic orientation of Saint-Domingue's economy and proposed a racial caste system that would unite "whites" across inequalities of wealth and class. The work antagonized multiple groups in France and Saint-Domingue, leading Versailles to suppress it in 1777, the year after its publication. Hilliard's advocacy of a more rational and market-driven colony based on free trade angered merchants in France's port cities, who depended on the colonial trade monopoly. His acknowledgment that slavery was indefensible except for its profitability incensed the planter class; administrators and intellectuals attacked his proposal to shift large numbers of people from slavery to freedom or back, based on their skin color. In addition, Hilliard's plan to replace social class distinctions with a three-part racial caste system helped politically motivate wealthy free men of color.[23] Yet *Considérations* captured the spirit of Saint-Domingue. Hilliard insisted that macroeconomic considerations should override every other constraint in fostering colonial growth. Thus, this law clerk was willing even to set aside due legal process in favor of a system that used skin color to determine status.

Commentators looking at these two colonies in the early 1770s might have thought that Saint-Domingue, rather than Jamaica, would have struggled most in that decade. As we have seen, Jamaica was at the peak of its prosperity and peacefulness in the early 1770s. Saint-Domingue, however, was hit by a succession of natural disasters. In 1770 an earthquake occurred in the area around Port-au-Prince. The Dominican priest and naturalist Jean-Barthélemy-Maximilien Nicolson estimated that two hundred people died in the capital and that at least forty died in Léogane, where he lived. He noted that "the plain of Cul-de-Sac especially [the chief plantation district outside Port-au-Prince], if one believes the talk, was unrecognizable the morning after that awful night. On several plantations the earth opened up as if a plow had passed there."[24] Hilliard remembered that "Port-au-Prince was turned upside down, the people and the leaders wandered over the ruins in clouds of dust and sulfur, crying out in despair. . . . A great number of prisoners , . . . escaped from death and set free, threw themselves at the feet of the governor and the intendant, the slaves surrounded their masters with signs and expressions of sadness. . . . We feared famine, not revolt."[25]

In 1772, within the space of thirty days, two hurricanes struck the colony.[26] Father Nicolson described in detail how the storms produced in Léogane "the horrors of a winter unknown in these burning climates. The banana trees were all broken; coffee bushes torn out or broken, cane fields flattened and uprooted, indigo gardens washed away by floods. No house remained intact, either for blacks or for whites."[27] On the rich sugar plain outside Les Cayes, on the southern coast, Moreau witnessed "a sickening spectacle" after the storm.[28] American merchants responded to the disasters of 1770 and 1772 by sending ships and supplies to Saint-Domingue. When French merchants protested that their commercial monopoly was being violated, the governor pointed out that France could not supply the livestock, lumber, or salt cod that the colony desperately needed.[29] The worsening political conflict in British North America, however, made it more difficult for British North American traders to provide goods for Saint-Domingue in their periodic natural disasters.

In August and September 1775 two more terrible storms occurred, badly damaging crops and warehouses.[30] At the same time, in 1775 and 1776, a serious drought struck the colony. That winter terrible weather froze French ports, driving up the price of already scarce provisions. Slaves suffered, particularly, and this time colonists did not attribute their deaths to a "poisoner" like Macandal. Administrators believed part of the problem was that planters counted too heavily on smugglers and had not planted enough provision crops. In August 1776, the governor and intendant lamented to the naval minister that "by an interest as inhuman as badly calculated, almost generally here planters sacrifice even the subsistence of their slaves for the temporary augmentation of their revenue." In response, officials set up new laws requiring planters to cultivate a minimum amount of land in manioc, yams, and plantains according to the number of slaves they had. The law required that militia officers inspect these provision grounds twice a year, but the administrators had to increase the number to three annual inspections. Still many planters did not comply. The intendant wrote to Versailles that the situation was "more critical than you can imagine," describing the devastation in a society where entrepreneurial planters lived close to the edge in order to extract maximum productivity from their operations.[31] By the second half of 1776 the drought lifted, and normal crop production resumed. This was fortunate, because the British blockade of New England was now more effective than during the Coercive Acts of 1774. Instead of trading with Saint-Domingue merchants, from 1776 North American captains began to attack British West Indian shipping. In 1776 alone, 250 British West Indies ships fell to these privateers.[32]

In 1777 the American Continental Congress proposed a treaty of commerce to Saint-Domingue's governor. At about the same time North American traders found ways to avoid the British blockade and began returning to Saint-Domingue. Moreover, in 1777, as France prepared to join the Revolutionary conflict, the Naval Ministry accelerated shipments of wheat for troops in Saint-Domingue, with New England merchants becoming the main suppliers.[33] This combination of North American, Dutch, and French shipping kept the colony relatively well supplied through the rest of the war. What would normally have been illegal shipping was plentiful enough that colonists were even able to export sugar, coffee, and other products. Overall, therefore, the American Revolution benefited Saint-Domingue colonists, because the war situation expanded their access to foreign ships, creating the kind of free trade they had been demanding. The natural disasters of the early 1770s were thus temporary blips in Saint-Domingue's ongoing growth and development.

For white Jamaicans, by contrast, the coming of the American Revolution signaled an end to years of glorious profits. While the American Revolution did not lead to inexorable decline—the slave plantation system was too strong to be substantially hindered by outside events in North America—it divorced its future historical trajectory from that of the United States of America. Jamaicans were instinctively loyal, but the advent of the Revolution made some whites sympathetic to American rebellion. It was probably a lively subject of discussion at planters' tables.

The dangers of talking about liberty in a slave society were illustrated in the Hanover slave plot of 1776, which many observers believed was intimately connected to the onset of the American Revolution.[34] The rumored plot was put down brutally, with seventeen slaves executed (some burnt by slow fire), forty-five transported, and eleven severely flogged. Governor Keith explained that Jamaican slaves knew that "the English were engaged in a desperate war, which would require all their force elsewhere," meaning that slaves in Jamaica "would not have a better chance [than in 1776] of seizing the country to themselves."[35] Nevertheless, the alleged plotters were not inspired by revolutionary ideas but rather by African objectives, modified to fit Jamaica. They wanted, according to white observers, to create a series of principalities corresponding to specific African ethnicities, such as Coromantee or Eboe, which would imitate and collaborate with Maroon communities.[36]

For loyalists such as the Reverend John Lindsay, rector of St. Catherine parish and a firm defender of empire and of slavery, the Hanover plot proved the folly of discussing rebellion in a slave country. He commented how "the topic of the American Revolution has been by the disaffected amongst us, dwelt upon and blandished off with strains of Virtuous Heroism" while refrains of "Dear Liberty [have] rung in the Heart of every House-bred slave, in one form or other, for these ten years past," at dinner tables "where by the bye every Person has his own waiting man behind him." The result, he argued, was to turn "even the creole Negroes, who were the savers of their Masters and Mistresses in the Rebellion of 1760" into rebellious slaves.[37] But it was not just the supporters of John Wilkes or Samuel Adams who used inflammatory rhetoric about how planters were to be soon reduced to slaves if misguided British policy was adopted.[38] Simon Taylor, for example, often employed metaphors of slavery in political discourse. In 1781, he fulminated that he might leave Jamaica "to some other Government where we might be able to make a shift to live, and not be held in Egyptian bondage."[39]

White Jamaicans, just like their counterparts in Saint-Domingue, had a complex relationship with the political discourse of enslavement. Fear of insurrection nagged at white Jamaican psyches. But nothing suggests that white Jamaicans were so paralyzed by fear of their slaves that they compromised their political beliefs. It was not fear of a slave uprising that stopped them from joining the rebellion of their North American compatriots. Instead, much of their behavior seems designed to provoke and terrify enslaved people. If a slave revolt did happen, Jamaicans thought they could deal with the consequences. Tacky's Revolt had shaken Jamaica. But, in retrospect, it was a singular event with limited long-term consequences. Indeed, recent Jamaican prosperity could almost be dated from the end of its suppression. Repression worked. The problem with repression, however, was how planter brutality played in the metropolis. There it played very badly.[40]

A more interesting historical question than why Jamaica did not join the American Revolution is why North America did not try harder to recruit it to its cause. For example, in the Articles of Confederation, the Continental Congress allowed colonies in what was to become Canada, but not any West Indian island, the option of later joining the rebels. As a writer in the *Danish*

American Gazette commented, the Americans declared independence from Britain without "once calling on the West Indians to declare their sentiments [and] ... without once proposing whether they would join with them in their ... race for liberty." Some North Americans, it was true, were concerned about how their actions affected West Indian commerce. Thomas Johnson Jr., for example, a delegate from Maryland to the Continental Congress, felt "troubled at the prospect of Wretchedness into which we are likely to plunge the West Indies" if America rebelled. Yet he was determined to proceed, because "any Relaxation of our System would be imputed by our Enemies to irresistible Interest." His concern about a West Indian alliance was that Britain would view any concessions to the islands as a breach in American solidarity.[41]

The voluminous records of the founding generation show little North American attention to Jamaica. A petition from the Jamaican Assembly in 1774, which argued that the colony was sympathetic to the patriot cause but could not join on account of its vulnerability to internal and external enemies, attracted some comment.[42] John Adams thought that "the Petition from Jamaica, and the behaviours of the other Islands, are great Events in our favour." It received a lengthy response from the Continental Congress, but that document accepted without question Jamaicans' explanation that "the peculiar situation of your island forbids your assistance." It acknowledged Jamaica's patriotism and apologized for including the West Indies in its nonimportation agreement. Finally it offered the lame, even insulting, advice that Jamaica might survive the American boycott "by converting your sugar plantations into fields of grain." "While the present unhappy struggle shall continue," it limply concludes, "we cannot do more." Some writers suggested that Jamaica could do more to support the patriot position. In one of the few American pamphlets to comment on the West Indies and their lack of support for the Revolution, West Indian–born Alexander Hamilton noted that while Jamaicans' passivity meant that they were "not chargeable with any actual crime towards America," their inaction might nevertheless "be esteemed criminal" for "it is the duty of each particular branch to promote, not only the good of the whole community, but the good of every other branch." Hamilton reminded West Indians that they could not "subsist independent of us ... [as] the necessaries they produce within themselves when compared with the consumption, are hardly worth Mentioning." Without America, he continued, "the West Indies might have the satisfaction of starving." Increasingly, Americans saw an impassable chasm between themselves and the islands. One writer in 1774 declared that "I would, as a Politician, divide our American settlements into two classes, the first and favourite one, the West India islands." Another argued that anyone who did not honor and obey the Continental Congress "is a slave." Such men should be sold "at a publick Vendue" and sent "to plant sugar with his fellow slaves in Jamaica" as someone unfitted to breathe the free air of America.[43]

What united Jamaican and mainland colonists in 1776 was anxiety about the future. Nevertheless, whereas mainland colonists were taking action to secure what that uncertain future might be, grounding their anxieties in grand republican sentiment, Jamaicans were tightly focused on present worries. In December 1776 the Jamaican Assembly assured the Crown that "it must be and is our desire as well as our interest, to support, maintain and defend your Majesty in your just rights, with the greatest loyalty and affection."[44] White Jamaicans feared that war would disrupt trade. They warned also that their slaves would take advantage of imperial weakness to continue and expand the slave rebellions begun in July in Hanover. And they wondered how they would fend off France and Spain, eager to take the island from the British. As early as 2 August 1776, Thistlewood noted that local gossip suggested that there were "Strong suspicions off the French in Hispaniola." Some feared with George III that American independence would bring with it the loss of his West Indian colonies and perhaps the disintegration of the whole empire.[45]

The Greater Antilles was a significant theater of military operations in the American Revolution. Following the narrative accounts of Piers Mackesy and Andrew O'Shaughnessy, we divide the war into two periods, the second period starting in 1778, when France and Spain entered the conflict.[46] The military narrative of events does not need, therefore, great explication in this book. The sole mention of Jamaica before 1778 by American military officers was in secret instructions to the naval commander John Paul Jones, from the financier and member of the American Marine and Maritime Committee Robert Morris. In early 1777, Morris encouraged Jones to harass shipping to the north of Jamaica and the south of Saint-Domingue after having taken the eastern Caribbean island of St. Kitts. Morris thought that destroying property in Jamaica would create such an alarm in Britain that it might encourage Britain to divide its forces so that Jamaica could be defended, to the benefit of the United States, which would have more "elbow room" to attack weakened British forces in North America. Morris's suggestions, however, came to nothing, as soon after this letter was sent Jones was ordered to sail to France.[47]

Jamaica was not part of Britain's war plans before France entered the conflict. In events leading up to the Declaration of Independence, London diverted troops from Jamaica to North America. Governor Keith lamented in a letter of 6 June 1776 that the departure of the Fiftieth Regiment for North America had "led to general consternation and alarm."[48] Later that year the Fiftieth Regiment was replaced by Dalrymple's regiment of 491 men. There were no more troops sent to Jamaica until two regiments arrived in March and November 1779 with 1,765 additional men and a further four regiments arriving between January and March 1780 with 2,730 men. Between 1776 and 1778, these troops did no fighting, except to help put down the Hanover slave plot.

The absence of any serious military threat to the island ensured that relations between the Assembly and the governor were generally harmonious.[49] One reason for this harmony was Jamaicans' support for British policy. In February 1777, for example, the parish of Westmoreland, where Thistlewood was a vestry member, sent a patriotic address to the governor and the king congratulating them on their success against "the Provincial Insurgents," whom they described as "that deluded people." It hoped that Americans "would soon be brought back to their Duty."[50] Opposition amounted to low-level grumbling over how little attention Britain was paying to the island's defense needs and concern over increased costs of trade. But Jamaicans had little reason to complain. Britain made multiple concessions to Jamaica from 1776 onward. It allowed West Indians the right to negotiate the dates of naval convoys protecting shipping from Jamaica to Britain; it relaxed trade restrictions between Ireland and the Caribbean; and it gave West Indian lobbyists in London easy access to important government ministers. The most momentous of these concessions came in 1782, after the Crown granted Ireland a new constitution. Jamaica was among the colonies that received the right to make laws for itself, subject to parliamentary approval, clarifying a controversial issue and acceding to long-standing Jamaican demands. Like Ireland, Jamaica was able to get through imperial edict constitutional gains that the American revolutionaries had believed they needed to go war in order to receive.[51]

Some Jamaicans took advantage of the island's peaceful status to trade with the enemy. The most prominent Jamaican smuggler was Eliphalet Fitch, a native of Boston who became a major merchant in Kingston in the early 1760s. He had close connections with merchants in Cuba and Saint-Domingue and was also willing to pass on information to North Americans about the movement of British warships. Fitch was not condemned for his activities. The advocate-general, Thomas Harrison, a man sympathetic to the positions held by Kingston merchants, advised Governor Keith that he should not prosecute Fitch and other merchants trading with the enemy under the rather specious argument that the government would have to prove not only that Fitch was trading with America but also that the individual Americans he was trading with were actually in rebellion.[52]

Most Jamaicans, however, did not trade illegally. They suffered from disruptions of trade, especially from North America, where they had previously obtained a large percentage of their supplies, but also in respect to the trade with Africa and with Britain, where ships had to sail in convoy to prevent attack.[53] Their aim throughout the conflict was a quickly negotiated peace between antagonists. This desire for peace was why the Society of West India Merchants in London eagerly welcomed the peace mission of the Earl of Carlisle to America in 1778, just after the signing of the Franco-American alliance. Britain proposed to give the thirteen colonies all the constitutional rights and privileges it had refused them in the decade before the Declaration of Independence if they agreed to return to the empire. By 1778, however, matters were too far gone for that to be a possibility.[54]

For white Jamaicans the entry of France and Spain into the conflict created an altogether more worrying situation than in 1776: a war that involved three European empires. Jamaica now moved, reluctantly, to center stage. Its defense became vital to British war aims. West Indians greatly increased their financial contributions to the war effort after 1778, when France entered the conflict. West Indians in London raised money to recruit troops and send them to the Caribbean.[55] Jamaica's Assembly dramatically raised the annual expenditure it granted to the governor, most of which was for military purposes. In 1780 this amounted to £153,000, or 3.2 million livres, up from £53,300 in 1772, when Jamaica was prosperous and peaceful. It meant that Jamaica's government debt rose to £282,260, or nearly 6 million livres, by the end of 1781.[56]

In London, George III saw the expansion of the war as an opportunity to complete Pitt's Seven Years' War plan to drive France out of the Caribbean.[57] In 1778 he wanted to "avenge the faithless and insolent conduct of France" by abandoning operations in North America for a West Indian offensive. For the monarch and many of his ministers such a war was less morally complex than a civil war with errant countrymen.[58] For white Jamaicans the fear of invasion outweighed the excitement of potential Caribbean conquests. After

1778 they complained a great deal more about their position than they had done before, and from 1779 the harmonious relationship between the Jamaican Assembly and Governor Dalling disintegrated. As one of Lord Shelburne's Jamaican correspondents commented, he went "from being one of the most popular characters a twelve month ago that perhaps ever was in this country . . . [to] the reverse."[59]

In 1780, Dalling seized the occasion afforded by the widening of the war to attempt to seize a significant chunk of present-day Nicaragua and thus compromise Spanish control over its Caribbean-focused American empire.[60] The expedition, predictably, was a colossal failure, because European troops were so vulnerable to tropical disease. Dalling's forces successfully besieged the fort of San Juan, but only 380 of his eighteen hundred men survived this "victory." A few soldiers died in battle, but disease was far and away the biggest killer. It was the single deadliest engagement in the entire Revolutionary conflict, more destructive to British forces than even Saratoga, the most decisive North American battle. This region of Central America, as soldiers bitterly lamented, was a malarial swamp. Moreover, the campaign was conducted in the rainy season, when deaths from disease were guaranteed. Thomas Dancer, a Jamaican doctor, concluded that "the late dreadful mortality of the troops . . . serve to evince the insalubrity of these climates and the difficulty of attending all military operations in this part of the world." As always in the West Indies, mosquitoes proved victorious over humans.[61]

Dalling's replacement as governor, Colonel Archibald Campbell, was a far more conciliatory figure than Dalling and much more liked by the Jamaican Assembly. But he also succeeded because by 1781 the Assembly was more convinced than it had been in 1779 and 1780 that it was in great danger from foreign attack.[62] The change of emphasis can be traced in Thomas Thistlewood's diary entries. Before 1778 he hardly mentioned the American conflict except to note that he was reading several books on the causes of the conflict. But in 1778 he became alarmed at the threat of a Spanish or French invasion. On 28 January 1778, he noted, "A report off Ten Thousand Spanish troops being arrived in Hispaniola." He repeated on 16 August 1780 an unfounded rumor conveyed to the residents of Savanna-la-Mar by a trooper "almost frightened out of his wits" that four hundred French soldiers had arrived at nearby Black River and were marching west. The rumor, "soon contradicted," caused "such a fright and confusion . . . as scarce be conceived especially among the few white women remaining and the negroes." He was worried in late 1781 when news of Count Admiral de Grasse's plans to attack Jamaica became widely known. He noted on 24 December 1781 that "people [were] terribly alarmed, for fear of Invasion." He followed events closely until April, when news of Rodney's victory at the Saintes reached Jamaica. The invasion made everyone "quite down in the mouth," especially in March 1782, when the French and the Spaniards "are Certainly expected, about the Fall off the Moon," or in less than two weeks from when he was writing. Significantly, he made no complaint, as he had done earlier in the war, about martial law being declared in 1781 and 1782.[63] Indeed, the period of maximum danger for Jamaica was in the months after the French and American victory at Yorktown in October 1781 when de Grasse turned his attention toward Jamaica.

In Saint-Domingue, the years from 1772 to 1778 were a largely peaceful period in which the major effect of the American Revolution was intensified smuggling with North America. Before the North American rebellion began, merchants and colonial administrators cited the hurricanes, the earthquake, and the droughts of 1770, 1772, and 1775 to justify trading with North America and Jamaica. One creole author even claimed that North American trade prevented the colony's ruin in the wake of the 1770 earthquake that destroyed Port-au-Prince. At the same time, French colonists also carefully watched political events in British America. The officially approved colonial broadside, the *Affiches américaines*, covered North American and British politics in detail, mostly drawing from English papers. It devoted long articles to events like the arrival of Stamp Act officials in New York in 1765 and to the Continental Congress's Declaration of Independence and its reception in 1776.[64]

Indeed, the beginning of the war was more than just a smuggling bonanza for Saint-Domingue merchants. It also made the colony an essential location for French secret aid to the rebels as well as military intelligence. The wrenching reforms of the 1760s, overseen by the duc de Choiseul, who was naval minister in the second half of the Seven Years' War until 1766, then minister of foreign affairs, had been driven by his notion that France would soon fight another war against Britain. He believed the Caribbean would be a major theater of that conflict, which was one key reason why he appointed so many West Indian creoles to ministerial offices.[65] These ideas were taken up again in 1774 when Louis XVI came to the throne and named the Count de Vergennes minister of foreign affairs. A new naval minister, Antoine de Sartine, reactivated Choiseul's policy of investing heavily in new ships of the line. When Sartine took office in 1774, only twenty-four such ships were ready to sail. By 1778, he had restored the French navy to fifty-eight ships of the line, compared to sixty-six in service in the British navy. Jonathan Dull notes that "not since the 1690s had the French come so close to equality" to Britain in naval strength.[66]

Foreign Minister Vergennes believed that whether Britain retained or lost its North American colonies, it would try to recoup its military expenses by taking Caribbean territories from France and Spain. Nevertheless, France remained officially neutral during the first years of the conflict, gauging the chances of the rebels and building its forces. Saint-Domingue, however, played an important role in strengthening relations between its metropolis and North American rebels. In February 1776 Governor d'Ennery wrote Versailles that the Continental Congress was asking to arrange a formal commercial treaty with the colony. Plombard, a merchant at Cap Français, proposed creating a depot for Chesapeake tobacco in Saint-Domingue, to facilitate the shipment of this product to Europe.[67] Nothing came of either of these suggestions, but in August d'Ennery welcomed the congressional agent Stephen Coronio to residence in Cap Français, assuring him full protection and guaranteeing the same for all United States of America ships on the colony's coasts and in its harbors. Saint-Domingue had so much commerce with North America that it was a logical center for French government intelligence. In July 1777, when Sartine heard that classified information about North America had become general knowledge in Paris, he immediately suspected that Saint-Domingue's governor and intendant had written indiscreet letters to friends and family in France.[68]

French neutrality in the Caribbean was thus a flimsy illusion. North American privateers sold their goods taken from British ships within easy distance of Cap Français. On 13 April 1777, for example, the *Hourdy Beggar* sailed into Fort Dauphin with a cargo of 260 slaves taken from the British ship *George*. A week later it sailed out again, now carrying no cargo. Instead, its captain had a certificate from the local customs house that he had sold the captives to a Spanish captain down the coast at Monte Cristo who planned to carry them to Havana. Some French captains entered this potentially lucrative business of illegal smuggling, gaining privateering commissions from the Continental Congress in Philadelphia. One condition of such commissions was that they had to hide their national identity, although one doubts that anyone was really fooled. In November 1777, fearing British reprisals, the colonial administration of Saint-Domingue auctioned off a prize taken by one of these "masked corsairs," informing Versailles that they were holding the proceeds in escrow against future claims by the British. In fact, the British navy in the course of 1777 captured close to two dozen ships leaving Saint-Domingue's ports, taking them to Jamaica on the grounds that they were sending supplies to the rebels.[69] Besselère, a young fortune seeker, had arrived from France around November 1777. He wrote his brother on 24 December, "We have been threatened with war since we arrived.... It is only in the last ten to twelve days that we have enjoyed the hope of peace."[70] The young overseer Regnaud de Beaumont feared war, for he had heard rumors that planters fired employees like himself in wartime. But in April 1778 he wrote to his mother that "sentiments are still divided here over the war. The affirmative side is often supported and right after that the negative." He hoped the wartime opening of the colony's ports would lift the near-famine conditions in Léogane.[71]

In 1775, even before France secretly agreed to help the Americans, two enterprising French merchants, Pierre Penet and Emmanuel de Pliarne, had sailed from Saint-Domingue to Rhode Island and then onto Philadelphia. They convinced the Secret Correspondence committee of the Congress that they represented Versailles and offered to supply the Continental army with war material. At about the same time a bona fide French agent arrived—also via Saint-Domingue; he was Julien Alexandre Achard de Bonvouloir, a minor nobleman from Normandy who had served in the Cap Français regiment.[72] France began to send gunpowder and other supplies to the rebels. Because its location made it the first stop for French ships crossing the Atlantic, Martinique was the most important site for this trade, but Saint-Domingue also played a key role.

The playwright and businessman Pierre Caron de Beaumarchais, a central figure in this operation, routed many of his shipments through Saint-Domingue. In 1776, using 2 million livres, or £86,580, in secret financing from the French and Spanish governments, Beaumarchais made Saint-Domingue a major node in the smuggling operations of his company, Rodrigue Hortalez.[73] He sent so much powder, artillery shot, tents, uniforms, and other equipment to the North American rebels via Saint-Domingue that his agent in Cap Français, Monsieur Carabasse, had to buy four small smuggling ships to carry these goods to the mainland. In 1777 he covertly arranged for one of his ships to be "captured" by an American privateer just before entering the Cap Français harbor. The plan was that while his Cap Français agent lodged a formal complaint, the privateers would sail the vessel to Philadelphia, where another of Beaumarchais's agents would complain to Congress. The agent would then secure control of the vessel, disembark the war materials and send the ship back to France loaded with tobacco.[74]

On 26 June 1778 Vergennes announced that France was at war with Britain, after receiving word that a British fleet had captured two French frigates.[75] Almost immediately, Saint-Domingue's governor officially opened all the colony's ports to North American commerce. After years of covert trade and intelligence, Saint-Domingue was openly involved in the War of American Independence. By October 1778 the British had blockaded its coast. Within two months they had seized two hundred ships coming out of the port of Cap Français.[76]

As in the Seven Years' War, the British blockade, at least at first, raised the price of provisions, especially wheat. The colony however, had suffered drought since 1776. Higher food prices and reduced shipping led to slaves dying of hunger because their gardens were too dry to produce food.[77] While France geared up for the war, Saint-Domingue received little naval protection, and even at the Naval Ministry colonial supporters spoke of the colony being "cruelly abandoned."[78] This changed in July 1779 when Count d'Estaing, the colony's controversial former governor, arrived in Cap Français. He was at the head of a squadron of twenty-five warships that had just recaptured Grenada, a former French possession lost to Britain in the Seven Years' War. Hoping to reinforce his troops for an expedition to the North American mainland, d'Estaing organized two volunteer colonial units in Saint-Domingue, one white and one free colored. These efforts were unsurprising given his previous attempts between 1764 and 1766 to drum up creole patriotism with a free colored Legion of Saint-Domingue. As before, colonial whites were unimpressed. The *Affiches américaines* sniffed that Saint-Domingue's planters would rather donate money to the Royal Navy. In the modern world, they argued, such financial sacrifices showed greater patriotism than the classical ideal of personal service. The paper mocked the scene of a "ferocious and barbarian mother" sacrificing her son to the fatherland."[79] Regnaud de Beaumont, who could barely feed and clothe himself, tried in vain to get out of monthly militia service, which started around September 1777. He was disparaging about the attempt to rally volunteers for d'Estaing's North American campaign.[80] The Grenadiers unit, slated to be six hundred strong, perhaps unsurprisingly, recruited only 156 white soldiers.[81]

Saint-Domingue's free men of color enrolled in the Chasseurs unit in far greater numbers, although Regnaud observed that "most muster against their will, and most are forced by generals and commanders."[82] Recruiting efforts began in March 1779, well before d'Estaing's late July arrival. When the ex-governor arrived, he visited the colony's most decorated free black veteran,

Vincent Olivier, who was said to be over one hundred years old and who had been freed from slavery after valorous combat in the 1697 French siege of Cartagena. After being presented at Versailles and fighting in Europe, Olivier had returned to Saint-Domingue, becoming the colony's first free black militia officer. Feted by d'Estaing and Governor d'Argout in August 1779, Olivier urged his own sons and free men of color throughout the colony to join the Chasseurs. Whether because of the respect shown to "Capitaine Vincent," as he was known, or pressure from their militia officers, within three weeks over nine hundred men of color volunteered. As also happened with the post-Seven Years' War free colored militia, these units were all commanded by whites. When d'Estaing sailed away in August, 545 Dominguan free men of color were on board his fleet.[83]

The rebels and their French allies failed to take Savannah, Georgia, from the British. D'Estaing, lightly wounded, returned to France with some Dominguan men of color. In May 1780 the colonial broadside reported that a detachment of free colored soldiers was returning to Cap Français from France: "It must have seemed extraordinary in Europe to see people of color armed, disciplined, and war-ready. This spectacle could become the subject of serious reflection. However we have been assured that this corps served very well and it seems that all the officers who served in the last campaign agree in their praise for it."[84] Other Chasseurs volontaires were not so fortunate. Sixty were sent to Charleston, and between 150 and 200 were stationed in Grenada, some of them remaining there until near the end of the war. Versailles wanted to spare its European troops from tropical disease, but as their commander, the Marquis de Rouvray, put it, "the Chasseurs are all property owners who have abandoned their fortune to serve the king."[85] As the free colored volunteers returned from France, Regnaud de Beaumont wrote his mother in terms that reveal his own frustrations at not finding an economic niche in the colony: "I have seen some of the whites who were part of d'Estaing's expedition. They are returning, so to speak, even more destitute than when they left. They earned only their food during the campaign and an escalin and a half per day, equivalent to 274 livres a year." Regnaud, by comparison, was barely surviving as an overseer on 1,500 livres.[86]

After d'Estaing's mixed record in 1779, 1780 was a year of unprecedented French success in the Atlantic theater. The allied French and Spanish navies outnumbered the British, allowing French convoys to reach the Caribbean, despite the blockade. On 1 January 1780, Saint-Domingue's governor and intendant observed that "in general, the colony is provided for.... This war will be counted as the first in which the government was powerfully occupied . . . with the well-being of the colonists."[87] Moreover, the allied navies captured nearly an entire British merchant convoy coming from the Caribbean, with some sixty-seven ships.[88] Saint-Domingue's broadside told of Jamaica's disastrous Honduras campaign and of various French successes, crowing that "the English (at least those of Jamaica), no longer believe themselves the masters of the sea. They are starting to fear for the land." In June 1780, it reprinted a letter said to be from the *Jamaica Gazette* a month earlier in which a group of planters addressed their governor informing him of their fear of a French invasion.[89]

In fact, an invasion of Jamaica was exactly what the French and Spanish were planning. Versailles had promised to help Madrid recover Gibraltar from the British: Jamaica was the only British possession worth trading for this important strategic location. After playing a critical role in the Battle of Yorktown (October 1781), de Grasse took his fleet from the Chesapeake to the Caribbean, where he planned to attack Jamaica. He needed, however, to wait for a French convoy that would restock his ships. Meanwhile, he attacked the Leeward Islands, taking St. Kitts, Nevis, and Montserrat in the eastern Caribbean. Whether de Grasse would have been able to conquer Jamaica is an open question. Experienced commanders doubted that any European army could survive the diseases that would strike its soldiers during a long tropical siege. Britain had discovered this in the 1762 attack on Havana. In 1779 the British generals Sir William Howe and Henry Clinton, both very experienced in American military operations, argued against sending more British troops to Jamaica, where commanders reported annual losses of 15 percent.[90]

One French response to these appalling annual losses was to try to extend the free colored military unit, the Chasseurs *volontaires*, that d'Estaing had raised to fight in Georgia. In 1780 the colony's governor, the Count d'Argout, authorized the creation of such a unit, tellingly renamed without the word "volunteer." He ordered French army officers to begin assembling men of color, prescribing harsh penalties for the families of those free colored militiamen who did not cooperate. The Marquis de Rouvray, who had commanded the Chasseurs at Savannah, assured his superiors that this new permanent unit would be the salvation of the colony, but he said little about foreign campaigns. Instead, he evoked the now quasi-mythical figure of Macandal as justification for forming the Chasseurs *royaux*.[91]

The difficulties that royal administrators faced in trying to establish a competent and well-functioning colored militia did not stop them from conceiving grand plans for an amphibious assault on Jamaica. In Jamaica, it was rumored that de Grasse's ships carried "50,000 pairs of handcuffs, and fetters... intended to confine the negroes."[92] The French and their Spanish allies devoted large amounts of men and resources to this project—a project that if successful would have sent shock waves throughout the Atlantic World, equivalent to Britain's capture of Havana in 1762. Surprisingly, it is nearly invisible in standard accounts of the war.[93] After conquering British St. Kitts, Nevis, and Montserrat, de Grasse then sailed his fleet of thirty ships of the line to Martinique to wait for a convoy carrying supplies and reinforcements from Europe. Jamaica was next on his horizon. Justin Girod-Chantrans, a traveler en route to Saint-Domingue, witnessed the rendezvous of the convoy with de Grasse's fleet. He described how "7,000 regular troops, pulled from all the [French] Windward islands, distributed on naval and convoy vessels, will join 4,000 other French, 10,000 Spaniards, and 14 ships of that same nation that are already at Cap Français. The general opinion is that they are going to besiege Kingston, in order to take Jamaica from the English. Such grand preparations!"[94] Even if Girod-Chantrans exaggerated the number of soldiers and sailors, the proposed invasion force was nearly double the number of soldiers that France committed to Yorktown in September and October 1781 and was perhaps equal in soldiers to the combined total of French and American soldiers at that decisive battle. The invasion force was larger than Jamaica's total white population.

But just as de Grasse attempted to take his men to Saint-Domingue his fleet was checked by Britain's Admiral Rodney on 12 April 1782, near the small Saintes islands that lie between Guadeloupe and Dominica. Rodney, reinforced by ships arriving from Britain, had thirty-six ships of the line and thus was able to engage de Grasse with near superiority in ships and guns, the first time in this war that any British admiral had such a tactical advantage. The Battle of the Saintes destroyed only a handful of French ships, but a number of major French vessels were so damaged they had to go to Boston for repairs.[95] Rodney seized five French ships, including de Grasse's flagship, and took his prizes to Jamaica, where Thistlewood saw them paraded by British sailors. Admiral Hood captured two more ships a week later.[96]

Despite this defeat, most of the French and Spanish invasion force made it to Saint-Domingue, as did Girod-Chantrans. He described the sight of twenty thousand French and Spanish troops stationed at Cap Français and its hinterland. Yet without de Grasse's fleet, this army suffered the fate of all other European armies in the Caribbean: "a menacing colossus, formed in Europe by two great powers... succumbs gradually, falls in ruins each day." As always with European soldiers in the West Indies, soldiers died in huge numbers from putrid fevers. Men died so quickly that mass graves in the Cap Français cemetery had to be reopened to make room for more bodies. According to Girod-Chantrans, the army lost one-third of its men in merely three months.[97] In short, Saint-Domingue suffered the greatest loss of men in any engagement of the whole Revolutionary War, without taking part in any battle. If indeed over seven thousand French troops lost their lives from disease in the abortive invasion of Jamaica, as is probable given death rates in other Caribbean campaigns in the eighteenth century, that loss was nearly as high as the eight thousand estimated battle deaths suffered by the Americans in North America throughout the Revolutionary War. It was over three times as many French soldiers killed in North America between 1778 and 1782.[98] Given the reality of tropical mortality, to conquer Jamaica, France would have needed to send at least fifty thousand men or rely on a rapid surrender.

Saint-Domingue's *Affiches américaines* said nothing about the Battle of the Saintes, although the news of the French defeat must have quickly swept through the colony. On 8 May 1782 the broadsheet recorded the arrival of the massive convoy that de Grasse was escorting but printed none of the political or military news those ships would have carried.[99] Nor did it discuss the army that was filling cemeteries in the colony's largest port city. Instead it drummed up patriotism by recounting French and allied victories in rich detail, like the Battle of Yorktown, and the French capture of Minorca. Writing one week later, in mid-May 1781, the overseer, Regnaud, warned his mother not to write him with political news: "These kinds of narrations have had a high cost for some people here. All we know from public rumors is that our affairs are not going well; it appears that the war will not be over soon."[100]

It was not until 31 July 1782 that the broadsheet mentioned the 12 April defeat and then only as unspecified "news of the disaster that touches the hearts of all true Frenchmen." The goal of this article was to rally support for a campaign to raise money to supply a naval vessel for the French navy. This idea had already taken root in France in the final years of the Seven Years' War; from 1763 to 1765 French port cities donated funds to build some fifteen ships of the line.[101] News of the Saintes produced a similar wave of patriotism in France, with provinces, port cities, and major corporate bodies pledging to donate funds for naval construction. In November 1782 the *Affiches* reported that the colony had responded to "the disadvantage we had last April 12" with a proposal to offer the king "a vessel of the premier rank." This sum was the equivalent of the 1 million to 1.5 million livres offered by the merchants of Marseilles, or the *fermiers généraux* (tax collecting corporation), to build a ship of the line with 110 cannons. Even after the Saintes, French officials still planned an invasion of Jamaica.[102] But as they delayed, negotiators forged a preliminary peace treaty. In September 1783 France ended the war by signing the Treaty of Versailles.

The size of the expeditionary force intended to capture Jamaica shows just how much importance the French placed on the Caribbean theater in the American Revolutionary War. If matters had turned out differently at the Saintes, Britain might have lost its wealthiest American possession. But the history of warfare in the Caribbean suggests that few conquests in the Caribbean were made easily. The massive loss of life suffered by the British in Honduras in 1780 and by the French in 1782 in Saint-Domingue prior to a potential invasion of Jamaica occurred with virtually no fighting. The American Revolutionary War in the Greater Antilles fits well-worn patterns, which might be why the aborted invasion of Jamaica has been so invisible to historians. These grand plans were yet another example of European ministers not understanding how difficult military expeditions were in the Caribbean and underappreciating how many men had to die to achieve even a limited objective.

The most significant events of this wartime period for the Caribbean—like the 1772 *Somerset* decision before war started—were those that changed European views about the culture of violence that marred the plantation system in the Greater Antilles. One such event occurred off the coast of Jamaica aboard a slave ship named the *Zong* in late November and early December 1781.[103] The *Zong* was a Dutch prize ship, purchased in West Africa by Richard Hanley, captain of the *William*, in February 1781, on behalf of his employers in Liverpool. The price of purchase included the ship, goods on board the ship, and 244 captives. Because it occurred in West Africa, the sale required the promotion of Luke Collingwood, the *William*'s ship surgeon to captain the *Zong*, despite never having commanded a slave ship previously. Between March and September 1781, Collingwood purchased more captives on the West African coast. On 6 September he left for Jamaica with 440 captives and one European passenger, Robert Stubbs, a former London ship broker who had recently been disgraced as the governor of the British fort at Anomabu.

It was a long and difficult journey. Near Tobago, the crew discovered that the ship's water casks were leaky, meaning that the *Zong* had only ten to thirteen days left of water at full rations. It was imperative that the voyage not be extended. But that is what happened. Either Collingwood or someone on the crew made a catastrophic navigational error, which meant that by 29 November the ship was days away from Jamaica with dwindling water supplies and a hold full of suffering Africans. Collingwood and his crew decided to jettison "cargo" under the pretext that a severe water shortage on an overcrowded ship would produce a slave insurrection. Collingwood or someone else onboard (Stubbs, as a former ship broker, is a likely suspect) knew his marine insurance law well. The owner of a ship that suffered a slave insurrection was entitled to insurance compensation under clauses relating to "perils of the sea."[104]

On 29 November 1781, members of the crew threw fifty-four women and children overboard. On 1 December 1781, they killed forty-two "stout healthy Men slaves" in the same way. Within a few days of these second killings, rainfall arrived, theoretically eliminating the need for more murders. Yet sometime after 6 December the crew threw twenty-six captives overboard, making a total of 122. Ten more slaves jumped overboard of their own accord: as suicides they could not be claimed for under insurance law. The *Zong* did not put into Jamaica for two more weeks, sailing for many more days than were plausibly needed to reach a friendly harbor. Finally on 22 December 1781 the ship sailed up the Black River in St. Elizabeth parish, a small port that had never before had a slave sale. On 9 January Collingwood sold the surviving cargo of 208 Africans.[105] Very shortly thereafter he died in Kingston, making him a convenient scapegoat for everyone else involved in the sordid business. The ship, quickly renamed the *Richard*, returned to Britain. The owners asked their insurers to compensate them for the value of 122 captives jettisoned as cargo. The underwriters objected. It went to trial on 6 March 1783. The court awarded the slave traders £3,660, or 84,546 livres for the loss of 122 captives. The underwriters appealed, and the appeal was heard by Chief Justice Lord Mansfield on 22–23 May 1783. Granville Sharp attended: his notes on the case are the main source for understanding the trial. Sharp spent much of the next few years publicizing the case. Mansfield overturned the earlier verdict and the underwriters therefore refused to pay the insurance claim.[106]

The *Zong* affair played a central role in making the abolition of the slave trade a matter of intense public interest in Britain. It served as a supreme example of the callous financial calculations on which the slave trade was based. It showed how slave traders had so effectively equated humans with property that "the taking away of the life of a black man is no more account than taking away the life of a beast." The black abolitionist Ottobah Cuguano spoke for abolitionists in general in 1787 when he described Captain Collingwood as "the inhuman monster" and his employers as "inhuman connivers of robbery, slavery, murder and fraud." For Cuguano the *Zong* case showed that "our lives are accounted of no value, we are hunted after as prey in the desert, and doomed to destruction as the beasts that perish."[107]

The *Zong* helped transform the antislavery movement into the most significant moral campaign in British history, a change that happened in a few short years in the mid-1780s.[108] Opponents of the slave trade never failed to express disgust at what had happened off the seas of Jamaica in late 1781 or to condemn the inhumanity of a legal system that saw the murder of 122 Africans solely as an interesting example of marine insurance law.[109] Of course, abolition was far in the future. The slave trade flourished in the twenty years following the *Zong* case. But after 1783 the debate over slavery, the slave trade, and the West Indian plantation system was no longer merely a commercial and political debate—these issues now had "a new ethical dimension."[110]

Fundamentally, the *Zong* case raised the question of what it meant to be British in a world of amoral commercial and imperial competition. Sharp hammered away to the reading public that British sailors in British-controlled waters had committed mass murder on a British ship under the direction of British merchants. Those British merchants then demanded compensation for the murder of innocent Africans who deserved British protection. A British court presided over by a British judge agreed with their claim and issued, he argued, a shameful British verdict. Sharp used the *Zong* to show that Britain was implicated in all aspects of slavery and the slave trade. Britons of all descriptions were concerned with what happened off the coast of western Jamaica. Increasingly, it became clear to the mass of British people that they should not accept the values of Jamaican planters regarding slavery. It was a short step from accepting this point to starting to regard planters as not conforming to British values and behavior.

The American Revolution in the Greater Antilles is a tale of events that did not happen. De Grasse did not invade Jamaica. Nor did Jamaicans join their North American brethren in rebellion, which would have transformed the political order of the region. The geopolitics of the region remained essentially as they had been before the Revolution. But the war did lead to some significant changes. Saint-Domingue's planters used wartime trade with North America to subvert official French trade policy; when the war was over they became more assertive about fashioning a new relationship with the metropole. The French attempt to use free colored soldiers without acknowledging their imperial patriotism foreshadowed what was to become a major theme in the history of Saint-Domingue up to the formation of Haiti in 1804. In the years after the war, Saint-Domingue continued to outstrip Jamaica economically, and only the revolutionary cataclysms of the 1790s ended its out-production of the rest of the Caribbean. Most important, the planters of Saint-Domingue retained the confidence of metropolitan officials and were increasingly recognized within French society as valuable partners in imperial expansion. That high position in metropolitan thinking was not the case in Jamaica. Jamaican planters and merchants may have demonstrated conspicuous loyalty to the British state and may even have done all that could have been asked of them in the War for American Independence. But, as the *Zong* case revealed, the more that Britons knew about Jamaican life and institutions the more they suspected that something was deeply amiss with planters and the plantation system in the British Atlantic World.

CHAPTER 9

Recovery and Consolidation in the 1780s

The end of the American Revolution in the Greater Antilles was greeted with great relief by Jamaicans. They saw Rodney's victory at the Battle of the Saintes as an almost providential deliverance and spent vast sums on an impressive monument to the admiral, dressed in classical costume, which graced the northern side of Government Square in Spanish Town.[1] This magnificent statue was not just a sign of Jamaicans' relief at having escaped a French invasion. It also announced their unbounded optimism about the colony's prospects. Jamaicans thought that they would continue to contribute their share to the growing grandeur of the British Empire in the last two decades of the eighteenth century, although they were aggrieved by shifts in British policies that they thought harmed Jamaica's future growth.[2]

The American Revolution was a good war for Saint-Domingue, unlike the Seven Years' War. Britain's naval blockade was less effective than it had been in the earlier conflict. As a result, planters had been able to sell their sugar and coffee directly to merchants from northern Europe and North America. The war temporarily cost Saint-Domingue access to new African slaves. But the government did not subject white colonists to the degree of militia "slavery" planters had complained of in the Seven Years' War. Instead, men of color handled a great share of defense work. After the war, while France faced a string of fiscal and political crises, partly as a result of massive wartime spending, Saint-Domingue entered into an unprecedented economic boom. With the postwar resumption of the slave trade from 1783, and the prospect of greater commerce with Britain's former mainland colonies, the colony was poised for great prosperity.[3]

Despite the optimism of contemporaries in 1783, when historians examine the postwar period in the West Indies they concentrate on whether the American Revolution precipitated the demise of the plantation system. But in 1783 neither Saint-Domingue nor Jamaica was at the start of an economic decline. Although the war and natural disasters had weakened some parts of these economies, these setbacks were small and temporary. By the middle of the 1780s in Saint-Domingue and by the start of the 1790s in Jamaica, plantations were at the peak of their prosperity. Their success was based on a resurgent Atlantic slave trade and a relentless push to wrest as much productive labor out of their slave force as possible. The plantation "machine" was beginning to be perfected: planters learned how best to exploit their human capital.[4]

The American War of Independence caused problems in Jamaica because the island was tied closely into a North Atlantic trading network that was badly disrupted in the mid-1770s. For an earlier generation of historians, concerned with explaining the rapid decline of the British West Indies in the nineteenth century, these postwar difficulties showed deep structural problems in the plantation economy. Lowell Ragatz and Eric Williams set the terms of the debate on the British West Indian side in 1928 and 1944, respectively. Ragatz traced West Indian decline back to the 1760s and to the rise of more efficient sugar producers, notably Saint-Domingue. Williams, conversely, saw the early 1780s as pivotal years. "Far from accentuating the value of the sugar islands," argued Williams, "American independence marked the beginning of their uninterrupted decline."[5] Only a few scholars today accept Williams's claims unreservedly. No one accepts Ragatz's. The prevailing modern consensus is that any economic difficulties caused by the American Revolution were temporary, not permanent.[6] Indeed, the point at which Jamaica seems to have been in clear economic decline gets pushed back closer and closer to the eve of emancipation, with major economic difficulties occurring only in the early 1830s.[7] Moreover, hurricanes were the principal difficulty Jamaica faced in the 1780s, not the conflict between Britain and North America. There is now a broad consensus among historians that British slave societies were increasing their output and their value as imperial trading partners in the 1790s; between 1792 and 1798 annual rates of return in the British West Indian plantations were of the order of 12.6 percent. In these years Britain expanded its Caribbean plantation empire to an unprecedented level as a result of Napoleonic Wars. This growing productivity probably continued in the ten years after 1800. Jamaica's productivity faltered a little after 1800 as its soil became less fertile and as some optimistic planters overpaid for land and slaves. But slavery in the Americas continued to be profitable in most places where it was established. David Ryden suggests that there was a short-term collapse of sugar prices in the British domestic market in 1806–7 but accepts J. R. Ward and Seymour Drescher's contention that Jamaican planters experienced no long term economic problems between the mid-1780s and the mid-1800s.[8]

Nevertheless, the War for American Independence caused Jamaican planters short-term economic distress. Their profits slipped dramatically, declining from an average of about 10 percent throughout the eighteenth century to 3 percent in the early 1780s. Sugar production fell from 51,218 casks shipped to Britain in 1775 to 30,282 in 1783. Lack of specie and increased freight charges harmed commerce. Between 1774 and 1781, charges on a hundredweight of sugar increased from 14 shillings to 37 shillings. Insurance rates rose from as low as 2 percent to as high as 28 percent. The price of food skyrocketed—if planters could find any.[9] The war made trade very difficult, as merchant Lowbridge Bright acknowledged to a potential migrant to the island in 1779. Bright attributed his tardy response to "the difficulties attending our shipping in wartime and the obstacles unexpected incidents which dayly occur."[10] Despite these difficulties, however, the colony's total wealth continued to rise, albeit at a smaller rate than previously. In 1754, total wealth was £10,338,236. It increased by 240 percent in the next twenty years, to £24,109,147 or 557 million livres by 1774. By 1778, it had increased a further 15 percent, to £27,860,517. The 1780s saw a slowing down of annual growth, but this was largely due to a string of devastating storms and the effects of war rather than to any long-term structural problems within the Jamaican economy. Between 1778 and 1787, total wealth grew by a modest 5 percent, to £29,299,918 or 677 million livres.[11]

In the 1970s, historians also advanced a decline thesis for Saint-Domingue after the American Revolution. Jean Tarrade's study of French colonial trade policies concluded that the kingdom's Caribbean commerce in the 1780s was in "a latent state of crisis under an apparent prosperity." He argued that the reliable profitability of the slave trade to Saint-Domingue in the 1780s obscured the fact that merchants were losing money on other colonial investments.[12] Françoise Thésée's 1972 history of the failed Bordeaux colonial merchant house of Romberg, Bapst et Cie supports Tarrade's analysis.[13] Her study shows how merchants' over-optimism about Saint-Domingue in the years after the American Revolutionary war led them into bad investments. Nevertheless, the 1780s in Saint-Domingue, we argue, were not the "beginning of the end" of prosperity in the colony but a speculative bubble that only a few firms failed to negotiate successfully. The fact that French merchants and creditors could not legally foreclose on debt-ridden colonial planters was a serious issue, but that was a political more than an economic dilemma. The economy of Saint-Domingue had more strengths than weaknesses right up until the events of 1791 changed the colony's future irrevocably.

The history of Romberg and Bapst illustrates the nature of that speculative bubble and the associated problem of debt collection. Based in the Austrian Netherlands (modern day Belgium), in 1781 Henry Romberg's merchant firm owned ninety-four ships. As neutrals, his captains could trade with Saint-Domingue during the American Revolutionary War because the British allowed them through the inevitable blockade.[14] Despite the conflict, they carried Saint-Domingue sugar, coffee, and other commodities directly to northern European markets. French merchants resented the competition but consoled themselves that their monopoly would return at the end of the conflict. Romberg too realized that peace would lock his ships out of French Caribbean ports. So in 1783 he moved from Brussels to Bordeaux and founded a new firm. A number of Swiss, German, and Dutch merchants did the same.[15] With substantial backing from investors in Brussels, Romberg and Georg-Christophe Bapst made slaving voyages and entered long-term contracts with planters in Saint-Domingue.[16]

Romberg and Bapst found planters eager to buy their African captives but had a harder time securing payment. As the colonist Lory wrote in 1780, "You often hear it said in France that good faith does not reign in Saint-Domingue; that phrase is repeated and will be repeated for a long time."[17] A colonial law prohibited creditors from foreclosing on plantations or any asset necessary for commodity production. Most colonial reformers believed this "no foreclosure" rule was essential to colonial prosperity, but it infuriated merchants.[18] Despite their expertise in Atlantic commerce, Romberg and Bapst badly mishandled their Saint-Domingue investments. Possibly they were too optimistic that their wartime profits would continue. Rather than merely selling slaves and buying colonial commodities, they signed contracts with dozens of planters. These agreements obligated the planter to ship all his produce on Romberg ships and to entrust the firm with selling these goods in Europe. In return, the firm provided these clients with guaranteed payments, including annual pensions to family members in France. In some cases the company paid substantial sums to consolidate a planter's other debts. In other cases they assumed complete responsibility for a plantation, installing their own managers and taking a commission before passing profits on to the absentee proprietor.

In 1789, after six years in business, Romberg and Bapst declared bankruptcy. They found new investors but went bankrupt again in 1793, with approximately 9 million livres or £390,000 in outstanding colonial loans. For Thésée, the Romberg failure illustrated that Saint-Domingue's economy was fundamentally flawed. The firm's inability to collect its colonial debts made its position untenable, but its clients also suffered from declining yields of cotton and indigo. She pointed to other prominent Bordeaux firms, like the Gradis and Bethmann houses, who gradually transferred their colonial investments into vineyards and other domestic businesses in the 1780s. Even during the French Revolution, in December 1791 at the height of its moderate phase, merchants and their representatives failed to change the laws preventing foreclosure on indebted planters.[19]

Bad business practices, however, not Saint-Domingue's economic conditions, doomed Romberg and Bapst. The firm specialized in indigo and cotton production, for example, commodities with prices that were especially volatile in the 1780s compared to sugar and coffee.[20] The firm's colonial managers were poor correspondents. Their letters were often late and contained few specifics about the problems of the estates they managed. The firm's primary problem was a kind of "irrational exuberance" about Saint-Domingue that may have been based on the profits Romberg saw during the American Revolution. In a typical contract the firm agreed to pay 200,000 colonial livres to resolve a planter's debts and advanced 10,000 French livres per year for living expenses. A contract giving the firm exclusive rights to carry and sell that planter's crops would, in theory, allow it to recoup its outlay. But many of Romberg's plantations delivered barely enough commodities to meet operating expenses. The firm's colonial representatives affiliated Romberg with sugar estates that they estimated could ship two million pounds of sugar every year. Yet over a decade Romberg received only 450,000 pounds of sugar. One cotton plantation that was expected to deliver twenty bales of cotton produced only three; another promised twenty but delivered only four. Most of the contracts contained no provisions for these kinds of shortfalls. Seventeen of the firm's fifty client plantations never shipped anything with the company. Some claimed they had nothing to ship, while others placed their goods with other merchants, sometimes because they owed more money to those firms.[21] Finally, the political chaos in France in the years 1789–93 disrupted the financial and commodity markets in the metropolis.[22]

Other French merchant firms, however, made money in the 1780s. Albane Forestier shows that the Chaurand merchant house of Nantes reaped strong profits from lending money to plantations in Saint-Domingue in the 1780s. Like Romberg, the Chaurands were shippers. They were commission agents for West Indian produce, slave traders, and direct investors in Saint-Domingue real estate. By the mid-1780s they had invested over 2 million livres in West Indian property. Like Romberg et Bapst, they set up contracts with thirty-eight large planters in the 1770s and 1780s, to take produce on consignment for a transaction fee of 2 percent.[23] Also like the Rombergs, the Chaurands made good profits during the American War, with privateering ventures generating 64 percent of profits between 1776 and 1784. Over the long term their best returns came from buying Saint-Domingue properties, especially in the western and southern provinces. They made these decisions quite carefully, despite "having never set foot in the colony." They evaluated plantations very comprehensively, paying special attention to the quality of management. This strategy was highly sensible given wide variations in estate profits in Saint-Domingue. Returns on plantations whose profits we know something about, such as the Fleuriau estate, in the Cul-de-Sac region in the Port-au-Prince hinterland, yielded net profits of more than 15 percent per annum between 1775 and 1784.[24] The Duplàa plantation was nearly as profitable, with net profits of about 10 percent per annum. Even the Chaurands' less profitable investments in Saint-Domingue yielded over 5 percent per annum return.[25]

These returns help explain why the Chaurands and Romburg increased their financial exposure to Saint-Domingue markedly after 1776. They made large initial loans and encouraged further borrowing to increase the productive capacity of estates. They aggressively sought new clients, even if these clients were heavily in debt, being confident that the contracts they required planters to sign reduced uncertainty and gave them a strong claim on plantation produce. The Chaurands found more profitable estates than Romberg and Bapst, and they managed their clients more effectively. They were also helped by rising prices for sugar, which kept their ratio of debt to revenues constant. The Chaurands, Forestier argues, readily accepted risky investments. They understood that their clients had taken on debt to achieve economies of scale and that these investments could maximize estate profits.[26]

The key problem with the decline thesis suggested by Tarrade and Thésée is that it confuses bad plantation management with the viability of Saint-Domingue's plantation complex. Like Jamaica with its hurricanes, the colony was struck more badly by drought and famine during the war than by any structural financial problem.[27] In 1776 and again in 1779 the entire colony suffered serious dry spells. The Fleuriau plantation outside Port-au-Prince saw its overall sugar production fall by close to 50 percent from 1778 to 1779.[28] The handful of plantations owned by the La Lorie heirs, with a total of more than five hundred slaves, experienced similar decline. Their sugar estates, managed by the Foäche merchant house at Le Havre, produced one-sixth less sugar by weight per acre in 1779 than in the previous year.[29]

The La Lorie heirs were concerned about this decline because they had contracted large debts, based on their confidence that their sugar production would grow. They owed over 900,000 livres, or £39,000, to various creditors, the largest being the Foäche firm, which, like Romberg, delivered them guaranteed incomes in exchange for a contract to manage their properties and sell their goods. Like many absentee owners, the La Lorie heirs devoted only a small portion of their annual incomes to paying down the debt, a procedure that was intended to take thirty years to complete.[30] Yet the very problem of indebtedness shows how much confidence French merchants had in the viability of the French plantation system; despite laws that blocked them from foreclosing on a plantation, they kept on lending money to planters.

Nor should this surprise us, for the economies of Jamaica and Saint-Domingue strengthened throughout the eighteenth century. Slave management techniques improved workers' productivity; planters began to use technology more effectively; the slave trade continued to flourish; and the market for plantation produce expanded along with the increase in European and North American populations. Nor did plantation economies elsewhere stop expanding after political events struck Jamaica and Saint-Domingue, slowing down their economies in the early nineteenth century. The profitability of plantations in nineteenth-century British Guiana, Cuba, Brazil, and above all the United States suggests that there was no obvious economic reason to expect that Jamaica or Saint-Domingue was spiraling into long-term decline.[31]

One reason an earlier generation of historians saw Caribbean plantation economies declining in the late eighteenth century is because of the powerful appeal of theories developed in this very period by Pierre Samuel Dupont de Nemours or Adam Smith, who made influential arguments for the superiority of free labor over slave labor. As Seymour Drescher has suggested, scholars have been tempted by Smith's chronology of historical change, and even more by Karl Marx's understanding of the place of slavery within the evolution of capitalism, "to look for a natural history of the institution in a sequence of youth, maturity, and senility."[32] The other reason is theological.

Slavery was an evil system, the "original sin," as Barry Higman puts it, of West Indian history. Abolitionists believed the sin of slavery would inevitably lead to the fall of the plantation system.[33] But the slave systems of the late eighteenth century did not unravel from within. Rather, it was planters' inability to insulate their estates from changes in metropolitan politics that opened the door to slave revolts in Saint-Domingue in 1791 and Jamaica in 1831. Until those external influences entered the picture, the plantation machine was improving as a wealth-generating institution, not declining. How did enslaved people fit into this picture? As workers, they were in constant negotiation with their masters, and in this way they shaped the nature of the plantation institution. But the planters' and merchants' pressure for profits, expressed in slaves' lives as borderline malnutrition and overwork, severely limited their ability to resist.[34]

Real, if temporary, economic hardship in Jamaica during and after the American Revolution came only after the entry of France and Spain into the war in 1778 and especially after hurricanes hit in 1780, 1781, 1784, 1785, and 1786. These years—a time of remarkable growth in Saint-Domingue—were probably the most troubling period for Jamaican planters and slaves alike since the Maroon Wars of the 1730s. The succession of storms caused considerable economic damage. The first hurricane of 3 October 1780 was especially devastating, although damage was localized to the western parishes. The 1781 and 1784 hurricanes affected the eastern regions of the island, while the 1785 and 1786 hurricanes were "more general." By the last hurricane of 20 October 1786 Jamaicans had become almost inured to hurricanes as a yearly occurrence. Alured Clarke, governor of Jamaica, wrote to London that "this unfortunate island has again been visited by one of those Public Calamities which seem to have become annual in this Quarter of the world."[35]

Those caught up in the direct path of a hurricane found the experience harrowing. Thomas Thistlewood suffered a direct hit on his property in 1780, losing by his estimation £1,000. He wrote about it at great length. Among the worst hurricanes to hit the Caribbean in recorded history, it left "sad havoc all through the countryside." Thistlewood described it in cosmic terms, as "most tremendous, awful & horrible. . . . The elements of fire, air, water and earth seemed to be blended together . . . [and] it seemed as if a dissolution of nature was at hand."[36] The same storm bankrupted William Beckford of Hertford, despite his being a member of the fabulously wealthy Beckford family. He had heavily mortgaged his three plantations in 1777, and the storm all but destroyed his estates. He wrote movingly, according to Thistlewood, who read his work in draft, about the effects of a hurricane on a landed proprietor: "a hurricane, particularly... one of so destructive a nature as that which happened in 1780, is... not only ruinous to the needy man, but more than the independent can support, and such as none but the truly affluent can repair."[37]

Poorer Jamaicans living in Westmoreland in 1780 faced destitution. During October 1782, fifty-nine residents of Westmoreland and Hanover parishes, the majority artisans, shopkeepers, or widows from Savanna-la-Mar, made sworn statements that were presented to the Jamaica Assembly on 12 November 1782. The petitioners were angry about what they thought was "a partial, inadequate and unjust distribution" of relief funds granted by the British parliament. In particular, they felt that sugar planters were being especially favorably treated. Mary Croll, for example, lamented that she was now "in danger of starving." Elizabeth Hamilton claimed that from being "in easy circumstances," she was now destitute. Hannah Fraser was "reduced to beggary." Isabella Bulman, aged seventy-two, was "reduced to the greatest extremity of misery and obliged to live in a hut on the charity of the benevolent."

But many inhabitants also stressed that without the hurricane they would have stayed in comfortable circumstances. Charles Payne, Elizabeth Hamilton, Robert Ross, Hugh Brodie, and Mary Costen all used the phrase "in easy circumstances" to describe their situation as of 2 October 1780. Catherine, the wife of merchant George Woodbine, described her family as being on that date "in easy circumstances, and in a flourishing and prosperous way of business." Henry Dickie was "in a fair way of trade and independent circumstances." Of course, the petitioners shaped their evidence to show that they were victims of a natural calamity. Their testimony also suggests, however, that without hurricanes the Jamaican economy would have suffered only marginally from the Revolutionary War. For example, the price of food rose dramatically in 1780 and 1781, but this owed as much to damage from local hurricanes as to wartime shipping problems. Indeed, hurricanes also affected shipping. The Kingston merchant Robert Mure, writing about high maritime insurance rates in September 1781, noted that "the underwriters are more afraid of the 3d of October than of all our enemies."[38]

Why did slaves not rise up against their masters in these times of hardship? Richard Sheridan describes 1776 as beginning a crisis in slave subsistence. He argued that increases in food prices following July 1776, the eventual cessation of supplies from North America, and the privations that resulted from a series of devastating hurricanes all led to famine and scarcity for enslaved Jamaicans.[39] White Jamaicans worried about slave rebellion, even if they were confident that they could overcome slave resistance. They conveyed that worry to politicians in Britain. In 1780, for example, the Marquess of Rockingham declared that Jamaica's military force was insufficient to quell a slave revolt.[40] Moreover, planters were greatly alarmed by declarations of martial law that took white employees away from plantations to do militia duty, stripping them of white supervisors. John Kelly, overseer on Golden Grove Estate, warned his employer in 1782 that "the continuance of Martial Law prevents the business of Plantations from going on with the usual regularity." In 1782, John Vanheelen, an overseer on Mesopotamia Estate in western Jamaica, claimed that during martial law "at times Estates are left without one white person," with business being "at the Mercy of the Negroes." In his opinion, "for the want of white people," during these periods, "estates have suffered very much" from slaves taking advantage of the absence of white people to pilfer large quantities of sugar and rum.[41]

Nevertheless, no slave rebellion happened in either Jamaica or Saint-Domingue during the American Revolutionary War or its immediate aftermath, unlike the Seven Years' War. Rebellious slaves in Jamaica tended to be brigands, like Three Finger'd Jack, who formed a gang of nearly sixty runaway slaves and created havoc in St. David and St. Thomas-in-the-East in the early 1780s.[42] David Geggus's conclusions for the enigma of slave quietude in Jamaica in the 1790s apply also to the period of the American Revolution. Geggus argues that the stability of a slave society depended less on ideas circulating in slave quarters than on the strength and unity of the planter class and the presence of regular troops on the island. Troop numbers in Jamaica during the 1770s and 1780s were sufficiently large, involving at least four regiments, to give enslaved people contemplating rebellion some pause.[43]

Perhaps slaves were too concerned by the challenges of hunger to think about rebellion. Simon Taylor recalled in 1789 how after the hurricane of 1781 he met slaves from a western sugar estate "hunting about in the Savannas for Sweet sops who told me they were starving." Hector McNeill recalled in 1788 how in 1780 he had been horrified by "the misery of beholding hundreds of wretched beings wasting around you, clamouring for food, and imploring that assistance which you cannot bestow." The hurricane of 1784 "committed horrid devastation," Governor Alured Clarke asserted. He noted the claim of one overseer that if relief was not forthcoming from opening up Jamaican ports to American ships "one half of the Negroes will not have a morsel to eat." A correspondent writing in 1785 informed a British friend that "it is a Truth indisputable as it is horrid . . . that many Thousands of Negroes perished this year by Famine." He estimated that "many hundreds have been killed in robbing provision grounds; many have died of actual hunger. And thousands are now bloated and in Dropsies from the same cause." The Jamaica House of Assembly summed up the effects of the recurring hurricanes, storms, and droughts: "the plantain walks, which furnish the chief article of support to the negroes, were generally rooted up, and the intense droughts which followed, destroyed those different species of ground-provisions which the hurricane had not reached. . . . The sufferings of the poor negroes, in consequence thereof, were extreme."[44]

Nevertheless, famine was a recurring feature of Jamaican and Saint-Domingue slave life. There was a subsistence crisis in Jamaica in the 1740s when "some of the poor Creatures pine away and are starved, others that have more spirits, go a stealing and are shot as they are caught in Provision Grounds; others are whipt or even hang'd for going into the Woods, into which Hunger and Necessity itself drives them to try and get Food to keep Life and Soul together."[45] When drought occurred, as in the late 1760s, slaves could not feed themselves. One subsistence crisis occurred in the mid-1790s when Jamaica suffered ten months with virtually no rain, the result being that "the Cattle, Mules & Horses are dying almost everywhere for lack of fodder, the Crops of Corn have failed, and the Plantain Trees are drying up, and weathering away" while every week "there are three or four Alarms in the Neighourhood of fire."[46]

The nearly annual catastrophes Jamaica suffered for more than a decade after the end of the American War added to the seasonal privations of the "hungry" times after every September, when slaves depended on yams and plantains for sustenance. Even in good times, planters' unrelenting style of slave management pushed workers to the edge of famine. The practice of making slaves grow their own food meant that those who were not ablebodied or who had numerous dependents could not provide for themselves, even in times of abundance.[47] As Gilbert Mathison explained in 1808, "If it should happen that, through idleness or sickness, or old age, or in consequence of too numerous a family of children, the provision-ground should be neglected, or become unproductive or insufficient, the Negro is not allowed to expect, nor in fact, does he obtain, assistance from the stores of the plantation."[48]

Planters in both colonies almost always chose profit maximization over amelioration. They accepted that slave populations were not self-sustaining. As long as the slave trade provided an ample supply of new Africans, planters accepted that a large proportion of their slaves would be afflicted by sickness, reproductive problems, and early death. Planters did little to make the material conditions of slavery any easier. On the contrary, they adopted more scientific methods of slave management during the eighteenth century that made plantations more efficient and slave work more difficult.[49] For example, rather than buy more workers for planting or harvesting, some planters hired extra labor—jobbing gangs—during these times. Their own slaves got little relief in times of lesser intensity than they might have gotten if plantation labor forces had been larger. Consequently, slaves were highly stressed, prone to disease, and likely to die. Their declining reproduction rates tell the story: when new arrivals from the slave trade are excluded, Jamaica's enslaved population declined by over 4 percent in 1700, by 3 percent between 1750 and 1775, and between 1 and 2 percent in the late eighteenth century.[50] Mortality rates, in other words, improved only slightly over time, even as the proportion of creoles (who should have lived longer than new arrivals from Africa) in the population increased. Mortality figures are harder to obtain for Saint-Domingue, but plantation records suggest annual decreases of over 6 percent in the 1770s and early 1780s.[51] These were periods of intensive migration from Africa.

One reason for natural decrease was that planters did little to secure healthy birthing conditions for pregnant slaves because they were reluctant to lose women's labor. As late as 1820, James Thomson, a Jamaican medic, complained how slave women were "difficult to manage." He insisted that pregnancy and work could go together. Pregnant women were perfectly capable of working, he thought, because "nature does not require such exemption from exercise during that state."[52]

Alexander Byrd and Barry Higman have shown how Simon Taylor of Golden Grove Estate, the preeminent Jamaican slave manager of his time, believed that the key to long-term plantation viability was careful management of labor supply and demand. Taylor continually pressed his employer to buy more slaves. He argued that if there was not a continual supply of new African laborers "it will be impossible to keep up your Estate at such great crops as it ought to make without it except by pushing your Negroes too much and killing them which will not be for your advantage." He warned against trying to plant and build at the same time. He explained to his employer's brother-in-law that "had I pushed both for a large plant and to build the house at the same time I should certainly have killed a good number of the Negroes which leaving humanity aside could never have been for the advantage of the gentle-man." To make his point, Taylor compared slaves to the hardened metal pivot pieces that held the rollers in a vertical sugar mill. He pointed out that "they [the slaves] are not steel or iron and we see neither gudgeons nor capooses last in this country." Because the plantation's productivity depended on an adequate supply of labor, he thought, slaves should not to be worked "above their ability." Moreover "without the addition of Negroes it will be impossible to keep the estate up." These new slaves would soon pay for themselves. He concluded that "it is a pity and for after having such good lands and works the estate should fall off for want of a sufficient strength of Negroes which if not put on must infallibly be the case."[53]

The result of not purchasing more slaves and continuing to work slaves very hard was seen at Golden Grove during the American War. Taylor was incensed at how his overseer, John Kelly, worked his slaves to death without caring about replacing them. He thought Kelly worked slaves so hard that "their hearts have been broke." Kelly gave them so few rations, and favored his own slaves so methodically over the slaves of his employer when allocating provisions, that the Golden Grove slaves "were Starving" and a "very feeble Set indeed." Working slaves as hard as Kelly did, in order to get an outstanding crop, was suicidal, Taylor thought, especially when the slave trade was in a state of severe contraction, as in the early 1780s. Golden Grove, Taylor argued, would not support "a Gang of Negroes while he [Kelly] had any thing to do with it," and soon the property, with slaves "killed by overwork & harassed to Death" would be a "Land without Negroes."[54]

The key problem planters faced in trying to maximize productivity while sustaining slave numbers was that the slave trade declined precipitously during the American Revolution. In the British West Indies the slave trade declined by one-quarter between 1776 and 1778: twelve leading Liverpool slave companies decided to get out of slave trading. In the three years before the war, 231 slaving ventures landing 60,480 slaves arrived in Jamaica. Between 1777 and 1782, however, only 95 ships with 34,526 slaves arrived in Jamaica. The year 1780 marked the nadir, with just 3,763 slaves arriving on the island. Trade conditions in Africa became more difficult.[55] In addition, the Middle Passage became increasingly dangerous. American privateers and French warships caused chaos. Insurance rates increased appreciably to deal with the problems of getting slave ships safely across the Atlantic.[56] The

American War forced slave traders to restructure their trade. They looked increasingly to the Royal Navy for support and tried, whenever possible, to travel in convoy. One solution to the problem of declining returns and growing costs was increasing the size of slave cargoes. Before the start of the American Revolution, very few ships carried more than six hundred slaves. The only ships carrying more than six hundred slaves to Jamaica before 1776 were the *Liberty* in 1763 (also a war year), which loaded 741 Africans and landed 605, and the *Britannia* in 1774, with 626 slaves embarked and 609 landed. But ships carrying enormous numbers of slaves became more frequent after 1778: two ships arrived with over six hundred slaves in 1779, one in 1780, two in 1781, four in 1782, three in 1783, and four in 1784.[57]

In addition, prices for African slaves in Jamaica were not encouraging. In 1775, the *Thames* arrived in Lucea in western Jamaica with 341 slaves from the Gold Coast and sold these slaves for an average price of £56.66 sterling. In 1779, 258 slaves from the Gold Coast onboard the *Rumbold* fetched £32.75 each. Prices improved a little in 1780—368 Biafran captives on the *Hawke* sold at Kingston for £37.85, and 295 slaves from Sierra Leone on the *Mars* sold for £43.59 apiece—but slave traders' profits were low, given the costs of providing maritime protection for slave ships.[58] Why slave prices did not increase when demand was high and supply low is not easy to understand and deserves further investigation. It seems likely that planters, no matter their need for new inputs of slave labor, were hesitant to buy many slaves given the political uncertainty of wartime. In addition, war made British imports markedly more expensive, straining planter finances. Credit may have also become harder to obtain during the American War.[59]

The crisis in slave subsistence that Sheridan thought was a consequence of the American War and successive hurricanes was in fact mainly a crisis in slave management: Jamaican planters worked their slaves so hard and gave such a small consideration to elements of risk (war and hurricanes) that even a limited decline in their ability to acquire food for slaves led quickly to destitution. It mirrored what happened in Saint-Domingue during the famines of 1775 and 1776 and during the Seven Years' War from 1757 to 1762. Slave losses from malnutrition and overwork were high even in years of prosperity, so that when adverse circumstances arrived, deaths quickly multiplied. With a sharply contracting Atlantic slave trade during the American War, Jamaican planters endangered their workforces by pushing slaves as hard as they did. Population figures tell their own story. David Ryden suggests that there were 194,721 slaves in Jamaica in 1768; 224,918 in 1774; 233,333 in 1775; 239,471 in 1778; and 246,043 in 1787. Whereas the slave population increased by 23,846 between 1754 and 1768 and by a further 38,612 between 1768 and 1775, the slave population increased by only 12,710 between 1777 and 1782. The major reason for the slowing of the growth rate was a decline in slave imports. After 1782, however, the slave trade quickly picked up. Between 1783 and 1785 37,046 slaves, or 12,348 per annum, were imported into Jamaica. After a dip in imports between 1786 and 1788, Jamaica entered into a surging Atlantic slave trade market. Between 1789 and 1795, Jamaica imported 113,569 Africans, or 16,224 per annum, peaking at a remarkable 27,650 in 1793.[60]

Planters and merchants tried to deny the harshness of Jamaican slavery by claiming that the majority of losses in slave population arose from war and natural calamity. The House of Assembly investigated "the Number of our Slaves, whose Destruction may be fairly attributed to these repeated Calamities and the unfortunate Measure of interdicting foreign Supplies." Using figures generally accepted as accurate, it estimated that the "Calamities" of the 1780s meant that fifteen thousand slaves "perished of Famine, or of Diseases contracted by scanty and unwholesome Diet, between the latter end of 1780 and the Beginning of 1787." Hard times clearly cost lives. But the Assembly's assertions were baldly self-serving. Slave numbers declined, they argued, for "causes not imputable to us, and which the people in Great Britain do not seem to understand." Their explanations were torturous, as they denied that prior to the hurricanes slave numbers were also declining. They invented complicated formulas inflating the numbers of runaways, Maroons, and free people of color in order to conclude that the slave population had decreased by the highly improbable figure of only 26,491 since 1655 from what it should have been, with fifteen thousand of these dying in the last few years from natural disasters. Simon Taylor applauded the Assembly's argument as showing "we are not such inhumane beings as those wicked Enthusiasts represent us to be & that people cannot murder or destroy Negroes as they do Dogs at home." Unlike Taylor, however, we should be suspicious of the Assembly's motives and skeptical about their reasoning.[61]

Some slave managers, the Assembly admitted, men like John Kelly, worked their slaves too hard in order to gain a short-term profit. Such men, it was argued, were few in number and not representative of the average humane Jamaican planter. Taylor was convinced that abolitionists seized on untypical cases like Kelly for polemical outbursts against wicked planters. "It is certainly owing to these sort of People," he fulminated, "that such stories are propagated among the People at home to our disadvantage."[62] There is no reason, however, why we should agree with Taylor. Even on their own testimony, given to the House of Commons in 1791 and published as *Abstract of the Evidence*, planters' callous disregard for black life and single-minded pursuit of profit maximization was everywhere evident. As Captain Thomas Lloyd of the Royal Navy noted in his evidence, slaves "were very generally considered as black cattle, and very often treated as post horses."[63] Slow slave population growth can be explained almost entirely by reference to fluctuations in the slave trade, rather than to wartime privations or natural disaster.

In Saint-Domingue, the plantation regime pressed onward to even greater productivity and even greater destructiveness for enslaved people than in Jamaica. Where the planters and administrators of Saint-Domingue excelled over their counterparts in Jamaica was in combining ruthless methods of slave management on sugar plantations with large cooperative irrigation works. By the 1780s the French state was involved in these projects, increasing their scale. The relentless demands of slavery in Saint-Domingue marked it out by the 1780s as not only among the most efficient but also among the most exploitative slave systems in history. Slaves worked under a regimented gang labor system, suffered the constant threat of famine, and endured high levels of violence from planters and from their employees. One response was spiritual: Africans flooding into the colony, as well as enslaved creole, or native-born, blacks, developed new spiritual communities, based on African or neo-African practices and concepts. Religious innovation and the transfer of African religious customs to Saint-Domingue gave enslaved people a sense of community, perhaps even power, as they worked with others to divine the future, ward off malevolence, and heal broken bodies and relationships.

French colonists did not understand slaves' religious practices. After Macandal, they often reduced Africans' spirituality to poisoning schemes designed to hurt planters by killing off their workers. Moreover, Enlightenment writers discussed African religions in ways that allowed elite colonists to dismiss slave practices as signs of their lower place in a hierarchy of civilizations. Others placed slave religion within established European contexts; Africans were adopting new European ideas, such as mesmerism, or were falling prey to swindlers, con men, and quacks. What they could not see is how Africans were using spirituality to build communities in conditions of enormous stress. These methods of creating social solidarity did not immediately lead to slave resistance, but it seems likely that they helped enslaved people better survive the deadly pressures of plantation life. There is little evidence of religious unity among slaves in 1780s Saint-Domingue. But the emergence of communities sharing some common religious assumptions would eventually allow slaves to take advantage of the French imperial political crisis of 1791, launching the Haitian Revolution.[64]

How planters interpreted slave religious practices changed somewhat over time. As we have seen, the colony's great mid-eighteenth-century slave conspiracy, the Macandal poisoning episode, was probably one of several severe wartime outbreaks of food-borne illness rather than a deliberate design to attack whites through the insidious use of poison. Planters continued to use torture to try to expose the poisoners they believed were killing their workers, as Chapter 10 describes and as they had done during the Seven Years' War. In the 1770s and 1780s, however, colonial administrators were willing to accept alternate explanations for mysterious waves of death. In 1773, 1774, and 1776, and again in 1779, Saint-Domingue's North Province experienced a marked rise in the mortality of livestock, enslaved workers, and white masters. Advocates for liberalizing the colony's food trade claimed that within a space of six weeks, fifteen thousand people died and another fifteen thousand died subsequently of hunger.[65] This kind of die-off, combined with Médor's confession, had launched the Macandal hysteria in 1757. But two decades later, officials explained what happened in less conspiratorial terms: deaths occurred because peddlers were selling bad food to starving plantation workers. That food was *tassou*, a form of dried beef made in Santo Domingo, apparently from the carcasses of animals that died in an extraordinary anthrax epidemic, which had also struck on the French side of the border.[66] It was only in 1769 that French doctors had identified anthrax as a specific livestock and human disease, different from smallpox, and other skin ailments. They eventually saw the Saint-Domingue outbreak as the largest anthrax outbreak ever recorded. Doctors believed it had been introduced into the colony by horses shipped from North America and noted that, initially, colonists suspected slave poison, noting that "this absurd opinion . . . gave rise to criminal prosecution and heinous punishments." But after seeing the deaths abate after rains ended the drought, then return with dry weather in 1774, 1775, and 1776, officials began to understand that this was an epizootic disease that could infect humans.[67] Planters were still worried about another Macandal, but this time no Médor came forward to confess that "a secret reigns among the blacks."[68] Instead colonists accepted the combination of deadly meat and the lack of imported food as an explanation for slave deaths.

Worries about slave poisoners remained, but they were tempered by rationalist explanations. In 1777, believing that the veterinary epidemic had passed, authorities in Fort Dauphin—the largest French town close to the Spanish border—tortured and executed Jean, an enslaved man, for poisoning hundreds of animals belonging to his master over a period of eight months. Jean was allegedly found with arsenic.[69] By December 1779, however, the Council of Cap Français was forced to admit that, in fact, anthrax had returned. In 1780 they repromulgated the law against selling tassou.[70] Authorities remained highly concerned about poison, but the surviving cases focus on medicinal rather than political or spiritual poisons. In 1781, the Cap Français Council levied a heavy fine against an apothecary for selling poison to a slave, who used it to kill himself.[71] In 1784 Zabeau, an enslaved woman in the North Province, was sentenced to be hanged by her armpits in the square of Cap Français, branded with the label "poisoner" on her chest, and then be jailed for the rest of her life. Her crime was to have regularly dosed her master with an emetic used to induce vomiting.[72] Nevertheless, elite colonists' attitudes about African and neo-African spirituality in the 1770s and 1780s had changed considerably since Macandal in 1757. They were much less ready than they were previously to immediately turn to accusations of poisoning when people died unexpectedly.

Given the success of enslaved people in securing freedom for themselves in the Haitian Revolution, it is not surprising that historians, like planters and their police forces, have long searched for rebels and conspirators in Saint-Domingue. Rather than accepting planters' unprovable assertion that slaves who possessed powerful toxins murdered each other in order to bankrupt their masters, historians might give more weight to slaves' claims. In the Macandal testimony, for example, slaves asserted that they used the potions, powders, and packets described as poisons to protect themselves, influence others, and heal relationships. Ripped from traditional and hierarchical societies, Saint-Domingue's Africans were likely familiar with slavery, but African slavery was very different from the quasi-anonymous proto-industrial labor of Caribbean sugar estates. In this environment, spiritual practices allowed enslaved men and women to demonstrate that the world they worked in was not the only world in which they lived. The difficulty, of course, is sources, as the religious world is one that slaves mostly kept secret from whites. It was also one that white observers seldom bothered to understand on the terms that slaves would have found sensible.

But we can get some small access to that world. James Sweet's microhistory of Domingo Álvarez, a vodun priest from Benin enslaved in Brazil in the 1740s, portrays this African religious specialist as a healer, intellectual, and creator of New World communities.[73] In Saint-Domingue we know little about the nature of slaves' spiritual communities, but it seems probable that they were fictive kinships, perhaps broadly similar to those that whites forged in the colony's Masonic lodges and Catholic baptismal fonts. These types of social bonds were likely not new to enslaved Africans, any more than they were to European immigrants. Like French colonists who found themselves in a society where the Church occupied a far less central place than it did in the metropolis, enslaved Africans were open to new ways of making meaning—new rituals, leaders, and concepts.

We can see this openness in slaves' enthusiasm for the missionary work of the Jesuit order, the one arm of the Catholic Church in Saint-Domingue that actively proselytized in the slave population. As we saw in Chapter 5, in 1761 the Council of Cap Français attacked Jesuit missionaries for being too close to slaves and free blacks, allowing them to create their own religious organization, choosing officers modeled on French parish officials.[74] This behavior is reminiscent of the confraternities that free and enslaved blacks established in Cuban and Brazilian cities with the help of Catholic authorities, who were willing to tolerate a certain amount of syncretism in order to strengthen the slave system.[75]

Things were different in the French West Indies, however. In Guadeloupe in 1752, the Jesuits attempted to create a free black confraternity, but local government prohibited this.[76] In Saint-Domingue's dynamic and materialistic society, colonists themselves had no such brotherhoods. When the colony expelled the Jesuit order in 1763, the charges against them included involvement with slave poisoners and profaners.[77] In 1785 Girod-Chantrans explained the Jesuit approach to converting slaves in a deeply ironic tone: "Eager to establish their domination, they focused on gaining the confidence of the slaves, by instructing them very specifically on their religion, revealing to them their own sublime nature, the majesty of man and his hopes for the future." This treatment, he believed, made "the slaves of that [earlier] period quite eager to attend masses, sermons, catechisms and all the ceremonies of the church."[78]

With the expulsion of the Jesuits from Saint-Domingue in 1763, Saint-Domingue's African-style religions had few European allies to urge syncretism with French traditions.[79] The huge influx of Africans after the Seven Years' War and the American Revolutionary War brought new religious traditions and leaders into the colony. We have little direct evidence about whether workers attracted to these new elements experienced them as a way to heal, destroy, or escape a society that was inflicting new extremes of hunger, work, and violence on them.

What scholars recognize as slave religion Saint-Domingue colonists saw as a collection of separate problems: disorderly dances, deadly poisons, and unauthorized healers, as well as breeding grounds for profaners and idolaters. While Jamaica outlawed "obeah" in 1760, it was only in 1788 that Moreau de Saint-Méry penned Saint-Domingue's first sustained discussion of "Vaudoux," and even then it was as part of his description of slave dance and music styles.[80] He recognized that "Vaudoux deserves to be considered not only as a dance, or at least it is accompanied by circumstances that make it one of the institutions in which superstition and bizarre practices play a large role." He described Vaudoux as "known for a long time, especially in the West Province" and noted that it had "principles and rules" that were watched over by Arada people, "the true members of the Vaudou sect" in the colony. Moreau's account focused on the role of a nonpoisonous snake used in the rituals he observed. He believed worshipers thought that the snake was an all-knowing all-powerful supernatural being, who communicated its wishes to a pair of male and female priests. They experienced a form of spiritual possession by the serpent, whose voice spoke from the mouth of the female priest, during "a convulsive state."[81]

Moreau described the scene: "Sometimes she flatters and promises happiness, sometimes she thunders and breaks out in reproaches; moved by her desires, her own interest or her caprice, she pronounces everything she wants to prescribe as laws without appeal, in the name of the serpent, to an imbecilic crowd who raises not the slightest doubt to the most monstrous absurdity, and who will simply obey whatever is despotically ordered." In return, worshippers asked the deity for favors; Moreau reported that most of them "ask for the ability to control the mind of their masters." Though his examples of worshipper's requests were all non-malevolent, Moreau saw the potential for danger: "there is no passion that is not represented in these wishes." He particularly condemned those slaves who prayed for the "success of their crimes." After participants made offerings, the ceremony moved to "the dance of the Vaudoux." In this rite the snake, from its box, communicated a kind of nervous trembling to the king and then the queen, which they communicated to the rest of the group, whose members danced, twirled, fainted, and tore their clothes.[82]

Moreau stressed that colonists never witnessed the "true Vaudoux, that which has lost the least of its original purity." He believed that members kept their most important rites secret and staged public performances only to convince colonists and magistrates that their dancing was harmless. He described how colonial police units had an ongoing "war" against the Vaudoux. Despite multiple references to "this system of domination . . . and blind obedience . . . a school where weak souls deliver themselves to domination that a thousand circumstances might render deadly," Moreau did not accuse Vaudoux devotees of revolt or rebellion, only noting that the actions of the group "do not consistently have good order and public tranquility as their goal." Nevertheless, his account describes the Vaudou community as closely knit. Participants swore "terrible oaths" and drank from a steaming pot of goats' blood—probably a secrecy oath from West Africa. The leaders, who described themselves as "father and mother," as well as "King and Queen" and "master and mistress," set aside worshipers' offerings, in part to pay expenses but also "to provide help for members who are absent or present who need it."[83]

Moreau also described something that "went beyond Vaudoux." This was the Dance of Dom Pèdre or simply Don Pedro, a similar rite that was especially agitated, convulsive, and rapid. Dancers experienced a kind of group *crise*, accelerated by drinking rum mixed with gunpowder. Today scholars see this as the origin of the Petwo/Petro rite within modern Haitian Vodou. Moreau suggested it was naïve to believe that Dom Pèdre was merely a dance. He saw it as potentially dangerous and as a means of fomenting rebellion. He undercut his claims that the dance was serious, however, by following his brief description with a mocking account of how blacks tried to copy white minuets, as if to deny the very possibility of slave creativity. Furthermore, he had little to say about the possible religious nature of Dom Pèdre. He described its origin in 1768 in the southern peninsula with a Spanish black who "abused the credulity of the slaves with superstitious practices" and noted that colonists prohibited the dance with severe penalties.[84] Nevertheless, his collection of colonial laws up to 1784 did not mention either Dom Pèdre or Vaudoux by name.

Unsympathetic and largely misinformed observers like Moreau do not provide us with much information about how enslaved people understood their world. Such reports do suggest, however, that in the 1780s the colony's growing West Central African population was shaping the culture of enslaved people. In May 1786 the Cap Français Council passed a law against a "class of macandals" holding large nighttime meetings in Marmelade parish in which they reduced individuals to "a stupor or convulsions, mixing sacred and profane" and then released them using holy water.[85] Authorities believed that three enslaved men led these nighttime assemblies. In August they arrested a maroon named Jean and tried to arrest Télémaque, a black *commandeur* (leader of a plantation work gang), and Jérôme, a mulatto slave known as Poteau. In Marmelade, which had been formed as a parish only in 1772, the slave population was largely from Central Africa, and the three men were conducting ceremonies and selling materials that seemed to be mostly derived from Congo cultural practices. Royal authorities in Cap Français were concerned enough to send the royal attorney, François de Neufchâteau, to investigate. He interviewed twenty-seven witnesses—white planters and overseers, as well as enslaved blacks—a lengthy investigation that suggests that authorities feared that another charismatic leader like Macandal was emerging.[86]

The investigator heard testimony that Jean, the maroon, was found carrying a hunting knife and a stick with iron at the end, perhaps the equivalent of the fighting sticks many slaves carried as canes, which were called *bangalas*.[87] He was leading gatherings called *mayombé* or *bila*, on two plantations. During one of these meetings a white overseer hidden in a nearby slave hut saw Jean kneeling on the ground atop two crossed machetes, before a table with lit candles, lifting up a fetish. He carried a small sack containing a crucifix, gunpowder, and nails. At his meetings participants drank rum mixed with "blanc d'Espagne," a kind of white clay used as a dye. Neufchâteau also investigated Télémaque and Jérôme Poteau, who were holding meetings with as many as two hundred participants in which they used and sold small sacks containing tiny limestone rocks called *bila*. Like Jean and perhaps like those leaders who called themselves Dom Pèdre, Don Pedro, or Dompete, Jérôme poured cannon powder into rum, mixed it with the crushed limestone, and gave it to his followers.[88] He claimed that he could help slaves detect poisoners and chicken thieves, with the help of small red and black seeds of an acacia plant, which he called *poto*. The most expensive items Jérôme and Télémaque sold their fellow slaves were fighting sticks called *mayombo*, which had crushed bila inside them and which were said to make them invincible to other stick fighters. Although the stick-fighting tradition was found in many parts of West Africa, Thomas Desch-Obi shows that in Martinique and Saint-Domingue it was strongly identified with Congolese slaves. Moreover, in nineteenth-century Cuba, the emergence of the religion Palo Mayombe was closely identified with Congos. We know little about what these ceremonies signified, though Desch-Obi hypothesizes that the bila were a kind of Congolese divination technology.⁸⁹ It seems likely that the gunpowder, crucifix, and iron used by the maroon Jean was also from Central African spiritual traditions. Colonial sources contradict each other about the fate of the men arrested; some say they were sentenced in absentia, while others claim they were sentenced to death.⁹⁰

Colonial jurists attempted to understand these practices in European terms. The Cap Français Council and Moreau both insisted that these ceremonies were a form of "magnetism," sometimes called "mesmerism," referring to the techniques of Anton Mesmer, which had arrived in Saint-Domingue from France in 1784. Immersion in magnetic tubs was said to have cured many people, including a female slave who had been paralyzed for fourteen years. One planter in Artibonite purchased a parcel of sick slaves and claimed to have restored them to health using magnetism. He then leased them to fellow planters at full price.⁹¹ Moreau de Saint-Méry's brother-in-law, Charles Arthaud, founded Saint-Domingue's scientific academy, the Cercle des Philadelphes, expressly to debunk the claims of mesmerists.⁹² By calling slave rituals "mesmerism," therefore, Moreau was labeling them as a hoax perpetuated on an ignorant public. As Neufchâteau, the royal attorney, put it, "This is just what happens elsewhere, a gullible and moronic multitude [is] seduced and deceived by skilled performers or clever charlatans, who take their stupid belief and use it for their own profit; but what elsewhere should only be considered an unscrupulous fraud and punished as such, is something very different because of the regime of the country in which we live."⁹³ The Council called Jérôme and Télémaque "false adepts [prodiges] of this so-called magnetism," which had been "usurped by the blacks and disguised by them under the name Bila." If there was a danger it was "to leave in the hands of the Negroes a tool that Physics [la Physique] itself only utilizes with care and whose abuse is so easy and so apt to be used by traveling entertainers [jongleurs], who are common among the Negroes, who venerate them."⁹⁴

If at least some Council judges considered magnetism to be a scientific tool, Moreau had no such illusion about Mesmer's magnetic cures. He described slaves' use of magnetism "matching as in Europe the beliefs of those who propagate them."⁹⁵ But Moreau cast around for other terms from recent French history, referring to Jérôme's followers as "convulsives," a reference to a controversial set of religious visionaries in Paris, and "Illuminés," a vague term that he perhaps meant to apply to the occult Freemasons of Martinès de Pasqually. Despite his scorn, Moreau was ahead of other colonists in being able to see the beginnings of syncretic religion, though it is not clear whether Jérôme was in fact using European ideas for his own purposes.[96]

In the 1780s the religious world of Saint-Domingue's slaves had not yet cohered to the point where we can talk about the existence of a common Vodou practice.[97] But communities like that which coalesced around Jérôme Poteau, Télémaque, and Jean seem likely to have been an important source of support for workers caught in the gears of Saint-Domingue's accelerating plantation machine.

The plantation system in Jamaica and Saint-Domingue did extraordinarily well in the 1780s. The two colonies did well despite the trauma of multiple hurricanes in Jamaica and some short-term setbacks during the American Revolution in both Saint-Domingue and, more seriously, in Jamaica. Profits from plantation production remained high and promised to become very high indeed. Moreover, the Atlantic slave trade reached its eighteenth-century peak in that decade, with imports to Saint-Domingue especially enormous. The short-term economic difficulties that Jamaica faced were largely over by 1787, and there was a new optimism about Jamaica's future development. In Kingston in 1787 there were more ships than ever before coming from foreign places, and Jamaican prosperity was so high that despite such a glut of traders "even indifferent negroes command very high prices." Between 1785 and 1789 both real sugar prices and slave prices were at their highest point in seventy-five years. As Lord Grenville summarized the West Indians' importance to Britain in 1787, the "commerce of our colonies was growing" and "promised a rapid increase in circumstance that could not but prove extremely beneficial to the merchants and planters of our West India Islands."[98]

In retrospect, and with the hindsight gained from the tumults in Saint-Domingue in the decade between 1794 and Haitian independence, the intolerable pressures that enslaved people were under might have led to significant slave rebellion in the British Caribbean, especially if a perspective is adopted that sees slaves as always interested in self-liberation and prepared to take opportunities as they arose.[99] Slaves revolted on some small British West Indian islands, notably in Grenada and Dominica, which had been French colonies until 1763. Martiniquan-born free men of color played a key role in both conflicts, which therefore cannot easily be separated from Britain's war against Revolutionary France.[100] Nevertheless, in most of the Caribbean, notably the larger islands of Barbados and Jamaica, slaves were quiescent in the

1780s, as they had been in Saint-Domingue throughout the whole period from the Seven Years' War and as they had mostly been in Jamaica after the great rebellion of 1760. In Jamaica, slaves were preoccupied just by staying alive as the plantation machine became more relentless in its demands and as famine threatened. In Saint-Domingue, slaves appear to have concentrated more on developing different value systems based on African religion rather than on resistance itself.

In the final analysis, the logic of the plantation system was the principal problem facing Jamaican and Saint-Dominguan blacks in the late 1780s, as well as the main source of whites' prosperity. It may have been a dynamic system of creative adaptation and capitalist enhancement, but it was also an intrinsically destructive method of production—destructive not only of black bodies but also of black spirits. As we describe in the following chapter, the perpetrators of violence and inhumanity were ultimately unwilling to face up to the implications of the vicious socioeconomic system they were shaping. Imperial officials, however, and other metropolitans interested in the West Indies were less willing than white colonists to accept the dehumanizing nature of plantation slavery. They tried to change it.

CHAPTER 10

The Ancien Régime in the Greater Antilles

In the years just before the French Revolution the relative positions of Jamaica and Saint-Domingue moved to the advantage of the latter. At midcentury, Jamaica was roughly the equal or somewhat ahead of Saint-Domingue in every measure of the plantation economy, including the size of enslaved and free populations, the number of plantations, and the volume of sugar production. By 1787 that situation was reversed in favor of Saint-Domingue. Given that the French colony had roughly three times as much land as Jamaica, this is not surprising. The one-to-three ratio in favor of Saint-Domingue prevails in many of the statistics compiled by the French planter Marc-Antoine Avalle in 1799. He reported that in 1787–88 Saint-Domingue exported to France at least 129 million francs' worth of goods while Jamaica produced 45 million francs' worth, roughly one-third as much. Avalle said that in 1788 Jamaica produced 85 million pounds of exports by weight while Saint-Domingue exported just over 277 million pounds in weight to France, again roughly one-to-three.[1] Indeed, by the 1780s, well-financed estates in both colonies produced roughly the same amount of sugar per acre.[2] Jamaica's Worthy Park in the center of the island made 3,770 kilograms of sugar per hectare while historical studies of Saint-Domingue plantations report a range of 3,031 to 3,789 kilograms of sugar produced per hectare.[3] By this measure the best Jamaican sugar estates were at least as productive as their counterparts in Saint-Domingue.

This parity in sugar production rates is surprising, given Saint-Domingue's massive irrigation systems. Moreover, heavy sugar cultivation had started decades earlier in Jamaica than in Saint-Domingue.[4] By rights, as had occurred in Barbados and Martinique in the Lesser Antilles, where planters had to work hard by the middle of the eighteenth century to grow sugar on land that was no longer naturally fertile, Jamaican sugar production should have been well below that in the fresher sugar lands of Saint-Domingue. Up to 1783, however, Jamaica had a far better supply of African slaves from a better resourced and more efficient slave trade than did Saint-Domingue. The results were seen in the amount of sugar produced per slave on Jamaican sugar estates. The best capitalized Jamaican plantations often had roughly 75 percent more slaves per hectare of sugarcane as their Saint-Domingue counterparts. Worthy Park, for example, had 3.58 slaves per hectare planted in cane.[5] By contrast, David Geggus's study of fifty-six Saint-Domingue sugar estates demonstrates that these estates had, on average, 2.06 slaves per hectare of cane.[6] In another analysis, Geggus found averages ranging from 1.8 slaves per hectare of cane in the 1770s to 2.0 in the 1780s.[7] It seems likely that Jamaican planters used their larger labor forces—larger relative to the size of their cane fields—to implement the techniques developed on smaller islands where soil exhaustion was an even greater problem. They used extra workers to fertilize or "dung" their fields; they were more likely to plant or "hole" new canes, instead of allowing cane stumps or "rattoons" to regrow after harvest. Some French planters, by contrast, used rattoons as many as five times in a field, which saved labor but successively diminished their sugar yields.[8]

Despite Saint-Domingue's extraordinary output—Avalle calculated that in 1787–88 it outproduced all the other colonies in the West Indies—French investors had higher expectations than ever in the boom that followed the American Revolutionary War. After the American Revolution ended in 1783, powerful figures at the French court and elsewhere poured capital into the colony. Their search for ever-higher profits encouraged Saint-Domingue's planters to try new techniques in sugar planting and refining. During the years that Britain was pioneering the Industrial Revolution in Europe, French Caribbean colonists powered ahead with technological improvements to plantation agriculture.[9]

Besides buying more slaves, which we discuss below, building new irrigation systems was the oldest investment strategy employed in Saint-Domingue. In 1776 Hilliard d'Auberteuil described how irrigation had transformed the colonial environment: "In Saint-Domingue's plains ... streams and rivers flow calmly where previously one saw only forests. ... Leaving the channels where artifice has carefully restrained it, [water] distributes itself into rivulets, covering the surface of the earth and helping the plants absorb a liquid gold which will soon be pressed out by mechanical means."[10] Since the 1740s planters had formed societies to build and maintain these systems. By the mid-1780s the French state had also become involved, as new larger projects required a level of technology and investment that planters by themselves could not provide. Irrigation systems were an intrinsic part of what McClellan and Regourd call "the colonial machine." Their use illustrates how by the 1780s the French government was committed to using science to improve commerce in the Antilles.[11] On 11 November 1786, for example, Saint-Domingue's governor Count La Luzerne visited the plantation of Bertrand de Saint-Ouen to observe the opening of a massive flood-control project on the banks of Saint-Domingue's largest river, the Artibonite. As dignitaries watched, an engineer sent from France started a steam engine built by the Perrier firm in France. This was a coal-powered single-condenser engine like that developed by Scottish inventor James Watt twenty years earlier. A recent geological survey of the colony identified local coal deposits that might eventually fuel the engine.[12] The French state paid 70,000 livres, or £3,030, for this machine that officials hoped could finally tame the Artibonite River. In 1761, for example, it took two thousand slaves to rebuild the damage caused by one particularly devastating flood. When planters in the Artibonite plain weren't waiting for waters to subside, they were often devastated by drought. Lack of water had cut indigo harvests in the 1770s by a third since the 1760s.[13]

In the 1780s Artibonite planters and the colonial government turned to new technology because they were unwilling to accept the limitations of conventional flood control. The new irrigation system not only incorporated the Perrier engine but also used a set of massive cast-iron siphon pipes. These devices promised to transform the lower Artibonite valley from a declining indigo zone into a rich sugar-producing area. According to some estimates, the project potentially increased the value of Saint-Domingue's exports by 10 percent per year. Early steam engines, however, like the steel and iron "gudgeons and capooses" in Simon Taylor's Jamaican sugar mills, did not hold up to the rigors of the tropical environment. The Perrier engine was so heavy that it gradually sank into the soil of the river bank. The planter Saint-Ouen, its main champion, died unexpectedly in 1787, and the government abandoned the coal-fired pump.[14]

In this same decade, entrepreneurs introduced new industrial technologies into other aspects of Saint-Domingue's sugar culture, including developments that were closely linked to embryonic industrialization in France.[15] By 1777, Louis XVI had awarded letters of nobility to Paul Belin de Villeneuve, a creole from Saint-Domingue, for his many improvements to the colony's sugar factories. Some of these improvements were relatively simple, like increasing the size of the two external rollers in a vertical sugar mill, or increasing the size of the forms in which raw sugar drained. But Belin also borrowed from industrial developments in Europe, like using the new English "reverberation furnace" to increase the efficiency of sugar-boiling houses. His design was published in 1786.[16] Coffee planters were undertaking similar mechanical experiments in the 1780s.[17]

Nevertheless, sugar refining remained the focus of private and government efforts. Unlike Jamaica, where improvements on sugar estates mainly revolved around better management of human resources, by the 1780s the French Naval Ministry was working closely with enterprising technologists to improve sugar production in Saint-Domingue.[18] In the 1770s, for example, the Boucherie brothers of Bordeaux developed new sugar refining techniques that they claimed would increase productivity by 40 percent. A massive refinery built in Martinique in the early 1780s employed these techniques but failed, supporters claimed, because Martinique's planters could not supply enough cane for the refinery to work profitably.[19] There was no shortage of cane in Saint-Domingue, however. The Boucheries' technology was installed on the plantation of Lafon de la Débat, a Bordeaux merchant who had invested in the colony after making rich profits during the American Revolutionary War.[20] From 1785 onward, the physician Jacques-François Dutrône La Coutûre, sponsored by the Naval Ministry, applied the Boucherie methods to Lafon's sugar estate near Cap Français and developed his own techniques as well. In his 1790 book, Dutrône claimed that by deploying new refining vessels and other innovations he eliminated nighttime refining, increased his sugar output by 6 to 8 percent, and removed all the molasses from the sugar.[21]

Productivity became an obsession for some of Saint-Domingue's sugar planters.[22] Perhaps the most notorious example of this mixture of rationality and irrationality is the case of Nicolas Caradeux, an important planter in the Cul-de-Sac plain outside Port-au-Prince. Like Dutrône, Caradeux was convinced about the value of the new Boucherie refining techniques. One of his friends, Le Doux, described the sugar Caradeux made as "astonishingly beautiful" in a letter to the merchant Fleuriau at La Rochelle. He assured Fleuriau, who owned a neighboring plantation, that he could "easily at least double the revenue that you receive every year without more than a little [additional] expense in your operation." Fleuriau's estate manager Arnaudeau confirmed to his employer that "since returning from France [Caradeux] follows the principles of M. Boucherie to which he adds his own special knowledge. His factory produces only very fine sugar." But Arnaudeau did not believe that Caradeux's devotion was helping him make the best profits that he could. He described how another neighbor, M. O'Gorman, had met the Boucheries in Paris and experimented with their techniques. But O'Gorman reverted to traditional methods, believing he had wasted money and time following the Boucheries' innovations. The Count de Vaudreuil, a debt-ridden creole who was a major patron of the arts at Versailles and Paris, had also tried to increase his sugar profits through the Boucherie method, to little avail.[23] Arnaudeau argued that "if it were really profitable, those who set up M. Caradeux's method would not have abandoned it to return to the old method after having spent as much money as they did.... Caradeux has the glory of making the best sugar in the colony, but he does not have the reputation of making much of it."[24]

Arnaudeau did not go so far as to question Caradeux's rationality. But he surely knew of Caradeux's other reputation—for extraordinary cruelty to his workers—which was remembered decades after the Haitian Revolution.[25] Charles Malenfant, another plantation manager in Cul-de-Sac, admired Caradeux as someone "of great foresight, full of ideas and plans that were useful and advantageous to the colony.... He had the passion to make the best sugar in the colony; that was what sometimes enraged him against his sugar makers and his drivers; it was also against them that he exercised the greatest cruelties."[26] Despite his admiration for Caradeux—"a man of merit, of prodigious strength"—Malenfant also described him "cold-bloodedly having slaves thrown into the furnaces, into the boiling cauldrons, or having them buried alive and standing with only their heads showing, and letting them die in this way." He told several stories of Caradeux's cruelty, including one in which he killed his enslaved sugar maker for twice making *vilain sucre* (ugly sugar). Caradeux reminds us that the obsession with improving methods was not clearly separable from the extraordinary violence of sugar plantation slavery.

If planters in Saint-Domingue pioneered new kinds of technology to increase their sugar output, planters in Jamaica tended to concentrate on improving their human property so as to sustain plantation profits. As we have argued above, the prosperity of Jamaica's plantation economy depended heavily on the slave trade. In the late 1780s, after the difficulties of the War of American Independence and the hurricanes of the mid-1780s, and especially as slave imports returned to normal levels, plantations began to return to higher profits. J. R. Ward estimates that in the eight years following the end of the

American War of Independence, from 1784 to 1791, the annual rate of profit for Jamaica as a whole more than doubled, from 3 percent to 6.4 percent. By the mid-1790s, with the Haitian Revolution driving up sugar prices, annual rates of profit reached the heady heights of 13.9 percent per annum, rates of profit even greater than those attained during the Seven Years' War period.[27] The large Penrhyn Estate in Clarendon, for example, entered into a period of prolonged prosperity from the late 1780s. In 1788 gross profits were £30,531, or 705,000 livres, up from an average gross profit of £19,068 for the previous years. After expenses were deducted, net profits amounted to £18,502—double the annual net profit that the enormously wealthy Dukes of Bedford cleared from their very profitable Bloomsbury lands in London. These numbers remained steady for the next few years before moving into bumper years following 1792, when net profit increased to a munificent £23,382, or 540,000 livres, in 1792.[28]

Moreover, Jamaica still contained opportunities for ambitious migrants. Jacob Graham arrived from England aged twenty-one in 1746. He worked as an overseer on a sugar estate, buying slaves as his money allowed. Sometime in the late 1760s he bought land in northwestern Jamaica. By 1774, he was a "jobber" with fifty slaves, and in 1787 made the move to becoming a sugar and coffee planter, buying Lapland plantation in the interior of St. James parish. By 1816, when he died, unusually old, at ninety-two, he owned 127 enslaved people and had non-landed assets of £18,750.[29]

We get some sense of Jamaica's prosperity from the details of a census taken in 1788.[30] It shows that the population had increased steadily since 1774. The hurricanes of the mid-1780s and the accompanying famine had taken their toll on the slave population, but it had still increased by over seventeen thousand, or 9 percent, in fourteen years. The free population had grown faster proportionately, with the white population increasing by 5,610 or 44 percent, and the free black and colored population increasing by 3,517 or a massive 86 percent. The colony's white population was still largely male. In Clarendon parish, in Jamaica's rural heart, men accounted for 70 percent of the white population. Most of them were between the ages of twenty-one and forty. Just 12 percent of its white population was aged ten or less and only fourteen people of six hundred were aged over sixty, with no women aged over seventy.[31] Enslaved blacks made up 95 percent of Jamaica's total population. As with whites, most of these were working-age adults. An analysis of three large sugar estates from the late 1770s suggests that the population was skewed toward productive field hands aged between fifteen and forty-four.

On these estates, there was a more equal gender balance among enslaved people than in the white population—males made up just over half of the enslaved population.[32]

The biggest change in Jamaica's demography was in Kingston, which had expanded rapidly in the 1780s. By 1788, Kingston had a population of 26,478, nearly double a mere fourteen years earlier. The white and free colored populations were 6,539 and 3,280 people respectively, or 38 percent of the free population of the island, which was enumerated as 25,957. What was even more remarkable was the growth of slavery in the town. It had 16,659 slaves in 1788—the largest number of enslaved people living in any town in the English-speaking world in the eighteenth century. Bryan Edwards thought highly of Kingston. It was, he thought, a "place of great trade and opulence" with "magnificent" houses and "markets ... inferior to none." It was even, he opined none too convincingly, "as healthful" a town "as any in Europe."[33]

Edwards reckoned Jamaica's total population in 1791 as having increased greatly from before the American Revolution, with forty thousand free people (black and white) and 291,000 slaves. He clearly overstated the number of free people, as the census of 1788 suggested a much lower population of whites and free coloreds. Nevertheless, his general point was sound: Jamaica was growing in both population and also in economic activity in the late 1780s and early 1790s. It was a place of flourishing commerce, with 473 ships (252 from Britain and Ireland; 133 from American States; 66 from British American Colonies, and 22 from the foreign West Indies) arriving in 1787, bringing in 8,845 sailors. It had exports in 1788 of £2,136,443, or 49.4 million livres, the great majority of which went to Britain. It had imports of £1,496,232, also largely from Britain but with Ireland, Africa, and the United States also important. Edwards concluded that Jamaica was a "delightful island, thus variegated by the hand of Nature, and improved by the industry of man."[34]

No surviving general census exists for Saint-Domingue after the American Revolutionary War, but estimates of the slave population for 1788 by the intendant Barbé-Marbois show the colony's extraordinary growth in the 1780s. In the eight years from 1780 to 1788, Saint-Domingue's enslaved population grew from 251,806 to 405,528, an increase of 60 percent achieved almost entirely through the slave trade.[35] The 1780s were an extraordinary moment for France's transatlantic trade in captive Africans. After 1783, ships from Africa unloaded between twenty-seven thousand and thirty-seven thousand captives a year in Saint-Domingue, easily double the numbers before the start of the American Revolution.[36] In 1767, following the Seven

Years' War, the Naval Ministry had thrown open the African trade to all French merchants, not just those in the major port cities. But it was only after the end of the War of American Independence in 1783 that the French-African trade to Saint-Domingue started on a period of continuous and record-breaking growth.[37] Between 1740 and 1770, the French trade in African captives delivered roughly ten thousand captives per year to Saint-Domingue. In 1776 this shot up to over nineteen thousand, although France's entry into the war in 1779 stopped most slave deliveries. In 1784, the year after the Peace of Paris, which brought an end to hostilities between France and Britain, traders brought over twenty-four thousand slaves. The number of arrivals grew until they exceeded forty-one thousand in 1790. By comparison, the annual number of Africans shipped to Jamaica in the 1780s was roughly nine thousand, with 15,809 captives disembarked in 1784.[38]

This explosion in the size of the slave trade to Saint-Domingue was due to a number of factors. After the War of American Independence, France made colonial growth a priority, maintaining its navy at wartime levels.[39] The French navy also supported attempts to extend the French trade into the Congo region. In 1784, after merchants complained that the Portuguese were interfering with their commerce at Cabinda, in Angola, Versailles sent a squadron to "reestablish the freedom of trade" in this region. The squadron went so far as to destroy a recently built Portuguese fort in this region. After this date, traders like Louis Drouin of Nantes, whose previous slaving voyages focused on the traditional French slave-procuring regions of coastal West Africa, began to send ships to Angola.[40]

A second factor was the Naval Ministry's determination to accelerate colonial growth while restraining smuggling. A monopoly reform in 1784 allowed foreigners to sell slaves in Martinique and Guadeloupe but kept them out of Saint-Domingue, despite considerable pressure from colonists to allow it. Although the administrators of Saint-Domingue's isolated southern peninsula had opened that region to foreign slave traders in 1783, Versailles ordered an end to this practice and instead provided subsidies for French merchants who brought captives there.[41] In 1785, Jean-Baptiste Dubuc, the Martiniquan absentee planter and free-trade advocate who had advised the Naval Ministry since the time of Choiseul, argued that Saint-Domingue needed at least eighteen thousand more slaves per year than French merchants were supplying. He declared that the colony could probably absorb another three hundred thousand workers, doubling its enslaved population, before reaching its full productive potential.[42] He argued that this expansion "could be achieved not just by clearing new land, although much of its arable land awaits cultivation, but also by more careful agriculture on established plantations, which require improvements that are impossible because of the lack of slaves."[43] Dubuc may have been thinking of the techniques used on Jamaican plantations, or he may have had in mind Dominguan estates like the Maré plantation, which in the years from 1785 to 1790 achieved yields per acre that were two to five times greater than equivalent plantations. Their secret appears to have been that they doubled or tripled the number of field slaves per acre of cane. Even if the yield per slave remained the same, a smaller plantation could dramatically increase its sugar output by buying more workers.[44]

Dubuc argued forcefully that the slave trade was essential for the future growth of the colony. "Slavery is the very existence of the colonies," he maintained. But he understood that there was a growing distaste in elite French society for this business. He proposed—quite unrealistically, given the way French merchants had flocked to participate in the profitable slave trade—that the kingdom should use captive Africans as a colonial workforce but have as little involvement with their transport as possible. It should turn this "odious and barbaric" trade, a "series of atrocious deeds" conducted by "pitiless calculators," over to other nations. He predicted that European slave traders would soon depopulate the African coast, but blamed the high death rates of Saint-Domingue's slaves, which required the ongoing slave trade to maintain population numbers, on France's refusal to allow planters to import North American provisions.[45] Dubuc's hand-wringing over the morality of the trade, therefore, was merely a way of arguing for the complete liberalization of colonial commerce. The English supplied slaves more efficiently than the French, and the North Americans grew food more cheaply. Versailles should allow its colonies to draw on those resources to make sugar as profitably as possible. The key point was that without more slaves, Saint-Domingue's remarkable growth would slow.

Unsurprisingly, the Naval Ministry did not accept this argument. The last thing French officials or port cities wanted was to give this lucrative trade over to the British or Portuguese. Merchants themselves also promoted the African slave trade, because trading slaves was more profitable than the complex business of dealing with colonists where laws on bankruptcy were tilted in favor of planters.[46] In the 1780s the slave-trading business was largely in the hands of experienced merchants. Of 741 French slave voyages in the 1780s for which we have the names of the owners, 397 were made by owners who made at least five voyages in that decade.⁴⁷ Only ninety-one voyages were made by owners whose name appeared only once. Yet it isn't accurate to separate French slave traders and colonial planters: at least 50 percent of the slave traders in port city of La Rochelle had kin who owned colonial plantations. It was not always as easy as Dubuc might have liked to separate the slave traders he called "pitiless calculators" from planters, who were just as calculating.⁴⁸

As David Geggus points out, 40 percent of the entire transatlantic slave trade in the years between 1784 and 1790 disembarked in Saint-Domingue.⁴⁹ In the eleven years from 1780 to 1790, Jamaica imported 98,395 Africans, compared to 215,477 recorded for Saint-Domingue.⁵⁰ This suggests that planters in Saint-Domingue had realized that increasing plantation output required more than technological improvements. The key ingredient had to be fresh inputs of labor, used more efficiently. It appears that, by 1788, both Jamaican and Saint-Domingue planters had adopted essentially the same strategy for managing their slave property, which was to accept large losses through excessive mortality from working slaves hard on sugar estates and to use the increasingly efficient Atlantic slave trade to make up for those losses. This strategy allowed planters to get the large short-term profits that they desired, even if that meant loading up their estates with debt. They were happy to go into debt to buy African captives because they considered them not only as workers but as highly liquid capital assets—they could always sell slaves—that often appreciated in value. Maintaining and increasing their estate work forces, while trying to improve slaves' value by deploying some men as tradesmen was the best way to make money from sugar and other tropical crops. Short-term imperatives made such strategies perfectly logical.

The examples of Caradeux and Dubuc illustrate that French colonists' relentless push to adopt new techniques for planting, labor management, and refining did not translate into enlightened attitudes toward enslaved workers, either by residents or by investors in France. Saint-Domingue's extraordinary prosperity after the American Revolution attracted not only new merchants to the slave trade, but also more absentee investors in plantations. These were powerful figures at court, and significant people involved in banking and commerce. They sought high investment returns from plantations, but had no intention of moving to Saint-Domingue. In 1784 and 1785 colonists fought bitterly against a new law that tried to regulate plantation management to protect the interests of metropolitan investors, and to restrain the cruelty of planters and managers toward their slaves. In December 1784 the Naval Ministry issued an order that addressed the concerns of this class, as well as those who were concerned about planters' mistreatment of colonial slaves.[51] It required plantation attorneys to register with local officials so wary authorities could prosecute and track fraudulent or incompetent managers; it forbade attorneys to manage more than two plantations simultaneously and required that they be within a certain distance of each other. The decree required managers and their staffs to maintain six separate estate registers, tracking crops, workers' health, sales, and other affairs. It laid out a system of record keeping for the sale and the shipment of plantation commodities.

At the heart of the decree, however, were the questions being raised in France since the late 1760s about plantation slavery. Enlightened defenders of the institution, like Hilliard d'Auberteuil, claimed the masters' self-interest motivated them to feed, clothe, and otherwise treat their chattel workers with humanity. When there were problems, proslavery writers often pointed to mismanagement by managers hired by absentee owners. Ministry officials wanted to protect not only wealthy absentee planters but the entire colony from the consequences of such mismanagement. The greed and short-term interests of attorneys could maim and kill valuable workers and, as the poet Saint-Lambert explained in his 1769 story "Ziméo," could provoke a slave revolt that would destroy an entire parish or colony. There has been no systematic study of this question in Saint-Domingue, but evidence from Jamaica—where there was a similar discourse blaming absentees for a host of problems—shows that attorneys were just as likely to be good managers as were resident planters.[52] It seems quite possible that in Saint-Domingue too, absentees and their employees were being scapegoated for problems that were inherent in this ruthless slave system.

Whatever the reality, the 1784 French law addressed this perceived problem by creating an incentive for managers to take care of an absentee's slaves and animals. It prohibited them from profiting by exploiting either of these kinds of the owner's property. Far more controversially, it made managers—or resident planters—legally liable for mistreating slaves, by mutilation, and even murder. If an overseer, or an owner, killed a slave, the state would prosecute him as a murderer, the law suggested. It repeated the Code Noir requirements that masters provide slaves with specific amounts of food and clothing. Moreover, it mandated that slaves be given their own garden plots, as was common practice in Jamaica. Plantations were all to grow their own provisions and not count the slave gardens as part of this. Most controversially, it gave the maréchaussée, the mounted constabulary, whose rank and file were largely composed of free men of color, the power to conduct plantation inspections and to interview slaves about the conditions they lived in.

In November 1785 the comments of Morange, the plantation attorney of a merchant and absentee owner, laid out planters' principal objection to these measures: "To believe the accusation of the slaves is to open the door to revolt and arm them against the whites. If the ordinance is maintained, the fate of a good slave will be preferable to that of a manager or an attorney, I would even say of an owner. You must know this country and the slaves to understand that I am correct. The more I go the more I see that they want to free the slaves and put the whites in chains."[53] An army officer stationed in Cap Français in May 1785 described the law in these terms: "In its substance this edict attacks the sacred rights of property and places a dagger in the hands of the slaves by submitting their discipline and their regimen to hands other than those of their masters. The constabulary of this country, composed in large part of low-born people, has the right to go on a plantation, question the Negroes about the manner in which they are treated, about the work that is demanded of them, about the punishments inflicted on them. God knows what disorders this could cause and what there is to fear from the unjust complaints of slaves and the vengeance of the leader of the constabulary who, by his very position, is in a natural state of war with the proprietors of this colony."[54]

The decree provoked a major backlash in the colony. The colonial Councils fought their obligation to officially register the law. The Port-au-Prince Council eventually ratified the law after delaying for a year. But the Cap Français Council refused. In 1785, responding to the showdown, the Naval Ministry produced another version of the law, demanding that both Councils approve each of the two laws.[55] The second one, the ministry claimed, would "remove any pretext for false interpretations." For example the revised law added new language underlining the importance of slaves' abject obedience. It also mandated that the constabulary give planters advance notice before they conducted any estate visit. Yet the revisions did not give much ground to planters. It listed the ways that overseers and managers could violate the rights of the enslaved people, and it charged parish militia commanders with watching over those rights. It contained a new section giving the governor and intendant explicit authority over all kinds of plantation affairs, including attorney fraud, problems with interlopers, and battles between slave forces on neighboring plantations. It provided an ad hoc tribunal to hear appeals by attorneys convicted of fraud or mismanagement. This new body included Council judges but remained under the control of administrative officials.

Versailles believed the new law protected metropolitan investors from the greed, incompetence or cruelty of their colonial managers. That cruelty deliberately included the tortures planters used to investigate and punish the widely feared but increasingly ill-defined slave crime of "poisoning." Even in 1757 and 1758, during the mysterious wave of mortality attributed to Macandal, administrators did not share planters' belief that Saint-Domingue faced a massive slave conspiracy. The tassou episode in 1776 and 1777 showed that spoiled food explained a surge of slave deaths much better than a network of African poisoners. As Malick Ghachem shows, the new estate laws of 1784 and 1785 were deeply pragmatic, intended to safeguard the institution of slavery by preventing abuses that would lead to a slave revolt. But planters did not see this as a mild reform. Their bitter opposition, centered in the Cap Français Council, shows how relentlessly they pushed for maximum productivity on their estates and how determined they were that no one interfere with what they saw as their control over their property. The Cap Français Council did not sign the legislation until March 1786.[56]

In January 1787, the Naval Ministry announced yet another controversial reform, effectively eliminating the Cap Français Council. It merged Saint-Domingue's two law courts and located the new body in Port-au-Prince. Many colonists saw this reform as a continuation of the 1784 struggle over the Crown's plantation regulations. While the merger of the two courts was not exactly a purge of the Cap Français magistrates, it nullified that body's pretentions to be a "sovereign" institution that could check the power of "despotic" royal governors, like d'Estaing and Rohan-Montbazon in the late 1760s. Moreau de Saint-Méry, a member of the Cap Français court, described the 1787 reform as instituting a "secondary despotism" over Saint-Domingue. He argued that it reduced the colonial judiciary to a "state of humiliation." He believed Versailles orchestrated the merger "in the hope that it would be easier to direct or even dominate one sovereign court instead of two." He predicted that, at some point, authorities would understand that "rule by fear, the least durable of all, cannot grow stronger in a colony this important."[57]

Guillaume Delamardelle, the royal attorney for the Port-au-Prince Council, was one of the authors of the 1787 reform.[58] He described the merger as a way to improve judicial efficiency. Combining the two courts in

Port-au-Prince, he argued, would save money and discourage interminable lawsuits.[59] But the question of slave treatment was at the heart of this change. Delamardelle pointed out that in the new system there would be no change in the number of local courts, because swift adjudication was especially necessary to bring slave owners into line with the long-term safety of the colony, an indirect reference to the 1784 slave law. While the planter Dubuc underlined the necessity of an ample slave trade for the prosperity of the colony, the attorney Delamardelle emphasized that another approach was to ensure the happiness and nutrition of slaves. The legal reform in that sense was part of the struggle between colonists and the metropole about whether royal laws would govern slavery or instead whether masters would remain all powerful. For Delamardelle, "It is essential that there be a counterweight, with the power and force to adjust [rétablir] relations between the free and slave population, relative to the different size of their populations. The establishment of this counterweight is all the more necessary so that the master will have more compassion [bonté] for his slaves, and that he will give them more freedom to procure their own well-being; this counterweight, we maintain, consists not only in a well-armed and active militia but also in an ever-present justice that can repress the power of the master when he abuses it and quickly make prompt example when the slave commits a crime."[60]

Although the colony's courts formally accepted these reforms, the Naval Ministry could not force colonists to embrace its ideas about the necessity of "repressing the power of the master when he abuses it." The 1788 case of Nicolas Lejeune was the most notorious example of this tendency. By the 1770s, many Saint-Domingue planters were deeply convinced that only complete sovereignty on their estates would save them from another Macandal. They insisted on the right to torture and execute suspected workers whom they thought to be "poisoners" without involving the state. Colonial administrators, on the other hand, believed that in the long term such cruelty would doom slavery by causing revolts. They pointed to Jamaica as evidence of how this assumption worked, with the memory of Tacky's Revolt in 1760, as ever, foremost in their minds. Some planters dismissed enslaved practices as mere "conjuring" and condemned the "excesses" of their neighbors. But colonists were generally united in their belief that planters' sovereignty over their estate workers had to be maintained at all costs or else slaves would rise up in revolt.[61]

In the late 1760s and early 1770s, colonial judges, many of them also planters, supported this position, which put them at odds with imperial desires to have a check on masters' power. In 1769, the Cap Français Council refused to bring charges against a planter named Cassarouy, who had tortured a number of his slaves, after the slaves ran away to protest his refusal to feed them. In June 1771 ten slaves belonging to the planter Dessources, presented themselves before the lower court [*sénéchaussée*] in Cap Français asking for protection from their master, who had burned and buried alive many members of his workforce after accusing the entire group of macandalism. The court returned Dessources's slaves to their master. Although the colony's slave laws prohibited torture, the judges claimed there was no need for the courts to oversee masters.[62] It seems likely, therefore, that most cases of planter brutality never reached court. The rumors about Caradeux's gruesome tortures and alleged murders of slaves do not seem to have ever resulted in an investigation, for example.[63]

While a small minority of slaves did bravely seek legal protection against abusive owners in the 1780s, some planters turned to African experts in order to reinforce their control over their workers.[64] In the early 1780s, in the coffee-growing Plaisance parish in the North Province, a planter named Caillon, known (perhaps ironically) as Belhumeur or "Good Humor," became convinced that a Congo woman named Marie Kingué, enslaved on his plantation, could detect poisoners.[65] According to a badly injured slave, who turned himself in to a judge in Cap Français, Caillon burned alive a number of workers, whom Kingué identified as "Macandals." When his friend Chaillau, the parish militia commander, became concerned that he was being poisoned, Caillon sent Kingué to examine him. She confirmed his fears and produced one frog, and then another, out of the commander's head, after which he reputedly felt much better. Chaillau persuaded Caillon to let her stay on his estate, where she identified seven or eight poisoners on his workforce. On the strength of her conviction Chaillau burned one of his slaves at the stake, had three or four whipped to death, and sold three others. Within a few years, Kingué had such wealth and prestige that she acted as a free woman. According to white authorities, blacks regarded her as "a god." They believed she could raise people from the dead. Whites and blacks purchased her talismans and wore them as protection against poisoning.

One planter with whom she may have consulted was Nicolas Lejeune *père*, a coffee planter living in the same parish as Chaillau. In 1781 he placed an advertisement in the colonial broadside informing the public that a creole slave named Jean-Baptiste had escaped from his plantation two weeks earlier. Lejeune labeled him "a thief and a macandal."[66] The second label does not tell us if Lejeune *père* thought Jean-Baptiste was a charismatic spiritual leader or a talented poisoner, in the chemical sense of that term. But the former characterization seems most likely, based on the little we know about Lejeune's estate. The intendant Barbé-Marbois later claimed that Lejeune's plantations since 1780 or 1781 had been a "theatre of masters' barbarity and slaves' vengeance." During those years, Lejeune's slaves had revolted against him, murdering his nephew. The courts had punished the slaves, breaking them on the rack, but the intendant later noted that "at no time did justice repress the inhumanity of the masters."[67] Lejeune's fevered reaction to the slaves he suspected were using witchcraft and poison makes it seem likely that he would have consulted Marie Kingué in 1781 for help finding his escaped "macandal."

In 1785, anonymous letters denouncing Kingué and calling her a "monster" came to the attention of authorities in Cap Français. As Cap Français's royal attorney described it, "The fanaticism about her finally reached a point where the greatest disturbances would have arisen among the plantation workers if the public authorities had not halted the course of events by arresting the sorceress."[68] Meanwhile, back in Plaisance, tensions on the Lejeune plantation grew so great that Nicolas Lejeune leased the plantation to his son, also named Nicolas. At this point the record becomes clearer. The younger Lejeune soon became convinced, like his father, that some of his slaves were poisoners. Within two years, forty-seven of his workers had died, and some reports say he also lost thirty mules, which were nearly as expensive to replace as humans. Lejeune did not feel these losses were just misfortune. Nor did he attribute these losses to his harsh management style. Something more sinister, he thought, explained his bad luck. Like his father and Caillon and the militia commander Chaillau, the younger Lejeune believed a slave was challenging his authority, and using poison to do so.[69]

In 1788, after another of his slaves died mysteriously, Lejeune decided to investigate. After her arrest Marie Kingué was presumably no longer available to assist him, but he enlisted the help of his white plantation surgeon, a man whom skeptical authorities later described as "profoundly ignorant." The surgeon dissected the body of the dead slave and claimed he had found evidence of poison. When the royal doctor and surgeon in Cap Français examined the woman's corpse they found no such evidence. But Lejeune was convinced. He continued his investigations, which meant torturing two women on the plantation. He claimed to have only heated their feet with hot coals, but their legs and thighs were burned so badly that they died of their wounds. Before dying they confessed to the crimes, though Barbé-Marbois argued "everything seemed to suggest that they were not guilty."[70]

The intendant wanted to make an example of Lejeune to demonstrate the power of the state to regulate the punishment of slaves. He believed that the Lejeunes' behavior was widespread in Saint-Domingue and insisted that such cruelty would eventually cause a slave revolt. He rejected planters' claims that their torture prevented slaves from rising up. He posed the question: "How many barbarities have the planters worked together to hide from the administration, whose vigilance on this matter is so hateful to them? What will happen if our common impotence in this affair is publicly demonstrated and if in the same time the tribunals join with the barbaric planters to oppress these unfortunates? For one hundred years these cruelties have been exercised with impunity; they are committed right before the slaves because it is known that their testimony will be rejected."[71]

Nevertheless, sometime in July or August 1788, Saint-Domingue's highest court absolved Lejeune of all guilt. The intendant and governor informed Versailles that public pressure on the judges had produced this verdict. The resolution of the Lejeune case illustrated that neither legal reforms nor arguments for the long-term benefits of humane standards would change colonists' idea that planters needed to have complete control over their workers. Lejeune's case shows how colonists and colonial judges, including those who condemned Lejeune's behavior, believed that the ability of planters to treat their slave property as they saw fit was fundamental to the continued prosperity of the plantation economy. Nevertheless, the fact that colonial judges brought Lejeune to trial, when they had not prosecuted earlier cases when white planters killed alleged poisoners, also proved a point: the allegation that a slave was a poisoner was not sufficient grounds for torture. Royal authorities believed that planters were to some extent answerable to public opinion for especially outrageous actions against enslaved people.[72]

The fact that colonists refused to sanction Lejeune for his sadistic actions just three years before the Haitian Revolution began makes it tempting to point to planter violence as a motivating factor for the great slave revolt of 1791. But this twin portrait shows that masters in Saint-Domingue and Jamaica exercised a similar level of brutality against their workers. The presence or absence of slave rebellion in these two societies, or indeed any slave society, does not indicate owners' cruelty or laxity toward their slaves.

Indeed, the similarities we have seen between Saint-Domingue and

Jamaica make it difficult to see elements in pre-Revolutionary society that account for the success of the 1791 slave conspiracy. Planters in both colonies believed they could control any rebellion through harsh estate discipline. They were wary of what slaves were doing but were convinced they could effectively repress slave discontent. Few believed there was a limit to the amount of repression slaves would endure. Nor did their attitudes about slave religion vary greatly. Jamaica outlawed obeah but permitted other kinds of slave religiosity, and there were few attempts to Christianize the slaves. In other words, both societies let slaves develop their own religions, repressing only selected practitioners. Planters and merchants were hugely optimistic about the future. Saint-Domingue's planters were, if anything, more optimistic because they believed that the national monopoly was the main thing holding them back. Liberalizing colonial trade, they argued, would make them more valuable to France, and giving them more local government powers, in a position more like Jamaica with its Assembly, would reduce absenteeism and increase colonists' loyalty to the France.

To be sure, there were differences underlying these similarities. The Crown's suppression of the Cap Français Council in 1785 aggravated French colonists' belief that they suffered under tyrannical royal control, which partially explains the enthusiasm of many about the French Revolution's political reforms. As we have seen, there were differences in the racial reforms that both Jamaica and Saint-Domingue enacted around midcentury, separating free people into more distinct "white" and "colored" castes. Saint-Domingue leaders consigned French-educated slave-owning families into the "ex-slave" (*affranchi*) class because they had one African grandparent or great-grandparent. To unify poor and rich colonists and build their attachment to the metropole, Saint-Domingue's elites did not allow wealth or education to override African genealogy in determining "whiteness." Jamaica had the Special Bills mechanism that allowed the richest families partial access to white status in voting and legal rights, while eliminating the hope that poor and middling Jamaican families of color could ever access those rights. Saint-Domingue, on the other hand, gave all free people full legal and property rights, allowing free coloreds to inherit even rich properties as a matter of course, in contrast with Jamaica's racial limits on inheritance.

Another difference was that Saint-Domingue experienced an extraordinary influx of enslaved Africans in the years after 1783, far beyond what they were accustomed to. While Jamaican slave owners welcomed royal troops, who played a central role in the Maroon War (1730–39) and in repressing

Tacky's Rebellion, Saint-Domingue's colonists worried that government interference, like the management reform of 1784 and 1785, would prevent them from dealing with internal threats.

In the context of Revolutionary developments in France, these distinctive elements—planters' frustration with French absolutism, greater frustration among wealthy free families of color, a massive influx of new Africans, and a greater insistence on planters' sovereignty on their estates—helped destabilize the French colony between August 1789 and the great slave uprising of August 1791. During those two years French colonists were far more worried about how to react to events in France than they were about their slave population. They disagreed with each other over whether to embrace the Revolution and whether to adopt metropolitan reforms that increased local autonomy from royal administrators. In 1790 the controversial Saint-Marc Assembly drafted a charter that went well beyond the local law-making powers claimed by Jamaica's contentious House of Assembly. Saint-Domingue's royal governor purged this body by force, and Parisian revolutionaries eventually approved his actions. But the question of how to structure imperial politics remained extremely contentious in Saint-Domingue. Similarly, in Saint-Domingue and Paris in 1789 and 1790 wealthy free men of color succeeded in making racial categories a central element in debates over colonial voting rights. When Parisian revolutionaries decided that such men could be full citizens, civil war broke out in Saint-Domingue during the summer of 1791, pitting whites against free men of color. As important as these destabilizing events were, changes in Saint-Domingue's enslaved population in the 1780s probably also played an important role in making the slave revolt of August 1791 possible. The availability of a rich repertoire of African and local religious practices may have helped unite enslaved creole men and women with newly arrived Africans in the quasi-mythical Bois Caiman ceremony of 21 August 1791, said to be the beginning of the Haitian Revolution.[73]

But by this time, the events of the Revolution as they played out in France and in Saint-Domingue had created a dynamic all their own. The courage and creativity of Saint-Domingue's slaves notwithstanding, there would not have been a 1791 slave revolt without the French Revolution. Saint-Domingue had wealthy free colored spokesmen that Jamaica did not possess, but in 1788 these men—Julien Raimond and Vincent Ogé—were getting no attention at Versailles for their complaints of racial injustice. Without the fall of the Bastille, we might never have heard of either. Similarly French colonists resented royal reforms, but by 1788 they had accepted the closing of the Cap Français

Council and had signed off on the 1785 estate management law. Without the French Revolution, if enslaved creoles and Africans had organized a revolt in 1791, they would have met the combined repressive force of white and free colored militia. The results might have been as devastating as Tacky's Revolt was in Jamaica thirty years earlier, but it seems likely that the master class would have prevailed, especially with military assistance from royal troops.

In retrospect, unresolved conflicts between Saint-Domingue and Versailles—like planter's sovereignty over their slaves and the dissolution of the Cap Français Council—foretold how heated French transatlantic politics would become in the early stages of the Revolution.[74] But such controversies, like the merger of Saint-Domingue's Councils or colonial judges' acquittal of Lejeune, caused few ripples of concern in France outside the Naval Ministry. This was not the case for Jamaican planters, however. In the aftermath of the American Revolution, the planters came under metropolitan scrutiny, mostly unfavorable, as never before.[75] The cause célèbre of the *Zong* focused attention on the inhumanity and financial amorality of the slave trade while encouraging Britons to take notice of the cruelties of plantation slavery. The many imperial scandals and issues that marked the 1780s also brought Jamaica into sharper focus.[76] Britons tended to accept the new principles of an imperialism that was beginning to bestride the globe but felt distinctly queasy about particular aspects of its commerce and governance.[77] In this way, British imperialism was considerably different from French imperialism, where concern over the actions of colonials was relatively muted.

British feelings about white settlers changed after the American Revolution. In particular, key thinkers began to doubt that white planters were as British as they claimed to be. The years after 1783 witnessed considerable changes in imperial thinking, as administrators responded to the ideological challenge of the American Revolution. More than ever before, a clarified sense of imperial purpose, based on the concept of "national honor," undergirded British imperialism. Ironically, by the 1780s it was Jamaican planters—always conspicuously patriotic in ways that French West Indian planters were not—who came under suspicion as being un-British and somehow not part of the British nation. By the late 1780s, Britain began attacks on the institution that West Indian planters held most dear—African chattel slavery. These attacks were combined with assaults on West Indians as people who deviated from an increasingly rigid understanding of what it meant to be British, both in Britain and abroad. The result was a devastatingly successful critique of

West Indian society that propelled abolitionism to sudden success as Britain's foremost humanitarian campaign.

British imperial policy was increasingly based on accommodating all people living under British rule, "guaranteeing protection to imperial subjects regardless of where they lived, what ethnicity they were, or what religion they espoused."[78] However, the British imperial state continued to refuse to agree with colonists that they had the right to govern themselves because they were freeborn Englishmen. Britain not only refused to accept such constitutional claims (thus continuing their stance from the American Revolution); they increasingly came to view white planters as fundamentally un-British, because of their moral behavior. Jamaican planters did not conform to emerging notions of "Englishness" based around bourgeois values of the family and a growing cult of moral improvement. Moreover, the American Revolutionary conflict made Britons more aware of colonial matters. They generally did not like what they read in their newspapers about Americans and West Indians. The American Revolution, Troy Bickham notes, was a war that "accelerated the creation of a nation of imperial critics." In this process, West Indian planters gradually got excluded from membership within a wider British family.[79] The West Indian planter generally cut a sad figure in British culture from the late 1780s onward. When artists depicted planters as a class they used the trope of the Oriental pasha, but when they portrayed individual planters they emphasized the pathetic rather than the tyrannical. Planters saw themselves as British gentlemen of upright character and firm morals, and as men capable of moderation, self-restraint, and refined gentility. But increasingly they were depicted in Britain as degenerating under the physical and moral pressures of tropical life. Even depictions of their wealth featured their luxury, effeminacy, gluttony, racial degeneracy, or sexual hybridity, all in highly negative terms.[80]

Such depictions helped abolitionists when they described slavery as an aspect of a degenerate and violent imperial polity, susceptible to all forms of corruption. Despite manifold evidence that Jamaican planters were capable innovators and dynamic capitalists, in the 1780s they were depicted as dogmatically fixed in their ideas, conservative in temperament, and out of step with modern times. Britons viewed planters as morally deficient, incapable of self-improvement, and in need of reformation from the outside. This critique, moreover, was highly gendered. The metropolitan opponents of planters were strongly masculinist and aggressively interventionist. Strangely, for men who had been symbols of aggressive masculinity in the mid-eighteenth century, the image of the planter contained significant degrees of effeminacy, reflecting the ways in which both ideas of race and ideas of gender were changing in the mid-eighteenth century.[81] In a period when symbols of masculinity suggested that proper men were healthy and virile with many children, colonists' notoriously poor health and lack of legitimate families raised doubts about their masculinity. Moreover, planter's "passion" and lack of self-control made them, in Orientalist discourse, more feminine than masculine.

The situation was substantially different in France. The Atlantic trade was the most dynamic element of the kingdom's economy, and the American Revolutionary War brought more favorable attention to the island colonies than ever before. French culture imagined the Antilles as an extraordinary engine of economic growth. If planters, as well as the vaguely defined category of "creoles," were associated with Oriental clichés—the sultan, the slave harem—these were clichés that tended to emphasize wealth, sexual libertinism, and masculine power, more than degeneracy. Although the Naval Ministry worried about how to maintain colonial loyalty while enforcing the national trade monopoly, by the 1780s antismuggling reforms and the extraordinary growth of French slave trading appeared to have resolved many of those concerns.

In the years after 1770, French and colonial intellectuals refined a biological conception of "race" and increasingly invoked it to justify the nature of plantation society. Moreover, French authorities drafted laws to reduce the numbers of enslaved and free people of color coming to the metropole. Yet these laws were not widely observed—especially not for wealthy free people of color who fit the metropolitan image of the rich American, as we have seen in the case of Julien Raimond. As for attacks on slavery, in 1787 Jacques-Pierre Brissot and Etienne Clavière founded the French antislavery society Les Amis des Noirs, with close ties to Thomas Clarkson and to other English activists. But French antislavery was a tiny and elite movement in the 1780s. Moreover, many of its supporters were attracted to the idea that slavery was economically inefficient rather than being concerned about the morality of the institution. Certainly, the French found it hard to contemplate abandoning slavery. As Madeleine Dobie has shown, French antislavery texts often included discussions of how profitable it would be to produce sugar with free workers. Conversely, many of those who wanted to lessen the inhumanity of Caribbean slavery did so for political reasons: eliminating torture would remove what they believed was the main cause of slave revolts.[82]

Nevertheless, by 1788 Saint-Domingue's planters had begun to worry about the threat of "abolitionists." They knew about Brissot's Amis des Noirs and its ties to Clarkson's Committee for the Abolition of the Slave Trade. What planters did not realize, however, was that their biggest vulnerability in metropolitan politics was not slavery but the colony's increasingly rigid racial laws. In 1790, two wealthy Dominguan planters of color—Julien Raimond and Vincent Ogé, men who fitted the vague and attractive French notion of the "rich American"—convinced Parisian revolutionaries to focus their colonial attention on Saint-Domingue's racial injustices, especially those injustices that affected wealthy free men of color.

We end this book deliberately in 1788 because we want to suggest that there was nothing inevitable about the later decline of Jamaica or Saint-Domingue. Contemporary commentators who thought about the future direction of the French and British Empires in the Greater Antilles saw both Jamaica and Saint-Domingue as places of extraordinary transformations. The land itself was transformed as West Indians developed the plantation as a new kind of social institution and economic engine. These were places of transplantation—new people, new fauna, new flora, and continual circulation of new ideas about how to make everything and everybody more profitable.

Personal transformations were especially important. The ongoing perfection of the integrated plantation model led to an increasingly complete commodification of the humans who worked in it. Planters transformed African people into West Indian slaves, and much of the growing wealth of the region was tied up in increasingly valuable enslaved people. Owners regarded these workers as human capital to be used ever more efficiently, often leaving them vulnerable to food scarcity and disease. Africans were expensive to buy, but profits gained from feeding Europe's growing sweet tooth allowed planters to use up and dispose of those expensive workers while making large amounts of money for themselves. The plantation machine operated smoothly, as long as the slave trade ran well. As we have seen, there were moments when that trade faltered, such as during the Seven Years' War in Saint-Domingue and during the War for American Independence. Even during such periods, however, planters continued their extreme demands on slaves so as to increase their productivity. Jamaica and Saint-Domingue were places where everything and everybody was turned into money and where, among the "chaos of money, Negroes and things," colonists defined themselves by a series of negotiable promises, calculations, and speculations.[83] These were cosmopolitan, profit-maximizing societies that measured progress by the twin standards of rapid economic growth and social mobility. Achieving either of these ideals depended on careful management of capital resources and entailed a close eye on the clock and calendar. On the sugar plantations of Jamaica and Saint-Domingue, more than almost any other place in the eighteenth-century Atlantic World, time was money. And although these societies were totally dependent on slavery, the slavery they practiced was a "modern" institution in which workers were units of capital, in theory barely distinguishable from a gear in a sugar mill, a boiling house cauldron, or any other expensive but fungible unit of production. Planters accepted—and created—high levels of mortality because they knew that overwork and malnutrition produced more sugar in less time. They also knew that the slave trade would supply more African bodies.

White Jamaicans and Saint-Dominguans were devoted to individualism, private property accumulation, and hedonistic pleasures to an extent that was unprecedented in late eighteenth-century Britain and France. These characteristics were allied to a determination to separate people out by color and attribute rights to other free people based not on status or class but on color and ethnicity. These social features made Jamaica and Saint-Domingue seem chaotic and immoral to an increasing legion of metropolitan critics. We can begin to see the outlines of this kind of criticism in how Jamaican and Saint-Dominguan planters were viewed in France and Britain on the eve of the French Revolution, even as some sectors of those metropolitan cultures embraced market mechanisms and racial categories. Thus, twin and contradictory views of Saint-Domingue and Jamaica were apparent by the end of our story.

But we would not like to end on such a note, a tone of disappointment, as if the two colonies and the particular social and economic structures they had created contributed little beyond the immiseration of enslaved people to the making of the French and British Atlantic Worlds. The abolitionists, who began in Britain from the 1780s and in France from the 1790s to demonize planters as irrevocably backward and as reverting in the heat of the tropics to sensual and languorous barbarism, may have won the historical argument about the representation of the settlers of Saint-Domingue and Jamaica and were, of course, correct in denouncing slavery as a particular sin that civilized countries needed to eradicate. But we do not need to accept all their arguments, especially the idea that the West Indies were incapable of any kind of creativity or cultural development. The Greater Antilles may not have been quite the places that might "be regarded as the principal cause of the rapid movement which stirs the universe," as the authors of the *Histoire des deux Indes* argued in 1770. But they were significant places, nonetheless. We cannot understand the development of either the British or the French Atlantic World without understanding that what happened in Saint-Domingue and in Jamaica between 1748 and 1788 profoundly influenced imperial policy. Events in these colonies also shaped the resolution of both the Seven Years' War and the American Revolution. Moreover, the Greater Antilles were forging grounds for new conceptions of how to define race and patriotism that continued to be important well after the Enlightened world of the eighteenth century turned into the imperial expansion of the nineteenth century. This book shows the relevance of both places for understanding the vitality of the plantation system, the importance of planters and merchants in that system, and the dynamism of slave societies in the making of the late eighteenth-century British and French Atlantic Worlds.

NOTES

The following abbreviations appear in the notes.

AHR	*American Historical Review*
AN	French National Archives, Paris
ANOM	French National Archives, Outre-Mer section, Aix-en-Provence
CO	Colonial Office Series, National Archives, Kew, London
CSP	*Calendar, State Papers: Colonial Series: America and the West Indies, 1574–1739*, 45 vols. (London: HMSO, 1860–1994)
FP	Fulham Papers, Lambeth Palace Library, London
HAHR	*Hispanic American Historical Review*
HCSP	Sheila Lambert, ed., *House of Commons Sessional Papers of the Eighteenth Century* (Wilmington, Del.: Scholarly Resources, 1975)
JA	Jamaica Archives, Spanish Town, Jamaica
JAJ	*Journals of the Assembly of Jamaica*, 14 vols. (Kingston: Alexander Aikman, 1792–1809), cited by date.
NA	National Archives, Kew, London
TSTDB	*Transatlantic Slave Trade Data Base*, http://slavevoyages.org/tast/index.faces
TT	Diaries of Thomas Thistlewood, cited by date, Beinecke Rare Book and Manuscript Library, Yale University
WMQ	*William and Mary Quarterly*

Chapter 1. A Comparative History of Jamaica and Saint-Domingue

1. For works that place the plantation system in global context, see Barbara L. Solow, "Capitalism and Slavery in the Exceedingly Long Run," *Journal of Interdisciplinary History* 17 (1987): 711–37; Philip D. Curtin, "The Environment Beyond Europe and the European Theory of Empire," *Journal of World History* 1 (1990): 131–50; and Giorgio Riello, *Cotton: The Fabric That Made the Modern World* (Cambridge: Cambridge University Press, 2013), 187–210. For later developments, see Dale M. Tomich, *Slavery in the Circuit*

of Sugar: Martinique and the World Economy (Baltimore: Johns Hopkins University Press, 1990).

2. Philip D. Morgan, "The Caribbean Islands in Atlantic Context, circa 1500–1800," in *The Global Eighteenth Century*, ed. Felicity A. Nussbaum (Baltimore: Johns Hopkins University Press, 2003), 64.

3. Abbé Raynal, as cited in Michael Duffy, *Soldiers, Sugar and Seapower: The British Expeditions to the West Indies and the War with Revolutionary France* (Oxford: Clarendon, 1987), 6.

4. This argument is made most forcefully by Sidney Mintz, *Caribbean Transformations* (New York: Aldine, 1974); and B. W. Higman, *Plantation Jamaica 1750–1850: Capital and Control in a Colonial Economy* (Kingston: University of the West Indies Press, 2005).

5. "Recensement général de la colonie de St. Domingue pour l'année 1753, joint à la lettre de Mr de Vaudreuil et de Lalanne du 3 may 1754," 1754, ANOM, G1/509, no. 27; Edward Long, *The History of Jamaica*, 3 vols. (London: T. Lowndes, 1774), 2:229.

6. Stuart B. Schwartz, "Introduction," in *Tropical Babylons: Sugar and the Making of the Atlantic World, 1450–1680* , ed. Schwartz (Chapel Hill: University of North Carolina Press, 2004), 18.

7. Schwartz, "A Commonwealth Within Itself: The Early Brazilian Sugar Industry, 1550–1670," *Tropical Babylons*, ed. Schwartz, 162–63.

8. Alfred Martineau and Louis May, *Trois siècles d'histoire antillaise, Martinique et Guadeloupe, de 1635 à nos jours* (Paris: Société de l'histoire des colonies françaises, 1935), 122–24; Richard S. Dunn, *Sugar and Slaves: The Rise of the Planter Class in the English West Indies, 1624–1713* (Chapel Hill: University of North Carolina Press, 1972), 60–65.

9. Philip P. Boucher, *France and the American Tropics to 1700: Tropics of Discontent?* (Baltimore: Johns Hopkins University Press, 2007), 230, 243.

10. Dunn, *Sugar and Slaves*, 94–95.

11. For the first references to gang labor, and for how little is known about the spread of this practice, see Russell R. Menard, *Sweet Negotiations: Sugar, Slavery, and Plantation Agriculture in Early Barbados* (Charlottesville: University of Virginia Press, 2006), 91; on the ubiquity of the "grand atelier" system among the large plantations in the French Caribbean, see Gabriel Debien, *Les esclaves aux Antilles françaises, XVIIe–XVIIIe siècles* (Basse-Terre: Société d'histoire de la Guadeloupe, 1974), 134–37.

12. Samuel Martin, *An Essay upon Plantership Humbly Inscribed to His Excellency George Thomas, Esq., Chief Governor of All the Leeward Islands, as a Monument to Antient Friendship*, 4th ed. (Antigua; London: Printed by S. Clapham; Re-printed for A. Millar, 1765), 2 and 36–37.

13. Schwartz, "A Commonwealth Within Itself," 176. Schwartz notes that in Brazil there were ten months in which sugar planters could harvest their canes; in the Caribbean, weather conditions only left four to six months for this activity.

14. Schwartz, "Introduction," 4 in *Tropical Babylons*, ed. Schwartz.

15. Douglas Hall, "Incalculability as a Feature of Sugar Production During the Eighteenth Century," *Social and Economic Studies* 10 (1961): 348.

16. Justin Roberts, "Uncertain Business: A Case Study of Barbadian Plantation Management, 1770–93," *Slavery and Abolition* 32 (2011): 254.

17. Frank Tannenbaum, *Slave and Citizen: The Negro in the Americas* (New York: A. A. Knopf, 1947); Carl N. Degler, *Neither Black nor White: Slavery and Race Relations in Brazil and the United States* (New York: Macmillan, 1971); Gwendolyn Midlo Hall, *Social Control in Slave Plantation Societies: A Comparison of St. Domingue and Cuba* (Baltimore: Johns Hopkins University Press, 1971).

18. B. W. Higman, *Slave Populations of the British Caribbean 1807–1834* (Baltimore: Johns Hopkins University Press, 1984), 1–5.

19. David Lowenthal, *West Indian Societies* (London: Oxford University Press, 1972), 40.

20. Higman, *Slave Populations of the British Caribbean*, 66–67, 377–78, 395–96. Important comparative works include Hall, *Social Control*; and Philip D. Morgan, *Slave Counterpoint: Black Culture in the Eighteenth Century Chesapeake and Lowcountry* (Chapel Hill: University of North Carolina Press, 1998). An exemplary comparative work in Atlantic history is J. H. Elliott, *Empires of the Atlantic World: Britain and Spain in America 1492–1830* (New Haven, Conn.: Yale University Press, 2006). Unfortunately, there is no explicitly comparative work in French historiography that compares Saint-Domingue to Martinique or Guadeloupe in the mid-eighteenth century.

21. Pierre-François-Xavier de Charlevoix and Jean-Baptiste Le Pers, *Histoire de l'isle espagnole ou de S. Domingue: Écrite particulièrement sur des mémoires manuscrits du P. Jean-Baptiste Le Pers, jésuite*, 3 vols. (Amsterdam: François L'Honoré, 1733), 3:342.

22. M. L. E. Moreau de Saint-Méry, *Description topographique, physique, civile, politique et historique de la partie française de l'isle Saint- Domingue* (1797; Paris: Société de l'histoire des colonies françaises, 1958), 38.

23. Long, *History of Jamaica*, 2:261–300; on white women in Saint-Domingue, see Moreau de Saint-Méry, *Description*, 39–44.

24. David William Cohen and Jack P. Greene, eds., *Neither Slave nor Free: The Freedman of African Descent in the Slave Societies of the New World* (Baltimore: Johns Hopkins University Press, 1974), 1–18.

25. Atkins to Board of Trade, 4 July 1676, CSP.

26. Boucher, *France and the American Tropics*, 242.

27. Thomas Craskell and James Simpson, *Maps of the Counties of Surrey, Cornwall and Middlesex, in the Island of Jamaica* (London, 1765).

28. R. H. Whitbeck, "The Agricultural Geography of Jamaica," *Annals of the Association of American Geographers* 22 (1932): 16–17.

29. Elmire M. Lobeck, "Haiti: A Brief Survey of Its Past and Present Agricultural Problems," *Journal of Geography* 53 (1954): 278.

30. "Carte Générale de La Partie Française de L'isle de St. Domingue: Relative Au Mémoire et Au Costier de Mr. d'Estaing," 1776, Library of Congress.

31. Guillaume Delisle, "Carte de L'isle de Saint-Domingue. Dressée en 1722 pour l'Usage du Roy, sur les Mémoires de Mr. Frezier, Ingénieur de S. M. et Autres, Assujetis aux Observations Astronomiques" (Paris: Lisle, Quai de l'Horloge, 1780).

32. Whitbeck, "Agricultural Geography of Jamaica," 17.

33. Nathan C. McClintock, "Agroforestry and Sustainable Resource Conservation in Haiti: A Case Study," Partners in Progress: Promoting and Advancing Sustainable Rural Development in Haiti, http://www.piphaiti.org/overview_of_haiti2.html.

34. David P. Geggus, "Urban Development in Eighteenth-Century Saint-Domingue," *Bulletin du centre d'histoire des espaces Atlantiques* 5 (1990): 216.

35. Franklin Knight, *The Caribbean, the Genesis of a Fragmented Nationalism*, 3rd ed. (New York: Oxford University Press, 2012), 63.

36. Jan Rogozinski, *A Brief History of the Caribbean: From the Arawak and Carib to the Present* (New York: Plume, 2000), 156.

37. Jane G. Landers, "The Central African Presence in Spanish Maroon Communities," in *Central Africans and Cultural Transformations in the American Diaspora*, ed. Linda M. Heywood (Cambridge: Cambridge University Press, 2002), 234.

38. Michael A. Gomez, *Reversing Sail: A History of the African Diaspora* (Cambridge: Cambridge University Press, 2005), 116.

39. Jonathan R. Dull, *The French Navy and the Seven Years' War* (Lincoln: University of Nebraska Press, 2005); N. A. M. Rodger, *The Command of the Ocean: A Naval History of Britain* (London: Allen Lane, 2004); and *TSTDB*.

40. Philip J. Stern and Carl Wennerlind, eds., *Mercantilism Reimagined: Political Economy in Early Modern Britain and Its Empire* (New York: Oxford University Press, 2013); and Sophus A. Reinart, *Translating Empire: Emulation and the Origins of Political Economy* (Cambridge, Mass.: Harvard University Press, 2011).

41. Jack P. Greene, *Evaluating Empire and Confronting Colonialism in Eighteenth-Century Britain* (New York: Cambridge University Press, 2013), 60–77.

42. 7–9 October 1762, JA 5:352–53; Nicholas Bourke, *The Privileges of the Island of Jamaica Vindicated* (Kingston, 1765).

43. Marilyn Lake and Henry Reynolds, *Drawing the Global Colour Line: White Men's Countries and the International Challenge of Racial Equality* (Melbourne: Melbourne University Press, 2008); Owen White, *Children of the French Empire: Miscegenation and Colonial Society in French West Africa, 1895–1960* (New York: Oxford University Press, 1999).

44. Ann McGrath, ed., *Contested Ground: A History of Australian Aborigines Under the British Crown* (Sydney: Allen and Unwin,1995); Timothy Keegan, *Colonial South Africa and the Origins of the Racial Order* (Charlottesville: University of Virginia Press, 1996); Saul Dubow, *A Commonwealth of Knowledge: Science, Sensibility and White South Africa, 1820–2000* (Oxford: Oxford University Press, 2006); Ann Laura Stoler, *Carnal Knowledge and Imperial Power: Race and the Intimate in Colonial Rule* (Berkeley: University of California Press, 2002); and Lisa Ford, *Settler Sovereignty: Jurisdiction and Indigenous People in America and Australia, 1788–1836* (Cambridge, Mass.: Harvard University Press, 2010).

45. Jean Tarrade, *Le commerce colonial de la France à la fin de l'ancien régime: L'évolution du régime de l'exclusif de 1763 à 1789*, vol. 2 (Paris: Presses universitaires de France, 1972), 776; Françoise Thésée, *Négociants bordelais et colons de Saint-Domingue: Liaisons d'habitations; La maison Henry Romberg, Bapst et Cie, 1783-1793* (Paris: Geuthner, 1972), 239-40.

46. For the history of Haiti after the Revolution, see David Nicholls, *From Dessalines to Duvalier: Race, Color, and National Independence in Haiti* (Cambridge: Cambridge University Press, 1979); and Laurent Dubois, *Haiti: Aftershocks of History* (New York: Metropolitan Books, 2011). For Jamaica after the abolition of the slave trade, see Mary Turner, *Slaves and Missionaries: The Disintegration of Jamaican Slave Society, 1787-1834* (Urbana: University of Illinois Press, 1982); Gad Heuman, *The Killing Time: The Morant Bay Rebellion in Jamaica* (Knoxville: University of Tennessee Press, 1995); Gale L. Kenny, *Contentious Liberties: American Abolitionists in Post-Emancipation Jamaica, 1834-1866* (Athens: University of Georgia Press, 2010); and Thomas Holt, *The Problem of Freedom: Race, Labor, and Politics in Jamaica and Britain, 1832-1938* (Baltimore: Johns Hopkins University Press, 1992). Important comparative works are Mimi Sheller, *Democracy After Slavery: Black Publics and Peasant Radicalism in Haiti and Jamaica* (Gainesville: University of Florida Press, 2000); and Edward B. Rugemer, *The Problem of Emancipation: The Caribbean Roots of the American Civil War* (Baton Rouge: Louisiana State University Press, 2008).

47. For the traditional view that the plantation contributed little to modern management methods, see Alfred D. Chandler, *The Visible Hand: The Managerial Revolution in American Business* (Cambridge, Mass.: Harvard University Press, 1977), 64-66. For a counterview, see Bill Cooke, "The Denial of Slavery in Management Studies," *Journal of Management Studies* 40 (2003): 1895-918. See also Higman, *Plantation Jamaica, 1750-1850*, 1-6 (quote, 5); and R. Keith Aufhauser, "Slavery and Scientific Management," *Journal of Economic History* 33 (1973): 811-24.

48. Andrew Jackson O'Shaughnessy, *An Empire Divided: The American Revolution and the British Caribbean* (Philadelphia: University of Pennsylvania Press, 2004).

49. Dubois, *Haiti*, 42, 99.

50. Vincent Brown, *The Reaper's Garden: Death and Power in the World of Atlantic Slavery* (Cambridge, Mass.: Harvard University Press, 2008); and Jeremy Popkin, *You Are All Free: The Haitian Revolution and the Abolition of Slavery* (New York: Cambridge University Press, 2010).

51. Saint-Domingue's impact on the French economy awaits a focused study, and historians debate why France lagged behind Britain in the path to industrialization. However, recent economic histories of France agree on the central importance of the colonial trade. Jeff Horn, *The Path Not Taken: French Industrialization in the Age of Revolution, 1750-1830* (Cambridge, Mass.: MIT Press, 2006), 76.

52. C. Knick Harley, "Slavery, the British Atlantic Economy and the Industrial Revolution," Oxford Discussion Papers in Economic and Social History 113, April 2013.

53. Perry Gauci, *William Beckford: First Prime Minister of the London Empire* (New Haven, Conn.: Yale University Press, 2013).

54. Catherine Hall et al., *Legacies of British Slave-Ownership: Colonial Slavery and the Formation of Victorian Britain* (Cambridge: Cambridge University Press, 2014), 1–33.

55. Kathleen Wilson, "Introduction: Histories, Empires, Modernities," in *A New Imperial History: Culture, Identity and Modernity in Britain and the Empire 1660–1840*, ed. Wilson (New York: Cambridge University Press, 2004), 1–28; Catherine Hall and Sonya O. Rose, eds., "Introduction," in *At Home with the Empire: Metropolitan Culture and the Imperial World*, ed. Hall and Rose (Cambridge: Cambridge University Press, 2007), 1–31.

56. Cécile Vidal, "The Reluctance of French Historians to Address Atlantic History," *Southern Quarterly* 43 (2006): 153–89. For the failure of Atlantic commerce after 1789, see Kevin O'Rourke, "The Worldwide Economic Impact of the French Revolutionary and Napoleonic Wars, 1793–1815," *Journal of Global History* 1 (2006): 123–49.

57. Albane Forestier, "A 'Considerable Credit' in the Late Eighteenth-Century French West Indian Trade: The Chaurands of Nantes," *French History* 25 (2011): 48–68.

58. Guillaume Daudin, *Commerce et prospérité: La France au XVIIIe siècle* (Paris: Presses de l'Université Paris-Sorbonne, 2005), 77–78.

59. François d'Ormesson and Jean-Pierre Thomas, *Jean-Joseph de Laborde: Banquier de Louis XV, mécène des lumières* (Paris: Perrin, 2002), 167–77.

60. Allan Potofsky, "Paris-on-the-Atlantic: From the Old Regime to the Revolution," *French History* 25 (2011): 89–107.

61. Christer Petley, "'Devoted Islands' and 'That Madman Wilberforce': Proslavery Patriotism During the Age of Abolition," *Journal of Imperial and Commonwealth History* 39 (2011): 393–415. For the Haitian Revolution in global context, see, inter alia, Laurent Dubois, *Avengers of the New World: The Story of the Haitian Revolution* (Cambridge, Mass.: Harvard University Press, 2005); Michel-Rolph Trouillot, *Silencing the Past: Power and the Production of History* (Boston: Beacon, 1997); and Ada Ferrer, *Freedom's Mirror: Cuba and Haiti in the Age of Revolution* (New York: Cambridge University Press, 2014).

62. For the rapid recovery of the West Indian economy after the American Revolution, see O'Shaughnessy, *Empire Divided*, 238–39; and Christer Petley, *Slaveholders in Jamaica: Colonial Society and Culture During the Era of Abolition* (London: Pickering and Chatto, 2009). For a more pessimistic view, see David Ryden, *West Indian Slavery and British Abolition, 1783–1807* (New York: Cambridge University Press, 2009). The West Indies as a whole was not in a state of economic decline at any time from the 1790s to the early 1820s. See Nick Draper, "The Rise of a New Planter Class? Some Counter-Currents from British Guiana and Trinidad, 1807–1834," *Atlantic Studies* 9 (2012): 65–83.

63. "La Méditerranée dans les circulations atlantiques," special issue of *Revue d'historie Maritime* 13 (2011); William Doyle, "Slavery and Serfdom," in *The Oxford Handbook of the Ancien Régime*, ed. Doyle (Oxford: Oxford University Press, 2012), 271. This

theme is expanded on in Paul Cheney, *Revolutionary Commerce: Globalization and the French Monarchy* (Cambridge, Mass.: Harvard University Press, 2010).

Chapter 2. The Plantation World

1. Vincent Brown, *The Reaper's Garden: Death and Power in the World of Atlantic Slavery* (Cambridge, Mass.: Harvard University Press, 2008).

2. [J. Hector St. John de Crèvecoeur], "Sketches of Jamaica and Bermudas and Other Subjects," in *More Letters from the American Farmer: An Edition of the Essays in English Left Unpublished by Crèvecoeur*, ed. Dennis D. Moore (Athens: University of Georgia Press, 1995), 106–13; Christopher Iannini, "'An Itinerant Man': Crèvecoeur's Caribbean, Raynal's Revolution, and the Fate of Atlantic Cosmopolitanism," *WMQ* 3rd ser., 61 (2004): 201–34.

3. "Sketches of Jamaica," 107–8.

4. Alexandre-Stanislas de Wimpffen, *Voyage à Saint-Domingue pendant les années 1788, 1789 et 1790* (Paris, 1797).

5. Desdorides, "Remarques sur la colonie de St Domingue, par Mr Desdorides, lieutenant-colonel du régiment provincial d'artillerie de Besançon, 1779, dédié au prince de Montbarey, ministre de la guerre," 1779, 13, Bibliothèque Mazarine, MS 3453 (we have been unable to find the first name of Desdorides).

6. Ibid., 16–17.

7. Thomas Craskell and James Simpson, *Maps of the Counties of Surrey, Cornwall and Middlesex, in the Island of Jamaica* (London, 1765); James Robertson, *Maps of the Counties of Surrey, Cornwall, and Middlesex, in the Island of Jamaica* (London, 1804).

8. See, for contrast, Charles Bochart and Humphrey Knollis, "A New and Exact Mapp of the Island of Jamaica," in *The Laws of Jamaica* (London, 1684), fig. 6, 173.

9. Edward Long, *History of Jamaica*, 3 vols. (London: T. Lowndes, 1774); Patrick Browne, *The Civil and Natural History of Jamaica* (London, 1756).

10. Michael Craton and James Walvin, *A Jamaican Plantation: The History of Worthy Park, 1670–1970* (London: W. H. Allen, 1970). See also Susan Dwyer Amussen, *Caribbean Exchanges: Slavery and the Transformation of English Society, 1640–1700* (Chapel Hill: University of North Carolina Press, 2007).

11. Adam Smith, *An Inquiry into the Nature and Causes of the Wealth of Nations* (London, 1776), ch. 7, pt. 3.

12. Sir Dalby Thomas, *An Historical Account of the Rise and Growth of the West Indies Collonies and the Great Advantages they are to England, in Respect of Trade* (London: J. Hindmarsh, 1690), 18. Thomas estimated that a hundred-acre plantation on the island of Barbados, with fifty enslaved Africans, seven white indentured servants, sugar mill, boiling works, equipment, and livestock would cost £5,625.

13. Jon Latimer, *Buccaneers of the Caribbean: How Piracy Forged an Empire* (Cambridge, Mass.: Harvard University Press, 2009).

14. David Buisseret, *Jamaica in 1687: The Taylor Manuscript at the National Library of Jamaica* (Kingston: University Press of the West Indies, 2006), 238–42, 247.

15. Thomas, *Historical Account of the Rise and Growth of the West India Collonies*, 14–15.

16. Both of these years are from Philip P. Boucher, *France and the American Tropics to 1700: Tropics of Discontent?* (Baltimore: Johns Hopkins University Press, 2007), 240.

17. Guillaume Delisle, "Carte de l'isle de Saint-Domingue. dressée en 1722 pour l'usage du roy, sur les mémoires de Mr. Frezier, ingénieur de S. M. et autres, assujetis aux observations astronomiques," *Atlas, map* (Paris: Lisle, Quai de l'Horloge, 1780), John Carter Brown Library, Rumsey Collection; Charles Frostin, "La piraterie américaine des années 1720, vue de Saint-Domingue (répression, environnement et recrutement)," *Cahiers d'histoire* 25 (1980): 178. See also James Pritchard, *In Search of Empire: The French in the Americas, 1670–1730* (New York: Cambridge University Press, 2004), 60–70; and Boucher, *France and the American Tropics to 1700*, 240–45.

18. Frostin, "La piraterie américaine des années 1720," 192.

19. Jean-Baptiste Labat, *Nouveau voyage aux isles de l'Amérique* (La Haye: Husson, 1724), 1:334, 2:363.

20. "Recensement général des dépendances des ressorts des conseils supérieurs du Petit Goave et du Cap pour l'année 1730," 1730, ANOM, G1/509, no. 20; "Recensement général de la colonie de St. Domingue pour l'année 1753, joint à la lettre de Mr de Vaudreuil et de Lalanne du 3 may 1754," 1754, ANOM, G1/509, no. 27; David P. Geggus, "Indigo and Slavery in Saint-Domingue," *Plantation Society in the Americas* 5 (1998): 189–204; John D. Garrigus, "Blue and Brown: Contraband Indigo and the Rise of a Free Colored Planter Class in French Saint-Domingue," *Americas* 50 (1993): 233–63.

21. "Saint-Domingue recensement général, 1730"; "Recensement général de l'île St. Domingue pour l'année 1739," 1739, ANOM, G1/509; the Jamaican slave numbers for both years come from Charles Frostin, *Les révoltes blanches à Saint-Domingue aux XVIIe et XVIIIe siècles (Haïti avant 1789)* (Paris: l'École, 1975), 30.

22. Gwynne Lewis, *France, 1715–1804: Power and the People* (New York: Pearson Longman, 2004), 108.

23. "Recensement général," 1753.

24. Jean-Laurent Rosenthal, *The Fruits of Revolution: Property Rights, Litigation and French Agriculture, 1700–1860* (New York: Cambridge University Press, 1992), 108–20.

25. James E. McClellan III, *Colonialism and Science: Saint-Domingue in the Old Regime* (Baltimore: Johns Hopkins University Press, 1992), 71–74.

26. "Recensement général," 1753; Frostin, *Les révoltes blanches*, 30.

27. Frostin, "La piraterie américaine des années 1720," 193; and Robin Blackburn, *The Making of New World Slavery: From the Baroque to the Modern, 1492–1800* (London: Verso, 1998), 431–35.

28. Lorena S. Walsh, *Motives of Honor, Pleasure, and Profit: Plantation Management in the Colonial Chesapeake, 1607–1763* (Chapel Hill: University of North Carolina Press,

2006); S. Max Edelson, *Plantation Enterprise in Colonial South Carolina* (Cambridge, Mass.: Harvard University Press, 2006).

29. J. R. Ward, "The Profitability of Sugar Planting in the British West Indies, 1650–1834," *Economic History Review* 31 (1978): 197–213.

30. B. W. Higman, "British West Indian Economic and Social Development," in *Cambridge Economic History: Colonial Period*, vol. 1, *The Colonial Period*, eds. Stanley L. Engerman and Robert E. Gallman (New York: Cambridge University Press, 1996), 326; Trevor Burnard, "'Prodigious Riches': The Wealth of Jamaica Before the American Revolution," *Economic History Review* 54 (2001): 506–24; David Eltis, Frank W. Lewis, and David Richardson, "Slave Prices, the African Slave Trade and Productivity in the Caribbean," *Economic History Review* 58 (2005): 673–700.

31. B. W. Higman, *Plantation Jamaica 1750–1850: Capital and Control in a Colonial Economy* (Kingston: University of the West Indies Press, 2005), 1–6; Burnard, "Prodigious Riches."

32. Richard Sheridan, *Sugar and Slavery: An Economic History of the British West Indies* (Bridgetown: Caribbean Universities Press, 1974), 216–17.

33. Lorena S. Walsh, *From Calabar to Carter's Grove: The History of a Virginia Slave Community* (Charlottesville: University Press of Virginia, 2001); Philip D. Morgan, *Slave Counterpoint: Black Culture in the Eighteenth Century Chesapeake and Lowcountry* (Chapel Hill: University of North Carolina Press, 1998).

34. Trevor Burnard, "'The Countrie Continues Sicklie': White Mortality in Jamaica, 1655–1780," *Social History of Medicine* 12 (1999): 45–72.

35. David Ryden argues that the recorded population figures underestimates slave numbers by one-seventh, making the actual number of slaves 86,946 in 1730 and 224,918 in 1774. Ryden, *West Indian Slavery and British Abolition, 1783–1807* (New York: Cambridge University Press, 2009), 201.

36. The inability of slave populations in Jamaica to grow naturally has provoked a large literature. See Kenneth Morgan, "Slave Women and Reproduction in Jamaica, c. 1776–1834," *History* 91 (2006): 231–53; B. W. Higman, *Montpelier, Jamaica*; and Katherine Paugh, "The Politics of Childbearing in the British Caribbean and the Atlantic World During the Age of Abolition, 1776–1838," *Past and Present* 221 (2013): 119–60.

37. Trevor Burnard, *Planters, Merchants, and Slaves: The Rise and Development of Plantation Societies in British America, 1650–1820* (Chicago: University Of Chicago Press, 2015); Brown, *The Reaper's Garden*; and Richard D. E. Burton, *Afro-Creole: Power, Opposition, and Play in the Caribbean* (Ithaca, N.Y.: Cornell University Press, 1997).

38. Buisseret, *Jamaica in 1687*, 276.

39. John Venn to Bishop Sherlock, 15 June 1751, Fulham Papers, 8:47. Lambeth Palace Library, London.

40. [Charles Leslie], *A New and Exact Account of Jamaica* ([Edinburgh, ca. 1740]), 41.

41. Henry Drax, "Instructions for the Management of Drax-Hall and the Irish-Hope

Plantation, to Archibald Johnson," in William Belgrove, *A Treatise upon Husbandry or Planting* (Boston: D. Fowle, 1755), 55.

42. Samuel Martin, *An Essay upon Plantership Humbly Inscribed to His Excellency George Thomas, Esq., Chief Governor of All the Leeward Islands, as a Monument to Antient Friendship*, 4th ed. (London: S. Clapham, 1765).

43. Long, *History of Jamaica*, 1:435.

44. Martin, *An Essay upon Plantership*, 9. Useful works include Richard S. Dunn, "Sugar Production and Slave Women in Jamaica," in *Cultivation and Culture: Labor and the Shaping of Slave Life in the Americas*, ed. Ira Berlin and Philip D. Morgan (Charlottesville: University of Virginia, 1993), 49–72; Dunn, "'Dreadful Idlers' in the Cane Fields: The Slave Labor Pattern on a Jamaica Sugar Estate," *Journal of Interdisciplinary History* 17 (1987): 795–822; J. R. Ward, *British West Indian Slavery, 1750–1834: The Process of Amelioration* (New York: Oxford University Press, 1988); Lucille Mathurin Mair, "Women Field Workers in Jamaica During Slavery," in *Slavery, Freedom and Gender: The Dynamics of Caribbean Society*, ed. Brian L. Moore (Kingston: University of the West Indies Press, 2001), 183–96; S. D. Smith, "An Introduction to the Plantation Journals of the Prospect Sugar Estate," in *Records of the Jamaican Prospect Estate* (microform) (Wakefield, Eng., 2003), 1–29; Philip D. Morgan, "Slaves and Livestock in Eighteenth Century Jamaica: Vineyard Pen, 1750–51," *WMQ* 3d ser. 52 (1995): 47–76; Brown, *Reaper's Garden*; Kenneth Morgan, "Slave Women and Reproduction in Jamaica"; and Justin Roberts, *Slavery and the Enlightenment in the British Atlantic, 1750–1807* (New York: Cambridge University Press, 2013). For British Guiana, see Trevor Burnard, *Hearing Slaves Speak* (Georgetown, Guyana: Caribbean Press, 2010); and John Lean, "The Secret Lives of Slaves: Berbice, 1819–1827" (Ph.D. diss., University of Canterbury, 2002).

45. Peter Thompson, "Henry Drax's Instructions on the Management of a Seventeenth-Century Barbadian Plantation," *WMQ*, 3d ser., 86 (2009): 565–604; Dunn, "Sugar Production and Slave Women in Jamaica"; David Collins, *Management and Medical Treatment of Negro Slaves in the Sugar Colonies* (London: J. Barfield, 1803), 176; and Heather Cateau, "The New 'Negro' Business: Hiring in the British West Indies, 1750–1810," in *In the Shadow of the Plantation: Caribbean History and Legacy*, ed. Alvin O. Thompson (Kingston: University of the West Indies Press, 2002), 100–120.

46. Morgan, "Slave Women and Reproduction."

47. Ibid.; Justin Roberts, "Working Between the Lines: Labor and Agriculture on Two Barbadian Sugar Plantations, 1796–1797," *WMQ*, 3d ser., 63 (2006): 551–86; William Dickson, *Letters on Slavery: To which are added, addresses to the whites, and to the free Negroes of Barbadoes: And accounts of some Negroes eminent for . . . their virtues and abilities* (London, 1789); Gilbert Mathison, *Notices Respecting Jamaica in 1808, 1809, 1810* (London: J. Stockdale, 1811), 34–35; Dunn, "Sugar Production and Slave Women in Jamaica."

48. Roberts, *Slavery and the Enlightenment*.

49. Pierre-François-Xavier de Charlevoix and Jean-Baptiste Le Pers, *Histoire de l'isle*

espagnole ou de S. Domingue: Écrite particulièrement sur des mémoires manuscrits du P. Jean-Baptiste Le Pers, jésuite 3 vols. (Amsterdam: François L'Honoré, 1733), 3:360–62.

50. Ibid., 3:364–65.

51. Labat, *Nouveau voyage*, 1:254.

52. Ibid.

53. "Recensement général de l'île St. Domingue pour l'année 1739," 1739, ANOM, G1/509; "Recensement général de la colonie de St. Domingue pour l'année 1771," ANOM, G1/509, no. 30; David P. Geggus, "Slave Society in the Sugar Plantation Zones of Saint Domingue and the Revolution of 1791–93," *Slavery and Abolition* 20 (1999): 35 (table 2); Jean Tarrade, *Le commerce colonial de la France à la fin de l'ancien régime: L'évolution du régime de l'exclusif de 1763 à 1789* (Paris: Presses universitaires de France, 1972), 2:739.

54. Louis de Beauvernet, "Plan de l'habitation de Févret de Saint-Mésmin à Saint-Domingue," n.d., Blérancourt; Musée national de la coopération franco-américaine, CFA C 187; Anthony Vidler, *Claude-Nicolas Ledoux: Architecture and Social Reform at the End of the Ancien Régime* (Cambridge, Mass.: MIT Press, 1990).

55. See Gabriel Debien and Pierre Pluchon, "L'Habitation Févret de Saint-Mesmin (1774–1790)," *Bulletin du Centre d'Histoire des Espaces Atlantiques* 3 (1987): 157–87. This study surprisingly does not mention this image.

56. Frank Weitenkampf, *Sketch of the Life of Charles Balthazar Julien Fevret de Saint-Mémin: Issued to Accompany an Exhibition of His Engraved Portraits at the Grolier Club, March 9–25, 1899* ([New York]: De Vinne, 1899), 3.

57. Gabriel Debien and Pierre Pluchon, "L'Habitation Févret de Saint-Mesmin (1774–1790)," *Bulletin du Centre d'Histoire des Espaces Atlantiques* 3 (1987): 158–59.

58. Ibid., 157.

59. Ibid., 168.

60. His personnel file makes it clear that Beauvernet was not a member of the military engineering corps but rather used court patronage to secure his position and was trained on the job. Louis de Beauvernet, "Beauvernet, Louis de, Prince, officier au corps royal de Génie, ingénieur du Roi à Saint-Domingue, et sa veuve (1774/1820)," n.d., ANOM, E 23, lettre B.

61. M. L. E. Moreau de Saint-Méry, *Description topographique, physique, civile, politique et historique de la partie française de l'isle Saint-Domingue* ([1797] Paris: Société de l'histoire des colonies françaises, 1958), 1137.

62. Debien and Pluchon, "L'Habitation Févret de Saint-Mesmin (1774–1790)," 161.

63. Desdorides, "Remarques sur la colonie de St Domingue," 7–8.

64. Richard S. Dunn, *Sugar and Slaves: The Rise of the Planter Class in the English West Indies, 1624–1713* (Chapel Hill: University of North Carolina Press, 1972).

65. V. S. Naipaul, *The Middle Passage: The Caribbean Revisited* (London: Macmillan, 1962), 27, 29.

66. Aimé Césaire, *Discours sur le colonialisme* (Paris: Présence Africaine, 1955); Edward [Kamau] Braithwaite, *The Development of Creole Society in Jamaica, 1770–1820* (Oxford: Oxford University Press, 1971). See Gordon K. Lewis, *Main Currents in*

Caribbean Thought: The Historical Evolution of Caribbean Society in its Ideological Aspects, 1492-2000 (Baltimore: Johns Hopkins University Press, 1983); Silvio Torres-Saillant, *An Intellectual History of the Caribbean* (London: Palgrave Macmillan, 2006); David Scott, *Conscripts of Modernity: The Tragedy of Colonial Enlightenment* (Durham, N.C.: Duke University Press, 2004); and Vincent Brown, "Social Death and Political Life in the Study of Slavery," *AHR* 114 (2009): 1231-49.

67. Haiti and horror have gone together from the early nineteenth century. See Leonora Sansay, *Secret History; or, The Horrors of Santo Domingo, in a Series of Letters* (Philadelphia, 1808). For the role that Haiti played in the development of philosophical thinking in the early nineteenth century, see Susan Buck-Morss, *Hegel, Haiti, and Universal History* (Pittsburgh: University of Pittsburgh Press, 2009). See also Brown, *Reaper's Garden*; and Trevor Burnard, "Jamaica as America, America as Jamaica: Hauntings in Vincent Brown's *The Reaper's Garden*," *Small Axe* 31 (2010): 200-211.

68. Brown, "Social Death," 1244.

Chapter 3. Urban Life

1. Philip D. Morgan, "The Caribbean Islands in Atlantic Context, circa 1500-1800," in *The Global Eighteenth Century*, ed. Felicity A. Nussbaum (Baltimore: Johns Hopkins University Press, 2003), 64.

2. Fernand Braudel, *The Structures of Everyday Life: The Limits of the Possible*, trans. Sian Reynolds (New York: Harper and Row, 1981), 479.

3. Jorge Cañizares-Esguerra, Matt D. Childs, and James Sidbury, eds., *The Black Urban Atlantic in the Age of the Slave Trade* (Philadelphia: University of Pennsylvania Press, 2013). For France, see Allan Potofsky, *Constructing Paris in the Age of Revolution* (London: Palgrave Macmillan, 2009). For Britain, see Peter Borsay, *The English Urban Renaissance: Culture and Society in the Provincial Town, 1660-1770* (Oxford: Oxford University Press, 1991). The pioneering work for the French Caribbean is Anne Pérotin-Dumon, *La ville aux îles, la ville dans l'île: Basse-Terre et Pointe-à-Pitre, Guadeloupe, 1650-1820* (Paris: Karthala, 2001); but see also Jeremy Popkin, *You Are All Free: The Haitian Revolution and the Abolition of Slavery* (New York: Cambridge University Press, 2010); and Franklin W. Knight and Peggy K. Liss, eds., *Atlantic Port Cities: Economy, Culture, and Society in the Atlantic World 1650-1850* (Knoxville: University of Tennessee Press, 1990).

4. René Phélipeau, "Plan de la ville du Cap François et de ses environs dans l'isle St. Domingue" (Paris: Phélipeau, 1786).

5. James E. McClellan, *Colonialism and Science: Saint Domingue in the Old Regime* (Baltimore: Johns Hopkins University Press, 1992), 94-97.

6. These numbers are from M. L. E. Moreau de Saint-Méry, *Description topographique, physique, civile, politique et historique de la partie française de l'isle Saint-Domingue* (Paris: Société de l'histoire des colonies françaises, 1958), 479. Errors in census numbers are discussed in David P. Geggus, "The Major Port Towns of Saint

Domingue in the Later Eighteenth Century," in *Atlantic Port Cities: Economy, Culture, and Society in the Atlantic World, 1650–1850*, ed. Franklin Knight and Peggy K. Liss (Knoxville: University of Tennessee Press, 1991), 102; see also Geggus, "The Slaves and Free People of Color of Cap Français," in *The Black Urban Atlantic in the Age of the Slave Trade*, ed. Cañizares-Esguerra, Childs, and Sidbury, 101–21.

7. Moreau de Saint-Méry, *Description*, 479, 1053, 1316.
8. Pérotin-Dumon, *La ville aux îles*, 343–44.
9. Ibid., 1037.
10. Moreau de Saint-Méry, *Description*, 982.
11. McClellan, *Colonialism and Science*, 95.
12. Edward Long, *History of Jamaica . . .* , 3 vols. (London: T. Lowndes), 2:281.
13. Emilien Petit, *Le patriotisme américain ou, Mémoires sur l'établissement de la partie française de l'îsle de Saint-Domingue, sous le vent de l'Amérique* (s.l.: s.n., 1750), 4.
14. McClellan, *Colonialism and Science*, 81.
15. Ibid., 188–89; and Jean Fouchard, *Le théâtre à Saint-Domingue* (Port-au-Prince: Impr. de l'État, 1955), 28–29, 53.
16. Philip P. Boucher, *France and the American Tropics to 1700: Tropics of Discontent?* (Baltimore: Johns Hopkins University Press, 2007), 254, 284.
17. Desdorides, "Remarques sur la colonie de St Domingue, par Mr Desdorides, lieutenant-colonel du régiment provincial d'artillerie de Besançon, 1779, dédié au prince de Montbarey, ministre de la guerre," 1779, 13, Bibliothèque Mazarine, MS 3453, 13–14.
18. Moreau de Saint-Méry, *Description*, 331–33, 370, the quote is from 331; Denné Dubuisson, undated letter ca. 1763, ANOM C9A, vols.118–19.
19. Ibid., 330.
20. Ibid., 319.
21. Ibid., 370–71.
22. Ibid., 981–82.
23. Ibid., 1053.
24. James Robertson, *Gone Is the Ancient Glory: Spanish Town, Jamaica, 1534–2000* (Kingston: Ian Randle, 2005).
25. Moreau de Saint-Méry, *Description*, 1030–32.
26. Boris Lesueur, "Les troupes coloniales aux Antilles sous l'Ancien Régime," *Histoire, économie & société* 28 (2009): 13.
27. Ibid., 6–8.
28. Charles Frostin, "Les 'enfants perdus de l'état' ou la condition militaire à Saint-Domingue au xviiie siècle," *Annales de Bretagne* 80 (1973): 327.
29. Jack P. Greene, "'Of Liberty and of the Colonies': A Case Study of Constitutional Conflict in the Mid-Eighteenth-Century British American Empire," in *Creating the British Atlantic: Essays on Transplantation, Adaptation, and Continuity*, ed. Greene (Charlottesville: University of Virginia Press, 2013), 140–207.
30. Michael Hay, *Plan of Kingston, 1745* (Kingston: s.n., 1745), Library of Congress, Map Division.

31. Long, *History of Jamaica*, 2:103.

32. Lord Adam Gordon, in *Travels in the American Colonies*, ed. Newton D. Mereness (New York: Macmillan, 1916), 377.

33. Long, *History of Jamaica*, 2: 103.

34. Wills, 10 June 1780, vol. 45, Island Record Office, Twickenham, Jamaica.

35. Moreau de Saint-Méry, *Description*, 479.

36. Ibid.

37. Ira Berlin has calculated that there were 12,444 blacks (both slave and free) in the port towns of Boston, Newport, Providence, New York, Philadelphia, and Charleston between 1770 and 1784. Ira Berlin, *Many Thousands Gone: The First Two Centuries of Slavery in North America* (Cambridge, Mass.: Harvard University Press, 1998), 374.

38. "Curtis Brett Journal," 11–12. We are grateful to Dr. Martin Brett of Cambridge University for allowing us access to this unpublished private journal.

39. Trevor Burnard and Kenneth Morgan, "The Dynamics of the Slave Market and Slave Purchasing Patterns in Jamaica, 1655–1788," *WMQ*, 3rd ser., 58 (2001): 205–28.

40. Henry Bright, Kingston, to Richard Meyler II, Bristol, 25 July 1750, in *The Bright-Meyler Papers: A Bristol-West India Connection, 1732–1837*, ed. Kenneth Morgan (Oxford: Oxford University Press, 2007), 226.

41. Kenneth Morgan, "Robert Dinwiddie's Reports on the British American Colonies," *WMQ*, 3rd ser., 65 (2008): 305–46.

42. Jack P. Greene, Charles F. Mullet, and Edward C. Papenfuse Jr., *Magna Charta for America: James Abercromby's "An Examination of the Acts of Parliament Relative to the Trade of our American Colonies" (1752)* (Philadelphia: American Philosophical Society, 1986), 14–15.

43. Gregory E. O'Malley, *Final Passages: The Intercolonial Slave Trade of British America, 1619–1807* (Chapel Hill: University of North Carolina Press, 2014), 309, 374–80.

44. Ronald Hoffman with Sally D. Mason, *Princes of Ireland, Planters of Maryland: A Carroll Saga, 1500–1782* (Chapel Hill: University of North Carolina Press, 2000).

45. TSTD.

46. David P. Geggus, "The Slaves of British-Occupied Saint Domingue: An Analysis of the Workforces of 197 Absentee Plantations, 1796–1797," *Caribbean Studies* 18 (1978): 19.

47. Moreau de Saint-Méry, *Description*, 479, 1053.

48. David Geggus, "The Slaves and Free People of Color of Cap Français." i, 101–22.

49. Ibid., 108–9; David Geggus, "On the Eve of the Haitian Revolution: Slave Runaways in Saint-Domingue in the Year 1790," *Slavery and Abolition* 6 (1985): 112–28; Jason Daniels, "Marronage in Saint-Domingue" (M.A. thesis, University of Florida, 2008), 78–79, 100.

50. [François Laplace], *Histoire des désastres de Saint-Domingue* (Paris, 1795), 112–13.

51. F3/126, fols. 408–9, Chambre d'Agriculture Report, 2 June 1785, ANOM.
52. Moreau de Saint-Méry, *Description*, 475, 544.
53. ANOM, F/3126, fols. 408–9.
54. Geggus, "Slaves and Free People of Color in Cap Français," 120–21.
55. Jerome S. Handler, *The Unappropriated People: Freedmen in the Slave Society of Barbados* (Baltimore: Johns Hopkins University Press, 1974), 16.
56. Léon Vignols, "Early French Colonial Policy: Land Appropriation in Haiti in the Seventeenth and Eighteenth Centuries," *Journal of Economic and Business History* 2 (1930): 135.
57. John D. Garrigus, *Before Haiti: Race and Citizenship in Saint-Domingue* (New York: Palgrave Macmillan, 2006), 64–65; Yvan Debbasch, *Couleur et liberté: Le jeu de critère ethnique dans un ordre juridique esclavagiste* (Paris: Dalloz, 1967), 83–85.
58. Moreau de Saint-Méry, *Description*, 479, 1053, 1316.
59. David Garrioch, *The Making of Revolutionary Paris* (Berkeley: University of California Press, 2002), 35.
60. John D. Garrigus, "Vincent Ogé Jeune (1757–91): Social Class and Free Colored Mobilization on the Eve of the Haitian Revolution," *Americas* 68 (2011): 45; Dominique Rogers, "Les libres de couleur dans les capitales de Saint-Domingue: Fortune, mentalités et intégration à la fin de l'Ancien Régime (1776–1789)" (Ph.D. diss., Université de Bordeaux III, 1999), 199.
61. Dominique Rogers and Stewart R. King, "Housekeepers, Merchants, Rentières: Free Women of Color in the Port Cities of Colonial Saint-Domingue, 1750–90," in *Women in Port: Gendering Communities, Economies, and Social Networks in Atlantic Port Cities, 1500–1800*, ed. Douglas Catterall and Jodi Campbell (Leiden: Brill, 2012), 375, 378–79.
62. Ibid., 370.
63. Moreau de Saint-Méry, *Description*, 426, 676, 873, 1906, 1341.
64. ANOM, G1/495, "Recensement des maisons de la ville de cap françois" (1776), 67.
65. These numbers do not include the soldiers and sailors in Moreau de Saint-Méry, *Description*, 479, 1053.
66. "Recensement des maisons de la ville de cap françois." See figures 2 and 3 in Zélie Navarro-Andraud, "La résidence urbaine des administrateurs coloniaux de Saint-Domingue dans la seconde moitié du XVIIIe siècle," *Articulo—revue de sciences humaines* [online], Hors-série 1 (27 May 2009): 11–12.
67. Richard Pares, "Merchants and Planters," supplement 4, *Economic History Review* (1960): 35–37.
68. Richard Mowery Andrews, *Law, Magistracy, and Crime in Old Regime Paris, 1735–1789* (Cambridge: Cambridge University Press, 1994), 1:552.
69. "Recensement des maisons de la ville de cap françois."
70. Moreau de Saint-Méry, *Description*, 321, 490, 493.
71. Maurice Begouën Demeaux, "Aspects de Saint-Domingue pendant la guerre de

Sept Ans, d'après les papiers Foäche," in *Mémorial d'une famille du Havre: Stanislas Foäche, 1737–1806; Négociant de Saint-Domingue* (Paris: Société française d'histoire d'outremer, 1982), 29.

72. Pierre-Louis de Saintard, *Essai sur les colonies françoises, ou Discours politiques sur la nature du gouvernement, de la population and du commerce de la colonie de S.D. . . .* (Paris: s.n., 1754), 135.

73. Pierre-Yves Beaurepaire, "Sociability," in *The Oxford Handbook of the Ancien Regime*, ed. William Doyle (Oxford: Oxford University Press, 2012), 383. Beaurepaire says there were between forty thousand and fifty thousand Freemasons in France in 1789, a figure that probably includes female lodges. Given a population of about twenty-six million in 1789, we calculate, overgenerously, that eligible adult men were one-quarter of that population, or 6.5 million.

74. McClellan, *Colonialism and Science*, 105–6. For a remarkable, though incomplete, list of French colonial Freemasons, see Elisabeth Escalle and Mariel Gouyon Guillaume, *Francs-maçons des loges françaises "aux Amériques," 1770–1850: Contribution à l'étude de la société créole* (Paris: s.n., 1993).

75. Kenneth Loiselle, "Living the Enlightenment in an Age of Revolution: Freemasonry in Bordeaux, 1788–1794," *French History* 23 (2009): 5–6; Éric Saunier, "L'espace caribéen: Un enjeu de pouvoir pour la franc-maçonnerie française," *Revista de Estudios Históricos de la Masonería* 1 (2009): 45.

76. Frederick-William Seal-Coon, *An Historical Account of Jamaican Freemasonry* ([Kingston: Golding Print. Service], 1976), 14.

77. Pierre-Yves Beaurepaire, "The Universal Republic of the Freemasons and the Culture of Mobility in the Enlightenment," *French Historical Studies* 29 (2006): 409–10. For Freemasonry in nineteenth-century Haiti, see Mimi Sheller, *Citizenship from Below: Erotic Agency and Caribbean Freedom* (Durham, N.C.: Duke University Press, 2012), 158–60.

78. Beaurepaire, "The Universal Republic," 425, 430.

79. Ibid., 417.

80. Seal-Coon, *Historical Account of Jamaican Freemasonry*, 14; James Robertson, "Eighteenth-Century Jamaica's Ambivalent Cosmopolitanism," *History* 99, no. 337 (October 2014), 626–28. On the 1780s, we are obliged to Dan Livesey for communicating his findings from the Jamaica Almanac. For the wider imperial context, see Jessica Harland-Jacobs, "'Hands Across the Sea': The Masonic Network, British Imperialism, and the North Atlantic World," *Geographical Review* 89 (1999): 237–53.

81. The facts of Morin's life are unclear and disputed. See the summary of the scholarship in Michel Taillefer, *La franc-maçonnerie toulousaine sous l'Ancien Régime et la Révolution, 1741–1799* (Paris: ENSB; CTHS, 1984), 77n39, 92–94. Taillefer notes that in the 1760s many French masonic lodges welcomed charismatic visitors who claimed they had secret masonic knowledge, which they were willing to share in exchange for money.

82. Seal-Coon, *Historical Account of Jamaican Freemasonry*, 23–26.

83. For the idea that Toussaint was a Freemason before the French Revolution, see Jacques de Cauna, "Autour de la thèse du complot: Franc-maçonnerie, révolution et contre-révolution à Saint-Domingue, 1789–1791," in *La Franc-maçonnerie au siècle des lumières: Europe-Amériques*, ed. Charles Porset and Cécile Révauger, *Lumières* 7 (Bordeaux 3, 2006), 289–310; Cauna's evidence is integrated into a larger portrait of Toussaint by Madison Smartt Bell, *Toussaint Louverture* (New York: Pantheon Books, 2008), 63, 77, 82; and Pierre-Yves Beaurepaire, "Fraternité universelle et pratiques discriminatoires dans la Franc-Maçonnerie des lumières," *Revue d'histoire moderne et contemporaine* 44 (1997): 195, 206.

84. Saunier, "L'espace caribéen," 42–55.

85. Éric Saunier, "Être confrère et franc-maçon à la fin du XVIIIe siècle," *Annales historiques de la Révolution française* 306 (1996): 617–34; Maurice Agulhon, *Pénitents et francs-maçons de l'ancienne Provence: Essai sur la sociabilité méridionale*, new ed. (Paris: Fayard, 1984).

86. For a brief overview of the scholarly literature on Martinés Pasqually, see Taillefer, *La franc-maçonnerie toulousaine*, 91.

87. Alfred Métraux, *Le vaudou haïtien* (Paris: Gallimard, 1958), 140.

88. Trevor Burnard, "A Failed Settler Society: Marriage and Demographic Failure in Early Jamaica," *Journal of Social History* 28 (1994): 63–82.

89. Moreau de Saint-Méry, *Description*, 41.

90. Daniel Livesey, "The Decline of Jamaica's Interracial Households and the Fall of the Planter Class, 1733–1823," *Atlantic Studies* 9 (2012): 107–23.

91. Garrigus, *Before Haiti*, 45–49.

92. Jacques Houdaille, "Trois paroisses de Saint-Domingue au XVIIIe siècle," *Population* 18 (1963): 100.

93. Stanhope to Board of Trade, 13 June 1763, CO 137/33/32. For racial mixing, see Brooke N. Newman, "Gender, Sexuality and the Formation of Racial Identity in the Eighteenth-Century Anglo-Caribbean World," *Gender and History* 22 (2010): 585–610; and Trevor Burnard, "'Rioting in Goatish Embraces': Marriage and Improvement in Early British Jamaica, 1660–1780," *History of the Family* 11 (2006): 185–97.

94. Martha Hodes, ed., *Sex, Love, Race: Crossing Boundaries in North American History* (New York: New York University Press, 1999); Trevor Burnard, *Mastery, Tyranny and Desire: Thomas Thistlewood and His Slaves in the Anglo-Jamaican World* (Chapel Hill: University of North Carolina Press, 2004), 228–40.

95. Trevor Burnard and Richard Follett, "Caribbean Slavery, British Antislavery and the Cultural Politics of Venereal Disease," *Historical Journal* 55 (2012): 427–52.

96. Sarah E. Yeh, "'A Sink of All Filthiness': Gender, Family, and Identity in the British Atlantic, 1688–1763," *Historian* 68 (2006): 66–83.

97. Pierre-François-Xavier Charlevoix, *Histoire de l'isle espagnole ou de S. Domingue*, 2 vols. (Paris: Didot, 1731), 2: 63; in 1750, Petit recommended requiring each colonial household to have a certain number of white female servants. Petit, *Patriotisme américain*, 109–10.

98. Kathleen Wilson, "Rethinking the Colonial State: Family, Gender, and Governmentality in Eighteenth-Century British Frontiers," *AHR* 116 (2011): 1317.

99. Kay Dian Kriz, *Slavery, Sugar, and the Culture of Refinement: Picturing the British West Indies, 1700–1840* (New Haven, Conn.: Yale University Press, 2008), 37–70. For Saint-Domingue, see Doris Garraway, *The Libertine Colony: Creolization in the Early French Caribbean* (Durham, N.C.: Duke University Press, 2005), 194–292.

100. Long, *History*, 2: 365–75 and Burnard, "Rioting in Goatish Embraces." For the wider context, see Colin Kidd, *The Forging of Races: Race and Scripture in the Protestant Atlantic World, 1600–2000* (Cambridge: Cambridge University Press, 2006), 79–120.

101. [Rev. Isaac Teale], "The Sable Venus—an Ode" (Jamaica, 1765), in *The History, Civil and Commercial, of the British West Indies*, 5 vols., ed. Bryan Edwards (London: Stockdale, 1794), 2:27–31; Long, *History of Jamaica*, 2: 327.

102. William Smith, *A New Voyage to Guinea* . . . (London, 1744), 146.

103. Garrigus, *Before Haiti*, 154–60.

104. Justin Girod-Chantrans, *Voyage d'un Suisse dans différentes colonies d'Amérique pendant la dernière guerre* (Neuchatel: Société typographique, 1785), 182.

105. Michel-René Hilliard d'Auberteuil, *Considérations sur l'état présent de la colonie française de Saint-Domingue: Ouvrage politique et législatif présenté au ministre de la marine*, vol. 2 (Paris: Grangé, 1776), 77.

106. David Buisseret, *Jamaica in 1687: The Taylor Manuscript at the National Library of Jamaica* (Kingston: University Press of the West Indies, 2006); [William Pittis], *The Jamaican Lady; or, The Life of Bavia* . . . (London, 1720), 35; William Beckford, *A descriptive account of the island of Jamaica* (London, 1788), 2:377.

107. Boucher, *France and the American Tropics*, 264.

108. Ibid., 211; see Yves Landry, *Les filles du roi au XVIIe siècle: Suivi d'un Répertoire bibliographique des filles du roi* (Montréal: Leméac, 1992).

109. Alexandre-Stanislas Wimpffen, *Voyage à Saint-Domingue, pendant les années 1788, 1789 et 1790* (Paris: Chez Cocheris, 1797), 2:108.

110. Letter from de Blénac and Mithon, 10 August 1713, cited in Pierre de Vaissière, *Saint-Domingue: La société et la vie créoles sous l'ancien régime (1629–1789)* (Paris: Perrin, 1909), 78.

111. Letter from Larnage and Maillart, 15 May 1742 and 22 April 1743, cited in Vaissière, *Saint-Domingue*, 78.

112. Petit, *Patriotisme américain*, 113–16.

113. Desdorides, "Remarques," 17.

114. Wimpffen, *Voyage à Saint-Domingue*, 2: 111–12; Justin Girod-Chantrans, *Voyage d'un suisse dans les colonies d'amérique*, ed. Pierre Pluchon (Paris: J. Tallandier, 1980), 242–43.

115. Moreau de Saint-Méry, *Description*, 19, 22.

116. Long, *History*, 2:278–81; Moreau de Saint-Méry, *Description*, 41–43.

117. Trevor Burnard, "'Gay and Agreeable Ladies': White Women in Mid-Eighteenth-Century Kingston, Jamaica," *Wadabagei* 9 (2006): 27–49.

118. Teresia Constantia Phillips, *An Apology for the Conduct of Mrs. T. C. Phillips*, 3 vols., 4th ed. (London: G. Smith, 1761). See also Lawrence Stone, *Uncertain Unions: Marriage in England 1660–1753* (Oxford: Oxford University Press, 1992), 236–74.

119. Joseph Roach, *Cities of the Dead: Circum-Atlantic Performance* (New York: Columbia University Press, 1996); Katherine Wilson, *Strolling Players of Empire: Theatre, Culture, and Modernity in the British Provinces, 1720–1820* (Cambridge: Cambridge University Press, forthcoming); and Wilson, "The Performance of Freedom: Maroons and the Colonial Order in Eighteenth-Century Jamaica and the Atlantic Sound," *WMQ*, 3rd ser., 66 (2009): 45–86.

120. Kathleen Wilson, *The Island Race: Englishness, Empire and Gender in the Eighteenth Century* (London: Routledge, 2003), 129–68.

121. Wilson, "Performance of Freedom"; Fernando Ortiz, *Cuban Counterpoint: Tobacco and Sugar* (Durham, N.C.: Duke University Press, 1995), 53, 97–103.

122. February 12, 1765, *JAJ*, 5:77–78

123. Phillips, *Apology*, 3:125–28.

124. Teresia Constantia Phillips, *A Letter Humbly Address'd to the Rt. Honourable the earl of Chesterfield* (London: Author, 1750), 20–21.

125. Wilson, *Island Race*, 143.

126. Ibid.

127. Paul Cheney, *A Colonial Cul-de-Sac: Plantation Life in Wartime Saint-Domingue, 1775–1782* (Chicago: University of Chicago Press, forthcoming).

128. Garraway, *Libertine Colony*, 210; Dian Kriz, *Slavery, Sugar and the Culture of Refinement: Picturing the British West Indies, 1700–1840* (New Haven, Conn.: Yale University Press, 2008).

129. Hilliard d'Auberteuil, *Considérations*, 1776, 2:43.

130. Rogers and King, "Housekeepers, Merchants, Rentières," 382–89; and Garrigus, *Before Haiti*, 57–66.

131. John D. Garrigus, "Redrawing the Color Line: Gender and the Social Construction of Race in Pre-Revolutionary Haiti," *Journal of Caribbean History* 30 (1996): 29-50; Rogers and King, "Housekeepers, Merchants, Rentières," 63.

132. Lauren R. Clay, *Stagestruck: The Business of Theater in Eighteenth-Century France and Its Colonies* (Ithaca N.Y.: Cornell University Press, 2013), 208.

133. Bernard Camier, "Les concerts dans les capitales de Saint-Domingue à la fin du XVIIIe siècle," *Revue de Musicologie* 93 (2007): 76.

134. Lauren Clay, "Identity and Race in France's Colonial Playhouses: Creating a 'Culture of Empire,'" ca. 2007, 17-19 (unpublished paper in possession of John Garrigus).

135. Camier, "Les concerts dans . . . Saint-Domingue," 88–89.

136. Moreau de Saint-Méry, *Description*, 367.

137. ANOM, Series T 210/3.

138. Moreau de Saint-Méry, *Description*, 362.

139. Camier, "Les concerts dans . . . Saint-Domingue," 88.

140. Girod-Chantrans, *Voyage d'un suisse dans les colonies d'amérique*, 152–54; Fouchard, *Le théâtre à Saint-Domingue*, 139; Emily Clark, *The Strange History of the American Quadroon: Free Women of Color in the Revolutionary Atlantic* (Chapel Hill: University of North Carolina Press, 2015).

141. Moreau de Saint-Méry, *Description*, 362.

142. Ibid., 97.

143. Ibid., 879.

144. By the early 1780s the girl's parents were both back in Port-au-Prince. In 1782 Minette's mother sold a house in that city for 12,000 livres or £520, a sum that placed her into the colony's free colored elite. That same year Minette's father married a white woman in Port-au-Prince. Bernard Camier, "Minette, situation sociale d'une artiste de couleur à Saint-Domingue," *Généalogie et histoire de la Caraïbe* 185 (2005): 4638–40.

145. Moreau de Saint-Méry, *Description*, 885, 989, 1054. Camier, points out that seventy out of 130 known instrumentalists in Saint-Domingue were enslaved. Camier, "Les concerts dans . . . Saint-Domingue," 80

146. Cited in Rogers and King, "Housekeepers, Merchants, Rentières," 382.

147. Clay, *Stagestruck*, 222.

148. Laurent Dubois and Bernard Camier, "Voltaire, Zaïre, Dessalines: Le théâtre des lumières dans l'atlantique français," *Revue d'histoire moderne et contemporaine* 54 (2007): 51–52.

149. Macandal, the alleged mastermind of a poisoning conspiracy, is discussed at length in Chapter 6.

150. Mirebalais was a remote mountain town with a large free colored population.

151. Dubois and Camier, "Voltaire, Zaïre, Dessalines," 51–52, 55; Marie-Christine Hazaël-Massieux, *Textes anciens en créole français de la Caraïbe: Histoire et analyse* (Paris: Editions Publibook, 2008), 127.

152. Clay, *Stagestruck*, 223–24.

Chapter 4. The Seven Years' War in the West Indies

1. Jonathan R. Dull, *The French Navy and the Seven Years' War* (Lincoln: University of Nebraska Press, 2005), 9–11.

2. P. J. Marshall, *The Making and Unmaking of Empires: Britain, India, and America c. 1750–1783* (Oxford: Oxford University Press, 2005), 137, 145–46.

3. Richard Harding, *Amphibious Warfare in the Eighteenth Century: The British Expedition to the West Indies 1740–1742* (Woodbridge, Suffolk: Boydell and Brewer, 1991); Richard Pares, *War and Trade in the West Indies, 1739–1763* (London: Cass, 1936), 181; Kathleen Wilson, "Empire, Trade and Popular Politics in Mid-Hanoverian Britain: The Case of Admiral Vernon," *Past and Present* 121 (1988): 74–109.

4. John McNeill, *Atlantic Empires of France and Spain: Louisbourg and Havana, 1700–1763* (Chapel Hill: University of North Carolina Press, 1985), 102–4.

5. Cited in Jeremy Black, *Natural and Necessary Enemies: Anglo-French Relations in the Eighteenth Century* (London: Duckworth 1986), 50.

6. Dull, *French Navy*, 4–5.

7. Brendan Simms, *Three Victories and a Defeat: The Rise and Fall of the First British Empire, 1714–1783* (London: Allen Lane, 2007), 350–52.

8. Dull, *French Navy*, 9–11.

9. *TSTDB*.

10. Pares, *War and Trade*, 311–24, 333; Dull, *French Navy*, 10.

11. Jamaica's exports fell from an average of £550,880 per annum between 1729 and 1738 to an average of £542,650 between 1739 and 1748; Pares, *War and Trade*, 472–73; Richard B. Sheridan, *Sugar and Slaves: An Economic History of the British West Indies, 1623–1775* (Bridgetown, Barb., University of the West Indies Press, 1974), 487–88.

12. Speech, 16 November 1749, in *Proceedings and Debates of the British Parliaments respecting North America*, ed. L. F. Stock (Washington, D.C.: Carnegie Institute, 1924–41), 5:369.

13. Trelawny to the Duke of Newcastle, 12 March 1748, CO 137/58.

14. Pares, *War and Trade*, 180–83; Charles Frostin, *Les révoltes blanches à Saint-Domingue aux XVIIe et XVIIIe siècles (Haïti avant 1789)* (Paris: l'École, 1975), 167–263.

15. Pares, *War and Trade*, 183; Charles Frostin, *Histoire de l'autonomisme colon de la partie française de St-Domingue aux XVIIe et XVIIIe siècles contribution à l'étude du sentiment américain d'indépendance* (Service de reproduction des thèses, Université de Lille III, 1973), 631–32.

16. Kenneth Banks, *Chasing Empire Across the Sea: Communications and the State in the French Atlantic 1713–1763* (Montreal: McGill-Queen's University Press, 2006), 37; Dull, *French Navy*, 16.

17. Fred Anderson, *Crucible of War: The Seven Years' War and the Fate of Empire in British North America, 1754–1766* (New York: Alfred A. Knopf, 2001), 11; Eric Hinderaker, *Elusive Empires: Constructing Colonialism in the Ohio Valley, 1673–1800* (New York: Oxford University Press, 1997), 135–44; and Patrice Louis-René Higgonet, "The Origins of the Seven Years' War," *Journal of Modern History* 40 (1968): 57–90. For useful overviews, see Daniel A. Baugh, "Withdrawing from Europe: Anglo-French Maritime Geopolitics, 1750–1800," *International History Review* 20 (1998): 11–16; and T. R. Clayton, "The Duke of Newcastle, the Earl of Halifax and the American Origins of the Seven Years' War," *Historical Journal* 24 (1981): 571–603.

18. Frostin, *Histoire de l'autonomisme colon*, 587; Dull, *French Navy*, 38.

19. McNeill, *Atlantic Empires*, 74–79; Gerald S. Graham, "Naval Defense of British North America," *Transactions of the Royal Historical Society*, 4th ser., 30 (1948): 95–110; B. R. Mitchell and Phyllis Deane, *Abstract of British Historical Statistics* (Cambridge: Cambridge University Press, 1971), 389–90; Richard Harding, *Sea Power and Naval Warfare, 1650–1830* (London: University College London Press 1999), 291.

20. Dull, *French Navy*, 11–12.

21. Cited in Bob Harris, *Politics and the Nation: Britain in the Mid-Eighteenth Century* (Oxford: Oxford University Press, 2002), 5.

22. Pares, *War and Trade*, 199–216.

23. Ibid., 289–316.

24. Maurice Begouën Demeaux, *Mémorial d'une famille du Havre: Stanislas Foäche, 1737–1806; Négociant de Saint-Domingue* (Paris: Société française d'histoire d'outremer, 1982), 17.

25. Jeremiah Meyler to Henry Bright, 9 May 1755; Robert Whatley to Henry Bright, 19 June 1755; Robert Stanton to Henry Bright, 24 June 1757; Richard Meyler II to Meyler and Hall, 18 January 1759, in *The Bright-Meyler Papers: A Bristol-West India Connection, 1732–1837*, ed. Kenneth Morgan (Oxford: Oxford University Press, 2007), 315, 317, 329, 345. For fear of invasion from Hispaniola, see George Metcalf, *Royal Government and Political Conflict in Jamaica, 1729–1783* (London: Longman, 1965), 140. For Bristol privateers, see David J. Starkey, *British Privateering Enterprise in the Eighteenth Century*, (Exeter: University of Exeter Press, 1990), 165, 181.

26. Naval Office Shipping Lists, 1763–64, C.O. 142/18.

27. Richard Middleton, *The Bells of Victory: The Pitt-Newcastle Ministry and the Conduct of the Seven Years' War 1757–1762* (Cambridge: Cambridge University Press, 1985); Anderson, *Crucible of War*, 212–13.

28. Dull, *French Navy*, 46, 99–103, 129–32; James C. Riley, *The Seven Years War and the Old Regime in France: The Economic and Financial Toll* (Princeton, N.J.: Princeton University Press, 1986), 80–81.

29. Dull, *French Navy*, 116.

30. Frank McLynn, *1759: The Year Britain Became Master of the World* (London: Jonathan Cape, 2004), 99–100.

31. Pares, *War and Trade*, 293–94.

32. McLynn, 101–3. One question is why Pitt did not focus his attentions on the greatest prize of all: Saint-Domingue. Pares explains it as follows: "Pitt feared offending Spain, then still neutral, by exciting her fears for Cuba and Santo Domingo. When Spain stopped her neutrality, then Saint-Domingue was under greater threat than in 1759. Of course, what happened was an assault in 1762 on Cuba rather than Saint-Domingue." Pares, *War and Trade*, 180n4.

33. J. R. McNeill, *Mosquito Empires: Ecology and War in the Greater Caribbean, 1620–1914* (New York: Cambridge University Press, 2010); Duncan Crewe, *Yellow Jack and the Worm: British Naval Administration in the West Indies, 1739–1748* (Liverpool: Liverpool University Press, 1993).

34. Dull, *French Navy*, 138–41; Frostin, *Histoire de l'autonomisme colon*, 633.

35. J. H. Parry, Philip Sherlock, and Anthony Maingot, *A Short History of the West Indies*, 4th. rev. ed. (Kingston: Macmillan Caribbean, 1987), 120; Pierre Pluchon, *Histoire de la colonisation française* ([Paris]: Fayard, 1991), 634.

36. TSTDB.

37. Ibid.; Banks, *Chasing Empire*, 42, puts the number at 27,500.

38. Banks, *Chasing Empire*, 42; Pluchon, *Histoire de la colonisation*, 635.
39. Dull, *French Navy*, 187, 198–99.
40. Pares, *War and Trade*, 195; Dull, *French Navy*, 209–11.
41. Cited in Pluchon, *Histoire de la colonisation*, 235.
42. Banks, *Chasing Empire*, 204–5.
43. Pluchon, *Histoire de la colonisation*, 235.
44. Frostin, *Histoire de l'autonomisme colon*, 588.
45. Pluchon, *Histoire de la colonisation*, 235–36; Dull, *French Navy*, 224–25; M. L. E. Moreau de Saint-Méry, *Description topographique, physique, civile, politique et historique de la partie française de l'isle Saint-Domingue* (Paris: Société de l'histoire des colonies françaises, 1958), 1455.
46. Moreau de Saint-Méry, *Description*, 266–67; Frostin, *Histoire de l'autonomisme colon*, 642–44.
47. 29 January; 1, 5 March, 1762, TT.
48. Dull, *French Navy*, 216, 226; Pluchon, *Histoire de la colonisation*, 236.
49. 1 December 1761; 29 January; 1, 5 March; 1, 2, 7 December 1762, TT.
50. McNeill, *Atlantic Empires*, 104.
51. Benjamin Franklin to Caleb Whiteford, 7 December 1762, British Library, Add. MSS, 36,593, fol. 53; Samuel Johnson, "Thoughts on the Late Transaction Respecting Falkland's Islands," in *The Yale Edition of the Works of Samuel Johnson*, ed. Donald Greene (New Haven, Conn.: Yale University Press, 1977), 10:374.
52. An earlier generation of scholars argued that Pitt had a deep understanding of the importance of the Americas in the global struggle of the Seven Years' War. See Kate Hotblack, *Chatham's Colonial Policy: A Study of the Fiscal and Economic Implications of the Colonial Policy of the Elder Pitt* (London: Routledge, 1917). Recent scholarship suggests that Pitt was motivated by domestic and European concerns rather than American prospects. Marie Peters, "The Myth of William Pitt, Earl of Chatham, Great Imperialist, Part I: Pitt and Imperial Expansion, 1738–1763," *Journal of Imperial and Commonwealth History* 21 (1993): 31–74; Richard Middleton, *Bells of Victory*, 7–9; Brendan Simms, *Three Victories and a Defeat*; Jeremy Black, "Exceptionalism, Structure and Contingency: Britain as a European State, 1688–1818," *Diplomacy and Statecraft* 8 (1997): 11–26; Black, *America or Europe? British Foreign Policy, 1739–63* (London: University College London Press, 1998). See also Hamish Scott, *British Foreign Policy in the Age of the American Revolution* (Oxford: Oxford University Press, 1990); and Stephen Conway, *War, State and Society in Mid-Eighteenth Century Britain and Ireland* (Oxford: Oxford University Press, 2006). For a counterview, see Daniel A. Baugh, "Great Britain's 'Blue-Water' Policy, 1689–1815," *International History Review* 10 (1988): 33–58.
53. Robert Louis Stein, *The French Sugar Business in the Eighteenth Century* (Baton Rouge: Louisiana State University Press, 1988), 30; Dull, *French Navy*, 232–34.
54. Pluchon, *Histoire de la colonisation*, 261–66; Dull, *French Navy*, 116.
55. Stein, *French Sugar Business in the Eighteenth Century*, 77.
56. Pares, *War and Trade*, 241.

57. Ibid., 184–85.

58. Pierre Pluchon, *Le premier empire colonial: des origins à la restauration* (Paris: Fayard, 1991), 234–35; Pierre H. Boulle. "Patterns of French Colonial Trade and the Seven Years' War," *Histoire Sociale-Social History* 7 (1974): 50–52.

59. Trevor Burnard, *Mastery, Tyranny, and Desire: Thomas Thistlewood and His Slaves in the Anglo-Jamaican World* (Chapel Hill: University of North Carolina Press, 2004), 93–94.

60. Stanley Ayling, *The Elder Pitt, Earl of Chatham* (London: Collins, 1976), 307–9. Pitt was less than honest when he gave this stirring speech. Pitt, when in power, had made up his mind very early on, in 1761, when negotiations between France and Britain began in earnest, that Canada rather than Guadeloupe had to be kept. It was only when Martinique was captured in 1762, raising the possibility that all the French West Indies except for Saint-Domingue might fall into British hands, that Pitt started to doubt his earlier decision. Pares, *War and Trade*, 224–26.

61. Anderson, *Crucible of War*, 505–6.

62. Cited in Zenab Esmat Rashed, *The Peace of Paris, 1763* (Liverpool: Liverpool University Press, 1951), 134.

63. Allan Christelow, "French Interest in the Spanish Empire During the Ministry of the Duc de Choiseul," *HAHR* 21 (1941): 530.

64. Stanley J. Stein and Barbara H. Stein, *Apogee of Empire: Spain and New Spain in the Age of Charles III, 1759–1789* (Baltimore: Johns Hopkins University Press, 2003), 11–12, 48–56; Christelow, "French Interest," 520; Christelow, "Contraband Trade Between Jamaica and the Spanish Main, and the Freeport Act of 1766," *HAHR* 22 (1942): 309–43; Hugh Thomas, *Cuba: The Pursuit of Freedom* (New York: Harper and Row, 1971); 50, 65; Simms, *Three Victories and a Defeat*, 481–82.

65. *Philosophical and Political History of the Settlements and Trade of the Europeans in the East and West Indies . . . by the Abbé Raynal*, trans. J. O. Justamond, 6 vols. (London, 1798; New York: Negro Universities Press, 1969), 5:485–86.

66. William Burke, "A Letter from a Gentleman in Guadeloupe to his Friend in London" (1760), cited in Frank W. Pitman, *The Development of the British West Indies 1700–1763* (New Haven, Conn.: Yale University Press, 1917), 346–47; and William Burke, *An Examination of the Commercial Principles of the Late Negotiations Between Great Britain and France in 1761* (London: R. and J. Dodsley, 1762), 29–30.

67. Of course, some of Pitt's outrage at British concessions to the French was disingenuous. Britain had to give something to the French given diplomatic realities of the time. To insist on winning every aspect of the peace negotiations, as Pitt did in 1763, was a luxury afforded to men out of office. Britain, as victor, made major gains from the peace negotiations: dominance in India, control of Senegal in India; Grenada, Dominica, St. Vincent, and Tobago in the West Indies; and the exclusion of the French from Canada and the Spanish from Florida on the North American mainland. North Americans were ecstatic. Benjamin Franklin spoke for many when he congratulated a London correspondent for "the glorious peace you have made, the most advantageous for all the

British nation, in my opinion, of any your annals have recorded." Cited in Gerald Stourzh, *Benjamin Franklin and American Foreign Policy* (Chicago: University of Chicago Press, 1954), 81.

68. *TSTDB*; Lucien-René Abénon, *La Guadeloupe de 1671 à 1759: Étude politique, économique et sociale*. 2 vols. (Paris: L'Harmattan, 1987), 2: 19–24.

69. *TSTDB*; Pares, *War and Trade*, 188–93; Richard Sheridan, *Sugar and Slavery: An Economic History of the British West Indies, 1623–1776* (University of the West Indies Press, 2000), 451; Anne Pérotin-Dumon, *La ville aux îles, la ville dans l'île: Basse-Terre et Pointe-à-Pitre, Guadeloupe, 1650–1820* (Paris: Karthala, 2001), 159–63.

70. John McNeill suggests that perhaps four thousand slaves went from Jamaica to Cuba, a considerable number probably arriving in the ships of John Kennion, a Liverpool merchant and owner of two plantations in Jamaica. In addition, merchants from Kingston and British North America sold the Spaniards great quantities of British manufactures. McNeill, *Atlantic Empires*, 166–70.

71. Elizabeth Donnan, *Documents Illustrative of the Slave Trade to America*, 4 vols. (Washington, D.C. : Library of Congress, 1930), 2:514–15.

72. Pluchon, *Histoire de la colonisation*, 63; "Recensement général de la colonie de St. Domingue pour l'année 1754, joint à la lettre de Mr de Vaudreuil et Delalanne du 10 juin 1755," 1755, ANOM, G1/509, no. 28; Extrait du récensement géneral de la population de St Domingue pour l'année 1788, ANOM, G1/509, no. 38; Trevor Burnard, "European Migration to Jamaica, 1655–1780," *WMQ*, 3d ser. 53 (1996): 772.

73. Emma Rothschild, "A Horrible Tragedy in the French Atlantic," *Past and Present* 192 (2006): 67–108 (quote on 72).

74. Ibid., 81–83.

75. Ibid., 81, 87. They concluded from a detailed analysis of the Kourou expedition that the disaster showed that "industrious and free people" could live only in the temperate North while the tropical lands of the South were "made, it seems, for despotism . . . a monstrous and weak alliance of a race of slaves with a nation of tyrants." *Philosophical and Political History* 5:19–23. For French Guyana and the Kourou expedition, see M.-L. Marchand-Thébault, "L'Esclavage en Guyane française sous l'Ancien Régime," *Revue française d'historie d'outre-mer* 47 (1960): 5–75; Jean Tarrade, "Alsaciens et Rhénans en Saintonge au XVIIIe siècle: Saint-Jean-d'Angély "entrepôt" des colons recrutés pour la Guyane sous la ministère de Choiseul," *Bulletin de la Société des Antiquaires de l'Ouest et des Musées de Poitiers*, 4th ser., 8 (1966): 163-82 and David Lowenthal, "Colonial Experiments in French Guiana, 1760–1800," *HAHR* 32 (1952): 22–43.

76. David Hancock, *Citizens of the World: London Merchants and the Integration of the British Atlantic Community, 1735–1785* (Cambridge: Cambridge University Press, 1995), 81–84, 119, 142, 146. 209, 216; S. D. Smith, *Slavery, Family and Gentry Capitalism in the British Atlantic World: The World of the Lascelles, 1648–1834* (Cambridge: Cambridge University Press, 2006).

77. D. H. Murdoch, "Land Policy in the Eighteenth-Century British Empire: The

Sale of Crown Lands in the Ceded Islands, 1763-1783," *Historical Journal* 27 (1984): 549-74.

78. Andrew Jackson O'Shaughnessy, *An Empire Divided: The American Revolution and the British Caribbean* (Philadelphia: University of Pennsylvania Press, 2000), 15, 88-89, 106-7.

79. Cited in ibid., 15, 17.

80. For George III, see Lowell Ragatz, *The Fall of the Planter Class in the British Caribbean, 1763-1833* (New York: Century Co.,1928), 50. For the West India lobby, see Andrew J. O'Shaughnessy, "The Formation of a Commercial Lobby: The West India Interest, British Colonial Policy, and the American Revolution," *Historical Journal* 40 (1997): 71-95 (quote 77); Lillian Penson, "The London West India Interest in the Eighteenth Century," *English Historical Review* 30 (1921): 373-92; David Ryden, *West Indian Slavery and British Abolition, 1783-1807* (New York: Cambridge University Press, 2009), 36-82. For the differing reaction to the Stamp Act in the West Indies, save St. Kitts and Nevis, see Andrew Jackson O'Shaughnessy, "The Stamp Act Crisis in the British Caribbean," *WMQ*, 3rd ser., 51 (1994): 203-26.

81. For contemporaneous reevaluations of empire and nationalism in France, see David A. Bell, "Jumonville's Death: War Propaganda and National Identity in Eighteenth-Century France," in *The Age of Cultural Revolutions: Britain and France, 1750-1820*, ed. Colin Jones and Dror Wahrman (Berkeley: University of California Press, 2002), 33-61.

82. Stephen Conway, *War, State and Society in Mid-Eighteenth Century Britain and Ireland* (Oxford: Oxford University Press, 2006), 237. The literature on changes in how Americans were viewed between the end of the Seven Years' War and on the entry of France into the American Revolution is large, but see Eliga H. Gould, *The Persistence of Empire: British Political Culture in the Age of the American Revolution* (Chapel Hill: University of North Carolina Press, 2000), 66-71, 106-47; H. V. Bowen, "British Conceptions of Global Empire, 1756-83," *Journal of Imperial and Commonwealth History* 26 (1998): 1-27; and Stephen Conway, "From Fellow-Nationals to Foreigners: British Perceptions of the Americans, Circa 1739-1783," *WMQ*, 3rd ser., 59 (2002): 65-100. For changes in how Britons conceived of nonwhite subjects of the Crown in this period, see C. A. Bayly, "The British and Indigenous Peoples, 1760-1860: Power, Perception and Identity," in *Empire and Others: British Encounters with Indigenous Peoples, 1600-1850*, ed. Martin Daunton and Rick Halpern (London: Routledge, 1999), 19-41. For changes in how British West Indians were viewed by Britons, see Trevor Burnard, "White West Indian Identity in the Eighteenth Century," in *Assumed Identities: Race and the National Imagination in the Atlantic World*, ed. John D. Garrigus and Christopher Morris (College Station: Texas A&M Press, 2010), 71-87. For earlier manifestations of British distaste for West Indian outlandishness, see Sarah E. Yeh, "'A Sink of All Filthiness': Gender, Family, and Identity in the British Atlantic, 1688-1763," *Historian* 68 (2006): 66-83.

83. *Philosophical and Political History*, 4:157-58.

84. For Britain, see Christopher L. Brown, "Empire Without Slaves: British

Concepts of Emancipation in the Age of the American Revolution," *WMQ*, 3rd ser., 56 (1999): 273–306.

85. [Anon.], *An Essay concerning Slavery and the Danger Jamaica is expos'd to from too great Number of Slaves . . .* (London: C. Corbett, 1746), 6, 38.

Chapter 5. Dangerous Internal Enemies

1. Michael Craton, *Testing the Chains: Resistance to Slavery in the British West Indies* (Ithaca, N.Y.: Cornell University Press, 1982), 336; Jean Fouchard, *Les marrons de la liberté*, ed. rev., corr. and augm. (Port-au-Prince: H. Deschamps, 1988), 365–85; for a more skeptical and nuanced view, see David P. Geggus, *Haitian Revolutionary Studies* (Bloomington: Indiana University Press, 2002), ch. 5.

2. Michèle Duchet, *Anthropologie et histoire au siècle des lumières: Buffon, Voltaire, Rousseau, Helvétius, Diderot*, 2nd ed. (Paris: Albin Michel, 1971), 142–43; and Madeleine Dobie, *Trading Places: Colonization and Slavery in Eighteenth-Century French Culture* (Ithaca, N.Y.: Cornell University Press, 2010), 155. Dobie cites Abbé Prévost's 1735 translation of the speech of a "purported Jamaican maroon, Moses Bom Saam"; the speech, which is probably apocryphal, appeared in *Gentleman's Quarterly* and other English publications in 1735. Thomas W. Krise, "The Speech of Moses Bon Sáam," in *Caribbeana: An Anthology of English Literature of the West Indies, 1657–1777* (Chicago: University of Chicago Press, 1999), 101–7.

3. [Charles Leslie], *A New and Exact Account of Jamaica* (Edinburgh, ca. 1739), 41.

4. Other historians are working on Tacky's Revolt. See Vincent Brown, *The Coromantee Wars: An Archipelago of Insurrection* (Cambridge, Mass.: Harvard University Press, forthcoming); and Maria Alessandra Bollettino, "Slavery, War, and Britain's Atlantic Empire: Black Soldiers, Sailors, and Rebels in the Seven Years' War" (Ph.D., University of Texas, Austin, 2009), 191–256.

5. John Hamilton to Rebecca Hamilton 12 June 1760, Ayrshire Record Office AA/DC/17/113.

6. This was an ironic reversal of what was to happen in the 1790s, when the real revolt was Saint-Domingue, while white Jamaicans experienced an unjustified scare about a possible slave conspiracy. David P. Geggus, "The Enigma of Jamaica in the 1790s: New Light on the Causes of Slave Rebellions," *WMQ*, 3rd ser., 44 (1987): 274–85.

7. Diana Paton, "Witchcraft, Poison, Law, and Atlantic Slavery," *WMQ*, 3rd ser., 69 (2012), 235–64. Poisoning trials in early nineteenth-century British Trinidad were almost completely instigated by planters who had moved to that island from Martinique and Saint-Domingue—see James Epstein, *Scandal of Colonial Rule, Power and Subversion in the British Atlantic During the Age of Revolution* (New York: Cambridge University Press, 2012), 245–258.

8. Pierre Pluchon, *Vaudou, sorciers, empoisonneurs: De Saint-Domingue à Haïti* (Paris: Karthala, 1987), 143–47.

9. M. L. E. Moreau de Saint-Méry, *Description topographique, physique, civile,*

politique et historique de la partie française de l'isle Saint-Domingue (Paris: Société de l'histoire des colonies françaises, 1958), 629

10. This was in 1801. Pierre Pluchon, *Toussaint Louverture: Un révolutionnaire noir d'ancien régime* (Paris: Fayard, 1989), 58.

11. For the assertion that he was born in Africa, see Moreau de Saint-Méry, *Description*, 629; Sylviane Diouf, *Servants of Allah: African Muslims Enslaved in the Americas* (New York: New York University Press, 1998), 152.

12. Léo Bittremieux, *Mayombsch idioticon* (Ghent: Erasmus, 1922), 352; David P. Geggus, "Haitian Voodoo in the Eighteenth Century: Language, Culture, Resistance," *Jahrbuch für Geschichte von Staat, Wirtschaft und Gesellschaft Lateinamerikas* 28 (1991): 33.

13. Moreau de Saint-Méry, *Description*, 629.

14. Carolyn E. Fick, *The Making of Haiti: The Saint Domingue Revolution from Below* (Knoxville: University of Tennessee Press, 1990), 71, says "it is probable that Médor had established formal links with Macandal and the documentary evidence certainly suggests this." The testimony shows that Médor was part of a community of poisoners, but I find no mention of Macandal or his associates.

15. Cited in Fick, *The Making of Haiti*, 68.

16. Sébastien Jacques Courtin, "Extrait [Inspection of Prison]," 9 September 1757, ANOM, C9A, vol. 102, microfilm 5345, reel 91.

17. Moreau de Saint-Méry, *Description*, 630–31. 20 January was the date of the legal death sentence.

18. [Anon.], *Rélation d'une conspiration tramée par les nègres dans l'Isle de S. Domingue, défense que fait le Jésuite confesseur, aux nègres qu'on suplicie, de révéler leur fauteurs & complices* (s.l.: s.n., 1758); M. de C. "Extrait du *Mercure de France*, 15 Septembre 1757, Makandal, histoire véritable," *Revue de la Société haïtienne d'histoire et de géographie* 20 (January 1949): 21.

19. "Recensement général de la colonie de St. Domingue pour l'année 1753, joint à la lettre de Mr de Vaudreuil et de Lalanne du 3 may 1754," 1754, ANOM, G1/509, no. 27.

20. Moreau de Saint-Méry, *Description*, 630.

21. The description of the plot is found in what appears to be a hastily added postscript to [Anon.], *Rélation d'une conspiration*, 8; the plot is described again in a 1779 memorandum on the need to reinforce the rural constabulary, reproduced in Pluchon, *Vaudou, sorciers*, 180.

22. Pluchon, *Vaudou, sorciers*, 168.

23. Delavaud's exact identity is not known. He was described in Médor's interrogation as one of the Lavaud brothers and so may have been Francois or Jean Lavaud, who owned several plantations in the North Province, or Bernard, a merchant in Cap Français in 1769; Moreau de Saint-Méry, *Description*, 627, 679, 681, 687, 1413.

24. The interrogation manuscript reads "faire face aux blancs."

25. John Thornton, "Cannibals, Witches, and Slave Traders in the Atlantic World," *WMQ*, 3rd ser., 60, no. 2 (April 2003): 278–80; Hein Vanhee, "Central African Popular Christianity and the Making of Haitian Vodou Religion," in *Central Africans and*

Cultural Transformations in the American Diaspora, ed. Linda M. Heywood (New York: Cambridge University Press, 2002), 253; writing of African healers accused of poisoning slaves in eighteenth-century Brazil, James Sweet notes that "the [West African] Fon word for powder (*atin*) translates as both 'medicine' and 'poison.'" Sweet, *Domingos Álvares, African Healing, and the Intellectual History of the Atlantic World* (Chapel Hill: University of North Carolina Press, 2011), 124; Sweet's subject, Álvarez, was also accused of conspiring to kill all the slaves on a plantation.

26. Paton, "Witchcraft, Poison, Law," 239.

27. Lynn Wood Mollenauer, *Strange Revelations: Magic, Poison, and Sacrilege in Louis XIV's France* (University Park: Pennsylvania State University Press, 2007).

28. "Arrêt du Conseil supérieur de la Martinique qui condamne Jeannot, nègre esclave, à être brûlé vif pour empoisonnement," May 1720, ANOM, Martinique C 251, fol. 1129; "Arrêt en règlement du Conseil souverain portant enregistrement de l'Ordonnance du Roi sur les vénéfices et poisons," 1724, ANOM, Martinique, C 253, fol. 215.

29. Pluchon, *Vaudou, sorciers*, 153; M. L. E. Moreau de Saint-Méry, *Loix et constitutions des colonies françoises de l'Amérique sous le Vent* (Paris: Quillau, 1784), 3:48–49.

30. Moreau de Saint-Méry, *Loix et constitutions*, 3:344 and 551; "Recensement général" ANOM G1509.

31. Moreau de Saint-Méry, *Loix et constitutions*, 3:492, 854–56.

32. Pluchon, *Vaudou, sorciers*, 267.

33. L'Huillier de Marigny, "Mémoire sur les poisons qui règnent à Saint-Domingue," 1762, ANOM, F3 88 210.

34. Sébastien Jacques Courtin, "Mémoire sommaire sur les prétendus pratiques magiques et empoisonnements prouvés au procès instruit et jugé au Cap contre plusieurs nègres et négresses dont le chef nommé François Macandal a été condamné au feu et exécuté le vingt janvier mille sept cents cinquante huit," 1758, ANOM, F3 88.

35. Sue Peabody, "'A Dangerous Zeal': Catholic Missions to Slaves in the French Antilles, 1635–1800," *French Historical Studies* 25, no. 1 (2002): 76n60.

36. Moreau de St.-Méry, *Description*, 33.

37. See the discussion of *minkisi* objects in the Congo, Haiti, and Cuba in Robert Farris Thompson, *Flash of the Spirit: African and Afro-American Art and Philosophy* (New York: Random House, 1983), 125–27; Geggus, "Haitian Voodoo," 32–33.

38. Hein Vanhee, "Central African Popular Christianity," 257; Terry Rey, "Kongolese Catholic Influences on Haitian Popular Catholicism: A Sociohistorical Exploration," in *Central Africans and Cultural Transformations in the American Diaspora*, ed. Linda Heywood (Cambridge: Cambridge University Press, 2002), 280–84.

39. The Jesuits Charlevoix and Le Pers wrote in 1733 that the Congo kings had been Christianized, but they noted that "one can barely find in some of them [Congos enslaved in Saint-Domingue] the slightest knowledge of our mysteries." Pierre-François-Xavier de Charlevoix and Jean-Baptiste Le Pers, *Histoire de l'isle espagnole ou de S. Domingue: Écrite particulièrement sur des mémoires manuscrits du P. Jean-Baptiste Le Pers, jésuite* 3 vols. (Amsterdam: François L'Honoré, 1733), 3:366; John Thornton and

Michelle Mauffette, "Les racines du Vaudou: Religion africaine et société haïtienne dans la Saint-Domingue prérévolutionnaire," *Anthropologie et Sociétés* 22, no. 1 (1998): 86–89; Cécile Fromont, *The Art of Conversion: Christian Visual Culture in the Kingdom of Kongo* (Chapel Hill, N.C.: Omohundro Institute of Early American History and Culture, 2014).

40. Courtin, "Mémoire sommaire."

41. Geggus, "Haitian Voodoo,"33n52.

42. *TSTDB*.

43. The documents do not identify Médor's nationality, but he seems to have spoken an African language, which makes it likely—though not completely certain—that he was born in Africa. His accomplice Venus testified that when the slave Gaou gave poison to Médor she told him to tell Médor "in their language that it was to finish off Mme Delavaud," who was their mistress.

44. Mollenauer, *Strange Revelations*, 130.

45. Courtin, "Mémoire sommaire."

46. L'Huillier de Marigny, "Mémoire sur les poisons qui règnent à Saint-Domingue," 1762, ANOM, F3 88 210.

47. Pierre-François-Xavier de Charlevoix and Jean-Baptiste Le Pers, *Histoire de l'isle espagnole ou de S. Domingue*, 3:364.

48. Adolphe Cabon, *Histoire d'Haïti*, 4 vols. (Port-au-Prince: Édition de La Petite revue, 1926–37), 1:166; Moreau de Saint-Méry, *Loix et constitutions*, 3:448.

49. Pierre H. Boulle, "Patterns of French Colonial Trade and the Seven Years' War," *Histoire Sociale/Social History* 7, no. 3 (1974): 53–54; Philippe Chassaigne, "L'économie des îles sucrières dans les conflits maritimes de la seconde moitié du XVIIIème siècle," *Histoire, économie et société* 7, no. 1 (1988): 94, 98.

50. Charles Frostin, *Histoire de l'autonomisme colon de la partie française de St-Domingue aux XVIIe et XVIIIe* (Université de Lille III, 1973), 592, cites ANOM C9A rec. 101.

51. Lambert, 15 July 1758, ANOM C9A, vol. 102, microfilm 5345, reel 91.

52. Ibid.

53. Thomas M. Truxes, *Defying Empire: Trading with the Enemy in Colonial New York* (New Haven, Conn.: Yale University Press, 2008), 56, 58–59.

54. Hilliard d'Aubertuil, *Considérations*, 1:167.

55. Truxes, *Defying Empire*, 66–69.

56. Bertie R. Mandelblatt, "'Beans from Rochel and Manioc from Prince's Island': West Africa, French Atlantic Commodity Circuits, and the Provisioning of the French Middle Passage," *History of European Ideas* 34 (2008): 422, describes the importance of wheat for French personnel stationed in West African slave factories, and the differential diets of officers and sailors.

57. McClellan, *Colonialism and Science*, 90; see Pares, *War and Trade in the West Indies*, 342, for unsuccessful attempts by French officials to get planters to eat local provision crops in wartime.

58. "Représentations que présente au Roi l'assemblée des deux conseils supérieurs à Saint-Domingue; 15 mars 1764"; ANOM, C9A, vol. 118–19, 120.

59. Jean-Barthélemy-Maximilen Nicolson, *Essai sur l'histoire naturelle de l'isle de Saint-Domingue avec des figures en taille-douce* (Paris: Gobreau, 1776), 54; McClellan, *Colonialism and Science*, 114.

60. Lambert, "Letter to naval minister," 21 November 1758, ANOM C9A, vol. 102, microfilm 5345, reel 91.

61. L'Huillier de Marigny, "Mémoire sur les poisons qui règnent à Saint-Domingue," 1762, ANOM F388, 210.

62. Steven L. Kaplan, *Provisioning Paris: Merchants and Millers in the Grain and Flour Trade During the Eighteenth Century* (Ithaca, N.Y.: Cornell University Press, 1984), 65.

63. Ibid., 394, 428.

64. Mary Matossian, *Poisons of the Past: Molds, Epidemics, and History* (New Haven:Yale University Press, 1991), 83.

65. See the doctor's report duplicated in Pluchon, *Vaudou, sorciers*, 264, and Courtin, "Mémoire sommaire." See the review of medical literature in Pieter S. Steyn, "Mycotoxins of Human Health Concern," in *Human Ochratoxicosis and Its Pathologies*, ed. Symposium international Ochratoxicose humaine et pathologies associées en Afrique et dans les pays en voie de développement et al. (Paris: Editions Inserm, 1993), 11–12.

66. Mary Matossian, "Ergot and the Salem Witchcraft Affair," *American Scientist* 70 (August 1982), 55–57; Matossian, *Poisons of the Past*; Linnda Caporael, "Ergotism: The Satan Loose in Salem," *Science* 192 (April 1976): 21; Nicholas Spanos and Jack Gottlieb, "Ergotism and the Salem Village Witch Trials," *Science* 194 (1976): 1390.

67. Alan Woolf, "Witchcraft or Mycotoxin? The Salem Witch Trials," *Clinical Toxicology* 38 (2000): 457–60.

68. Martin Howard, *Napoleon's Poisoned Chalice: The Emperor and His Doctors on St. Helena* (Stroud, Gloucestershire: History Press, 2009).

69. Steyn, "Mycotoxins of Human Health Concern," 12.

70. Ibid., 11–12.

71. Fournier de la Chapelle, 15 October 1758, ANOM C9A, vol. 102, microfilm 5345, reel 91.

72. Louis Auguste Aymar, François Borel de Neuilly, Jacques Hamelin, and Augustin Riches, "Extrait des pièces déposées en le procès criminel instruite au siège royale du Fort Dauphin contre les nommés Daouin et Venus," 26 May 1757, ANOM C9A, vol. 102, microfilm 5345, reel 91.

73. Delaye, "Déclaration du Sr Delaye au sujet d'un nègre à lui appartenant accusé d'empoisonnement" (Extrait des Registres du greffe du siège royale du Cap), 8 June 1757, ANOM C9A, vol. 102, microfilm 5345, reel 91.

74. Philibert Le Blondain, "Extrait des registres du Greffe de la Chambre criminelle du Fort Dauphin," 25 September 1757, ANOM C9A, vol. 102, microfilm 5345, reel 91.

75. "Interrogatoire de la négresse Marie Jeanne. Extrait des minutes déposés au

Greffe du siège royal du Cap," 26 October 1757, ANOM F388; "Interrogatoire de la nommée Nanon: Extrait des minutes déposés au Greffe du siège royal du Cap," 26 October 1758, ANOM F388; Delabrillon et al., "Interrogation du nègre Hauron. Extraits des minuttes déposés au siège royal du Cap," 1 November 1757, ANOM F388.

76. Regnier DuTillet, [Letter dated Port-de-Paix, 6 April 1758] ANOM, C9A, vol. 102, microfilm 5345, reel 91.

77. M. de C. "Extrait du *Mercure de France*, 15 Septembre 1757, Makandal, histoire véritable," *Revue de la Société haïtienne d'histoire et de géographie* 20 (January 1949): 24 ;"Interrogatoire . . . du nommé Joseph Pierre LaCoste navigateur," 6 December 1758, ANOM C9A, microfilm 5345, reel 91.

78. Moreau de Saint-Méry, *Loix et constitutions*, 4:451 and 463.

79. "Extrait d'un lettre de Port-au-Prince," *Les Affiches Américaines*, 12 February 1766, 57; Moreau de Saint-Méry, *Description*, 1067.

80. Mandelblatt, "Beans from Rochel," 418.

81. Pluchon, *Vaudou, sorciers*, 177, cites a 26 December 1759 letter of Saint-Domingue's governor.

82. Thomas M. Truxes, *Defying Empire: Trading with the Enemy in Colonial New York* (New Haven: Yale University Press, 2008), 74; Pares, *War and Trade*, 602; Brooke Hunter, "Wheat, War, and the American Economy During the Age of Revolution," *WMQ* 62, no. 3 (July 2005): 505–26.

83. M. de C., "Makandal, histoire véritable," *Revue de la Société haïtienne d'histoire et de géographie* 20 (January 1949), 21.

84. Pluchon, *Vaudou, sorciers*, 167.

85. Lambert. "Letter to naval minister de Massiac," 15 July 1758, ANOM C9A, vol. 102, microfilm 5345, reel 91.

86. L'Huiller de Marigny, "Mémoire sur les poisons qui règnent à Saint-Domingue."

87. Pluchon, *Histoire de la colonisation*, 631, 634.

88. Pluchon, *Vaudou, sorciers*, 196.

89. Moreau de Saint-Méry, *Loix et constitutions*, 4:222–23.

90. Ibid., 4:225–29.

91. Ibid., *Loix et constitutions*, 4:234.

92. Ibid., 4:352–56.

93. For a different view, see Kate Ramsey, "Legislating 'Civilization' in Postrevolutionary Haiti," in *Race, Nation, and Religion in the Americas*, ed. Henry Goldsmidt and Elizabeth McAlister (New York: Oxford University Press, 2004), 233.

94. Pluchon, *Vaudou, sorciers*, 198.

95. Moreau de Saint-Méry, *Description*, 631.

96. Sue Peabody, "'A Dangerous Zeal': Catholic Missions to Slaves in the French Antilles, 1635–1800," *French Historical Studies* 25 (2002): 79–81.

97. Pluchon, *Vaudou, sorciers*, 193.

98. David P. Geggus, "Saint-Domingue on the Eve of the Haitian Revolution," in *The*

World of the Haitian Revolution, ed. David P. Geggus and Norman Fiering (Bloomington: Indiana University Press, 2009), 3–20.

99. Edward Long, *History of Jamaica* . . . , 3 vols. (London, 1774; T. Lowndes), 2:447, 470; and Craton, *Testing the Chains*, 138. For the 1641 rebellion in Ireland, see M. Perceval-Maxwell, *The Outbreak of the Irish Rebellion of 1641* (Montreal: McGill-Queen's University Press, 1994).

100. The ability of the rebels to keep their plans secret is impressive, given divisions within slave communities about whether to support rebellion—rebellions in which if enslaved people did not lose their lives they would likely lose property. Not all the enslaved people on Forrest's estate, for example, followed Wager into battle. Nero, Congo Molly, and Beckford "made their escapes from the rebels, and alarmed sundry estate in the neighbourhood; by means whereof several white peoples' lives were saved, and the negroes of three estates prevented from rising." Cato, a slave of Forrest's, and Jemmy, a slave of Mr. Smith, an overseer killed at Forrest's estate, were granted their freedom for preserving the lives of white people. Reports to the Assembly, 5, 8, 20 December 1760, JJAJ. It is possible, also, that Maroons knew about the rebellion. Thistlewood noted after the rebellion had ended in Westmoreland that "it is said Col. Cudjoe went to Col. Barclay [the leading magistrate of the parish] & the gentlemen of this Parish a good while ago to warn them of this which has happened." 20 July 1760, TT.

101. 29 July, 18 October 1760, TT

102. 22, 25 May; 19 October 1760, TT.

103. 2 June; 30 September; 6 December 1760, TT.

104. Long, *History of Jamaica*, 2:452, 456; 26 May; 7 June; 24 October 1760, TT.

105. Edwards arrived in Jamaica in 1759, but he did not publish his poem on Alico's death, written in 1760, until just before the publication of his major history of the British West Indies. It coincided with visits made by Edwards to Saint-Domingue during the early days of the Haitian Revolution. Bryan Edwards, "Stanzas, Occasioned by the Death of Alico, an African Slave, Condemned for Rebellion in Jamaica, 1760," in Edwards, *Poems Written Chiefly in the West Indies* (Kingston, 1792), 38. Edwards was careful to explain to his readers that this poem arose from youthful ignorance of Africans' true nature. Thirty years on from Tacky's Revolt, he now understood, he declared, the necessity for exemplary executions such as that suffered by Alico, now seen less as a hero than as a unrepentant villain. The story of the slave burnt by fire is in Bryan Edwards, *The History, Civil and Commercial, of the British West Indies*, 5 vols. (London: Stockdale, 1794), 2:270–71.

106. The slave revolts in Jamaica were extensively reported in the British and North American press. See, inter alia, *British Magazine* 7 (1760): 443–44, 501, 504, 557; *Gentleman's Magazine* 30 (1760): 179–81, 294, 307–8, 392; 31 (1761): 321, 329, 377; *London Chronicle*, 21 June–1 July, 24–26 July, 16–26 August, 2–4 and 25–27 September 1760, 5–7 February 1761; *London Evening Post*, 24 June–1 July, 31 July–2 August, 26–28 August, 2–9 and 20–27 September 1760; *New York Mercury*, 16 June, 7, 21, and 28 July, 11 August 1760; *Pennsylvania Gazette*, 5 June, 17 and 24 July, 4 September 1760, 9 June 1763; *Boston*

Evening Post, 9 and 16 June, 7, 14, 21, and 28 July, 20 October 1760, 9 February, 31 and 20 April 1761; *New Hampshire Gazette*, 13 and 30 June, 1, 15 August 1760, 13 February, 17 April 1761.

107. In a confirmation of Thistlewood's doubts about the effectiveness of local troops, Moore had considerable trouble getting horse and foot militia to act against the rebels. He told the Board of Trade that "many of the Private Men had the Insolence to tell their Officers that they would not Appear under Arms, and should be ready to pay the fine of Ten Shillings which the Officers were Impowered to lay upon them by Virtue of the Militia Act for their non-appearance." Moore responded to such "Unnatural Behaviour in a time of Public Calamity" by declaring martial law. Henry Moore to Board of Trade, CO 137/32/3–6.

108. Moore to William Pitt, 21 May 1760, CO 137/60/300.

109. Long, *History of Jamaica*, 2:455; Moore to William Pitt, 21 May 1760, CO 137/60/300.

110. Long, *History of Jamaica*, 2:447, 455–58, 465–71; 7 June 1760, TT; Moore to Board of Trade, 24 July 1760, CO 137/32/23–24.

111. James Knight, "The Natural, Moral, and Political History of Jamaica and the Territories thereon depending," Long MSS, Add. MSS,12, 418, fol. 79–80, British Library, London.

112. Marjoleine Kars, "'Cleansing the Land': Dutch-Amerindian Cooperation in the Suppression of the 1763 Slave Rebellion in Dutch Guiana," in *Empires and Indigenes: Intercultural Alliance, Imperial Expansion, and Warfare in the Early Modern World*, ed. Wayne F. Lee (New York: New York University Press, 2011), 251–76.

113. For Akan influences, see Monica Schuler, "Ethnic Slave Rebellions in the Caribbean and the Guianas," *Journal of Social History* 3 (1970): 374–85. For African leadership, see Craton, *Testing the Chains*, 125–39; and Michael Mullin, *Africa in America: Slave Acculturation and Resistance in the American South and the British Caribbean, 1736–1831* (Urbana: University of Illinois Press, 1992), 40–43. For obeah, see Vincent Brown, *The Reaper's Garden: Death and Power in the World of Atlantic Slavery* (Cambridge, Mass.: Harvard University Press, 2008), 144–52. For Maroons, see Alvin O. Thompson, *Flight to Freedom: African Runaways and Maroons in the Americas* (Kingston: University of West Indies Press, 2006), 307–9.

114. Brown, *Reaper's Garden*, 148, 153.

115. Tacky is mentioned by neither Fred Anderson, *The Crucible of War: The Seven Years' War and the Fate of Empire in British North America, 1754–1766* (New York: Alfred A. Knopf, 2000), nor Brendan Simms, *Three Victories and a Defeat: The Rise and Fall of the First British Empire, 1714–1783* (London: Allen Lane, 2007).

116. Anderson, *Crucible of War*, 198, 246, 363, 760. It is briefly mentioned as an event in the Seven Years' War in Peter Linebaugh and Marcus Rediker, *The Many-Headed Hydra: The Hidden History of the Revolutionary Atlantic* (London: Verso, 2002), 221–24. We have worked on this interpretation of Tacky's Revolt simultaneously to Maria Alessandra Bollettino, who will deal with the revolt as part of movements in the Seven Years'

War to use black people more extensively as soldiers and sailors in a forthcoming manuscript called "Slavery, War and Britain's Atlantic Empire: Black Soldiers, Sailors and Rebels in the Seven Years' War." We are grateful to Dr. Bollettino for sharing some preliminary research with us and alerting us to some newspaper entries that we had missed.

117. Long, *History of Jamaica*, 2:452–53.

118. Admiral Holmes to John Cleveland, 25 July 1760, ADM 1/236, fols. 60–61, NA.

119. Thomas Cotes to John Cleveland, 19 April 1760, ADM 1/235, NA.

120. "Address to the King concerning the need for more troops, as evinced by the late rebellion," 15 December 1760, *JAJ*.

121. See also a report that "the prayer of the people of Jamaica, admit the confusion occasioned by the Negroes is, that the inhabitants of Martineco may not think of making an attempt on their island." *Boston Evening-Post*, 20 October, 1760; 21 July 1760, TT, 12 July 1760; Henry Moore to the Assembly of Jamaica, 18 September 1760, *JAJ*.

122. Lyttleton to the Board of Trade, 26 January 1762, CO 137/32/98–99; William L. Grant and James Munro, eds., *Acts of Privy Council of England: Colonial Series*, 6 vols. (London: Public Record Office, 1880–1912), 4:548. Jamaicans were happy to receive British troops on the island, thus making them willing to compromise their civil liberties to an extent unimaginable in North America. They were also willing to pay handsomely for the support of British troops. Andrew O'Shaughnessy estimates that they spent between £50,000 and £60,000 Jamaica currency on inland barracks between 1760 and 1763, reaching £100,000 by 1770. Such expenditure was sufficient for 2,572 troops. The annual cost of maintaining the army was £18,000 Jamaica currency in 1773. Nevertheless, what Jamaicans really wanted was not the army but a greater naval presence. Andrew Jackson O'Shaughnessy, *An Empire Divided: The American Revolution and the British Caribbean* (Philadelphia: University of Pennsylvania Press, 2000), 46–47, 49–50.

123. Vincent Brown, "Spiritual Terror and Sacred Authority in Jamaican Slave Society," *Slavery and Abolition* 24 (2003): 27–29; Brown, *Reaper's Garden*, 140–41. See also Kathleen Verdery, *The Political Lives of Dead Bodies; Re-Burial and Socialist Change* (New York: Cambridge University Press, 1999).

124. 4 June 1760, TT; Long, *History of Jamaica*, 2:458.

125. 17 July 1760, TT.

126. 17 July, 27 September, 6 October 1760, TT.

127. 20 December 1760, TT. The all-encompassing legislation passed by the Jamaican Assembly did the following: it banned irregular assemblies of slaves; it stopped slaves from having access to arms; it insisted that slaves could travel only with a pass; it stopped obeah; it restricted overseers from leaving their estates on certain specified occasions; it obliged all free coloreds to register their names in vestry books; and it stopped mariners from returning transported slaves to the island. It also appointed guards to monitor the Kingston slave market. 6, 16 December 1760; 14 October 1761, *JAJ*. For the less than enthusiastic response, see "Sir Mathew Lamb's Reports on Jamaican Laws," 4 November 1761, CO 137/32/78–81, and 8 November 1762, CO 33/17–19.

128. Jerome S. Handler and Kenneth M. Bilby, "On the Early Use and Origin of the

Term 'Obeah' in Barbados and the Anglophone Caribbean," *Slavery and Abolition* 22 (2002): 88. The first connection between obeah and slave rebellion came in the Antigua slave conspiracy of 1736. David Barry Gaspar, *Bondmen and Rebels: A Study of Master-Slave Relations in Antigua with Implications for Colonial British America* (Baltimore: Johns Hopkins University Press, 1985), 246–47. For the legal "creation" of obeah, see Diana Paton, "Obeah Acts: Producing and Policing the Boundaries of Religion in the Caribbean," *Small Axe* 28 (2009): 1–18. See also Diane M. Stewart, *Three Eyes for the Journey: African Dimensions of the Jamaican Religious Experience* (Oxford: Oxford University Press, 2005), 36–46. For the connections between obeah, magic, and medicine, see Juanita De Barros, "'Setting Things Right': Medicine and Magic in British Guiana, 1803–38," *Slavery and Abolition* 25 (2004): 28–50; De Barros, "Dispensers, Obeah, and Quackery: Medical Rivalries in Post-Slavery Guiana," *Social History of Medicine* 20 (2007): 243–61; and Randy Browne, "The 'Bad Business' of Obeah: Power, Authority, and the Politics of Slave Culture in the British Caribbean," *WMQ*, 3rd ser., 68 (2011): 451–80.

129. 25 April 1753, TT; R. R. Madden, *A Twelvemonth's Residence in the West Indies*, 2 vols. (Philadelphia: Carey, Lea and Blanchard, 1835), 2:69. Jerome S. Handler and Kenneth M. Bilby, *Enacting Power: The Criminalization of Obeah in the Anglophone Caribbean 1760–2011* (Kingston: University of the West Indies Press, 2012), 46–47.

130. Handler and Bilby, *Enacting Power*.

131. "An Act to Remedy the Evils Arising from Irregular Assemblies of Slaves, and to Prevent their Possessing Arms and Ammunition, and Going from Place to Place Without Tickets; and for Preventing the Practice of Obeah; and to Restrain Overseers from Leaving the Estates under their Care on Certain Days (act 24)" (1 January 1761), in *Acts of Assembly Passed in the Island of Jamaica: From the Year 1681 to the Year 1768* (Assembly of Jamaica: St. Jago de la Vega, 1771); Handler and Bilby, *Enacting Power*, 5, 46–49.

132. Brown, *Reaper's Garden*.

133. Long, *History of Jamaica*, 2:416–18, 451, 473; Diana Paton, "Punishment, Crime and the Bodies of Slaves in Eighteenth-Century Jamaica," *Journal of Social History* 34 (2001): 931, 938. See also 25 April 1753, 6 January 1754, 22 July 1768, TT.

134. Jerome S. Handler, "Slave Medicine and Obeah in Barbados, ca. 1650 to 1834," *New West Indian Guide* 74 (2000): 57–90; Handler and Bilby, "On the Early Use and Origin of the Term 'Obeah.'"

135. *HCSP*, 69:219.

136. Kate Ramsey, *The Spirits and the Law: Vodou and Power in Haiti* (Chicago: University of Chicago Press, 2011), 40–43.

Chapter 6. Racial Reconfigurations Before the American Revolution

1. Mark Harrison, *Climates and Constitutions: Health, Race, Environment and British Imperialism in India, 1600–1850* (Oxford: Oxford University Press, 2002); Colin Kidd, *The Forging of Races* (Cambridge: Cambridge University Press, 2006).

2. For terror in backcountry Pennsylvania contemporaneous to Tacky, see Peter

Silver, *Our Savage Neighbors: How Indian War Transformed Early America* (New York: W.W. Norton, 2008); Gregory Evans Dowd, *War Under Heaven: Pontiac, the Indian Nations, and the British Empire* (Baltimore: Johns Hopkins University Press, 2002); and Krista Camenzind, "Violence, Race, and the Paxton Boys," in *Friends and Enemies in Penn's Woods: Indians, Colonists, and the Racial Construction of Pennsylvania,* ed. William Pencak and Daniel K. Richter (University Park: Pennsylvania State University Press, 2004), 201–37.

3. Robert Martin Lesuire, *The Savages of Europe,* trans. James Pettit Andrews ([London: T. Davies, 1764]); David A. Bell, *The Cult of the Nation in France: Inventing Nationalism, 1680–1800* (Cambridge, Mass.: Harvard University Press, 2001), 84–106. It is noticeable that in his extensive reply to Lesuire's book, written at the height of the American Revolution, Edward Long, the defender of the Jamaican plantocracy, was able to defend the British from most charges of cruelty but was unable to defend Britons acting as "patrons and abettors of Wanton Homicide," through their association with "real" savages, such as Native Americans. Long, *English Humanity No Paradox* (London: T. Lowndes, 1778), 84.

4. *London Evening Post,* 31 July–2 August, 26–28 August, 2–4, 20–27 September 1760. See also *London Chronicle,* 24–26 July, 2–4, 20–23, and 25–27 September 1760; *Boston Evening Post,* 11 August 1760; *Pennsylvania Gazette,* 4 September 1760.

5. *London Chronicle,* 4–6 March 1762.

6. For ideas on how imperial growth could occur without slavery emanating from imperial reorientations after 1763, see Christopher Leslie Brown, "Empire Without Slaves: British Concepts of Emancipation in the Age of the American Revolution," *WMQ,* 3rd ser., 56 (1999): 273–306. A classic text is Edmund S. Morgan, "Slavery and Freedom: The American Paradox," *Journal of American History* 59 (1972): 5–29.

7. Johnson's quip is in [James Boswell], *Boswell's Life of Johnson,* ed. G. B. Hill, rev. L. F. Powell, 6 vols. (Oxford: Clarendon Press, 1934–64), 3:200. For the *Gentleman's Magazine,* Savage's poem, and the essay in the *Idler* 81 (3 November 1759), see G. Basker, "'The Next Insurrection': Johnson, Race, and Rebellion," *Age of Johnson* 11 (2000): 39, 45, 47.

8. [Sarah Scott], *The History of Sir George Ellison,* ed. Betty Rizzo (Lexington: University Press of Kentucky, 1996), 10; George Boulukos, *The Grateful Slave: The Emergence of Race in Eighteenth-Century British and American Culture* (New York: Cambridge University Press, 2008), 75–81, 116–39; Markman Ellis, *The Politics of Sensibility: Race, Gender and Commerce in the Sentimental Novel* (Cambridge: Cambridge University Press, 1999).

9. Vincent Brown, *The Reaper's Garden: Death and Power in the World of Atlantic Slavery* (Cambridge, Mass.: Harvard University Press, 2008), 154; Brad S. Gregory, *Salvation at Stake: Christian Martyrdom in Early Modern Europe* (Cambridge, Mass.: Harvard University Press, 1999); Thomas Day and John Bicknell, *The Dying Negro, a Poetical Epistle . . . ,* 3rd ed. (London: W. Flexney, 1775).

10. [Charles Leslie], *A New and Exact Account of Jamaica* ([Edinburgh, ca. 1739]), 41.

11. Trevor Burnard, *Mastery, Tyranny, and Desire: Thomas Thistlewood and His*

Slaves in the Anglo-Jamaican World (Chapel Hill: University of North Carolina Press, 2004), 104.

12. Ibid.

13. 26 May 1760; 6–8 January and 8 February 1761; 31 December 1765, TT.

14. Bryan Edwards, *The History, Civil and Commercial, of the British Colonies in the West Indies*, 2nd ed., 3 vols. (London: J. Stockdale, 1793), 2:7.

15. Ibid.

16. Ibid.; Patrick Browne, *The Civil and Natural History of Jamaica* (London: T. Osborne, 1756), 22; Jack P. Greene, *Imperatives, Behaviors, and Identities: Essays in Early American Cultural History* (Charlottesville: University of Virginia Press, 1992), 2–12.

17. William Lyttleton to the Board of Trade, 13 and 24 October 1762, CO 137/32/207–11, 212–15.

18. Trevor Burnard, "Harvest Years? Reconfigurations of Empire in Jamaica," *Journal of Imperial and Commonwealth History* 40 (2012): 537–38.

19. Jack P. Greene, "The Jamaica Privilege Controversy, 1764–1766: An Episode in the Process of Constitutional Definition in the Early Modern British Empire," *Journal of Imperial and Commonwealth History* 22 (1994): 16–54.

20. Nicholas Bourke, *The Privileges of the Island of Jamaica Vindicated* (Kingston, 1765).

21. 8, 13–15 December 1766 *JAJ*.

22. Edwards, *History, Civil and Commercial*, 2:428

23. Ibid., 53.

24. Burnard, "Harvest Years?" 549.

25. Malick W. Ghachem, *The Old Regime and the Haitian Revolution* (New York: Cambridge University Press, 2012), 130–38.

26. Jean-Baptiste Labat, *Nouveau voyage aux isles de l'Amérique*, vol. 2 ([The Hague]: P. Husson, 1724), 132, 133. See also Andrew S. Curran, *The Anatomy of Blackness: Science and Slavery in an Age of Enlightenment* (Baltimore: Johns Hopkins University Press, 2011), 61; and Madeleine Dobie, *Trading Places: Colonization and Slavery in Eighteenth-Century French Culture* (Ithaca, N.Y.: Cornell University Press, 2010), 139.

27. Dobie, *Trading Places*, 136–38.

28. Aphra Behn, *Oronoko, Traduit de l'Anglois de Madame Behn* (Amsterdam: s.n., 1745) ; Hans Sloane, *Histoire de la Jamaïque* (London: Nourse, 1751) ; Charles Leslie, *Histoire de la Jamaïque, traduite de l'anglois. Par M.***, ancien officier de dragons. . . .* 2 vols. (London: Nourse, 1751).

29. Ghachem, *Old Regime*, 64, 122; on Montesquieu's work with the translation of Sloane, see Jean Ehrard, *Lumières et esclavage: L'esclavage colonial et l'opinion publique en France au XVIIIe siècle* (Paris: André Versaille, 2008), 141.

30. Duchet, *Anthropologie et histoire*, 138–43.

31. Jean Baptiste Targe, *Histoire d'Angleterre, depuis le Traité d'Aix-la-Chapelle en 1748, jusqu'au Traité de Paris en 1763: Pour servir de continuation aux histoires de mm. Smollett et Hume*, 4 vols. (Paris: Desaint, 1768), 4:216.

32. Jean-François de Saint-Lambert, "Ziméo," in *Les Saisons, poème* (Amsterdam, 1769), 226–59.

33. Ghachem, *Old Regime*, 128, 131, 140.

34. Emma Rothschild, "A Horrible Tragedy in the French Atlantic," *Past and Present* 192 (2006): 67–108; J. R. McNeill, *Mosquito Empires: Ecology and War in the Greater Caribbean, 1620–1914* (New York: Cambridge University Press, 2010).

35. Trevor Burnard, "A Failed Settler Society," *Journal of Social History* 28 (1994): 63–82.

36. Edwards, *History, Civil and Commercial*, 2:169–70.

37. For a counter view, see Daniel Livesey, "The Decline of Jamaica's Interracial Households and the Fall of the Planter Class," *Atlantic Studies* 9 (2012): 107–23.

38. Nicholas Bourke, *The Privileges of the Island of Jamaica Vindicated* (Kingston, 1765), 44–47.

39. Charles Knowles to Board of Trade, 3 December 1754, CO 137/28/43; 13 November 1758 *JAJ*.

40. Trevor Burnard, "'Rioting in Goatish Embraces: Marriage and Improvement in Early British Jamaica, 1660–1780," *History of the Family* 11 (2006): 185–97.

41. [Anon.], *An Essay Concerning Slavery and the Danger Jamaica is expos'd to from too great Number of Slaves* . . . (London, 1746), 43.

42. Ibid., 42–53.

43. Ibid.

44. Ibid.

45. Ibid.

46. William Beckford, *A Descriptive Account of the Island of Jamaica* (London: T. and J. Egerton, 1790), 2:348.

47. Cited in Samuel J. Hurwitz and Edith F. Hurwitz, "A Token of Freedom: Private Bill Legislation for Free Negroes in Eighteenth-Century Jamaica," *WMQ*, 3rd ser., 24 (1967): 429.

48. "Act to prevent the Inconveniences arising from Exorbitant Grants," 23 October, 17 December 1761, 30 September, 16 November 1762, *JAJ*; Lovell Stanhope to the Board of Trade, 13 June 1762, CO 137/33/34–43; Hurwitz and Hurwitz, "Token of Freedom," 429–30.

49. Stanhope to Board of Trade, 13 June 1762, CO 137/33/34–43.

50. 16 December 1768, *JAJ*; Livesey, "Decline of Interracial Households."

51. 15 September 1777, 25 March 1779, Minute Book of the Trustees of Wolmer's Free School, MS 97a, fol.119, National Library of Jamaica, Kingston.

52. Curran, *Anatomy of Blackness*, 118–28, 134; Ehrard, *Lumières et esclavage*, 113.

53. Andrew Wells, "Race Fixing: Improvement and Race in Eighteenth-Century Britain," *History of European Ideas* 36 (2010): 134–38; Long, *History of Jamaica*, 356.

54. Peter Fryer, *Staying Power: The History of Black People in Britain* (London: Pluto, 1984), 161; Roxann Wheeler, *The Complexion of Race: Categories of Difference in Eighteenth-Century British Culture* (Philadelphia: University of Pennsylvania Press,

2000), 141–43, 209–33; Boulukos, *Grateful Slave*, 113–14; Dror Wahrman, *The Making of the Modern Self: Identity and Culture in Eighteenth Century England* (New Haven, Conn.: Yale University Press, 2004), 113–14, 116. See also James Delbourgo, "The Newtonian Slave Body: Racial Enlightenment in the Atlantic World," *Atlantic Studies* 9 (2012): 185–207.

55. Edwards, *History, Civil and Commercial*, 2:310.

56. James Ramsay, *An Essay on the Treatment and Conversion of Slaves in the British Sugar Colonies* (London: James Phillips, 1784), 288–89.

57. Burnard, *Mastery, Tyranny, and Desire*, 310; Winthrop D. Jordan, *White over Black: American Attitudes Toward the Negro, 1550–1812* (Chapel Hill: University of North Carolina Press, 1968), 177; Hurwitz and Hurwitz, "Token of Freedom"; Kingston Burial Register, 1722–75, JA. Daniel Livesay follows Mary Augier's daughter Elizabeth, who had five children with the wealthy white merchant John Morse. All five children established themselves in England, where their identities were largely that of wealthy white West Indians. However in a complex and drawn-out inheritance dispute, a white cousin evoked the Jamaican inheritance restriction in an unsuccessful bid to keep them from inheriting their father's estate. See Daniel Livesay, "Children of Uncertain Fortune: Mixed-Race Migration from the West Indies to Britain, 1750–1820" (Ph.D. diss., University of Michigan, 2010), 178–98.

58. Trevor Burnard, *Mastery, Tyranny, and Desire*, ch. 3; See also Trevor Burnard, "'Gay and Agreeable Ladies': White Women in Mid-Eighteenth-Century Kingston, Jamaica," *Wadabagei* 9 (2006): 27–49.

59. Garrigus, *Before Haiti*, chs. 4 and 5.

60. Frostin, *Les révoltes blanches*, 293; Moreau de Saint-Méry, *Description*, 1325.

61. M. L. E. Moreau de Saint-Méry, *Loix et constitutions des colonies françoises de l'Amérique sous le Vent*, vol. 2 (Paris: Quillau, 1784), 96, 344–47, 553–60, 754–59.

62. John D. Garrigus, "Le Patriotisme Américain: Emilien Petit and the Dilemma of French-Caribbean Identity Before and After the Seven Years' War," in *Proceedings of the Western Society for French History: Selected Papers of the 2002 Annual Meeting* (Lansing: Michigan State University Press, 2004), 18–29.

63. Claude Desmé-Dubussion, "Etat des services militaires," 1764, 300–01, ANOM, E 126, Moreau de Saint-Méry, *Description*, 1472.

64. Claude Desmé-Dubussion, "Lettre au ministre," 1752, 297, ANOM, E 126.

65. Jean-Pierre Desmé-Dubussion, "Mémoire pour le Sieur Desmé Dubuisson," 1764, 6, ANOM, E 126; and Yves de Tarade, "Claude IV Desmé Branche Desmé Dubuisson, Généalogie Yves de Tarade," *Généalogie Yves de Tarade*, 2000, http://detarade.chez.com/Genealogie_SuzanneDesme2.html.

66. Begouën Demeaux, "Aspects de Saint-Domingue pendant la guerre de Sept Ans, d'après les papiers Foäche," 28; Gabriel Debien, "Gouverneurs, magistrats et colons: L'opposition parlementaire et coloniale a Saint-Domingue (1763-1769)," *Revue de la société haïtienne d'histoire, de géographie, et de géologie*, October 1958, 3–4.

67. Moreau de Saint-Méry, *Loix et constitutions*, 4:281, 362, 380.

68. Ibid., 4:439–44, 464.
69. Ibid., 4:418, 514; Garrigus, *Before Haiti*, 115–16.
70. See the comparison of Bory's racial ideas with those of Hilliard d'Auberteuil in William Max Nelson, "Making Men: Enlightenment Ideas of Racial Engineering," *AHR* 115 (2010): 1376–77.
71. Moreau de Saint-Méry, *Description*, 180.
72. [Hardy], "Histoire des derniers mouvements qui ont agité la colonie francise etablie dans l'isle Saint-Domingue," trans. Gabriel Debien (1770s), 5, Collection Chatillon 61J 15, piece 7, Archives départementales de la Gironde.
73. Moreau de Saint-Méry, *Loix et constitutions*, 4:539.
74. Ibid., 4:552.
75. Ibid., 4:594–600.
76. Ibid., 4:566, 616.
77. Lewis, *France, 1715–1804: Power and the People*, 187–94.
78. Moreau de Saint-Méry, *Loix et constitutions*, 4:383–84.
79. Ibid., 4:584.
80. Ibid., 4:617–18.
81. Debien, "Gouverneurs, magistrats et colons," 5–7.
82. Moreau de Saint-Méry, *Loix et constitutions*, 4:648–50.
83. Michel Vergé-Franceschi, *Les officiers généraux de la marine royale, 1715–1774: Origines, conditions, services*, 4 vols. (Paris: Librairie de l'Inde, 1990), 4:1897–920.
84. Moreau de Saint-Méry, *Loix et constitutions*, 4:273; Vergé-Franceschi, *Les officiers généraux*, 4:1910, 1913.
85. Charles d'Estaing, "Discours de Mr d'Estaing au Conseil supérieur le 7 may 1764," ANOM, C9A, microfilm 5345, reel 104.
86. Moreau de Saint-Méry, *Loix et constitutions*, 4:731–32.
87. Ibid., 4:839.
88. Ibid., 4:644–705.
89. Ibid., 4:839–41.
90. "Nouvelles de France, Politique, mars 1765," in *Mercure historique et politique*, vol. 148 (A la Haye: chez Frederic Staatman, 1765), 552; Rousseau, "Nouvelles politiques, mars 1765," in *Journal encyclopédique* . . . , [ed. Pierre Rousseau] (De l'Imprimerie du Journal, 1765), 167.
91. Desmé-Dubussion, "Mémoire pour le Sieur Desmé Dubuisson," 15.
92. Ibid., 12–15; Moreau de Saint-Méry, *Description*, 291, 293, 383.
93. As Sebastian Conrad argues, the Enlightenment's global impact was not energized solely by the ideas of Parisian philosophers but was the work of historical actors around the world who invoked Enlightenment principles for their own specific purposes. Conrad, "Enlightenment in Global History: A Historiographical Critique," *AHR* 117 (2012): 1001.

Chapter 7. The Golden Age of the Plantocracy

1. Bernard Foubert, "Les habitations Laborde à Saint-Domingue dans la seconde moitié du XVIIIe siècle: Contribution à l'histoire d'Haïti (Plaine des Cayes)" (Ph.D. diss., Université de Paris IV, Sorbonne, 1990).

2. M. L. E. Moreau de Saint-Méry, *Description topographique, physique, civile, politique et historique de la partie française de l'isle Saint-Domingue* (Paris: Société de l'histoire des colonies françaises, 1958), 697–98, 1158, 1281, 1325–26; Léon Vignols, "Early French Colonial Policy: Land Appropriation in Haiti in the Seventeenth and Eighteenth Centuries," *Journal of Economic and Business History* 2 (1930), 135; The Praslin-Choiseul family received a very large indemnity payment of 139,971 francs for its colonial properties after the Haitian Revolution. Jean-François Brière, *Haïti et La France, 1804–1848: Le Rêve Brisé* (Paris: Karthala, 2008), 142.

3. Michel Vergé-Franceschi, *Les officiers généraux de la marine royale, 1715–1774: origines, conditions, services,* 4 vols. (Paris: Librairie de l'Inde, 1990), 4:2365.

4. Jean Tarrade, *Le commerce colonial de la France à la fin de l'ancien régime: L'évolution du régime de l'exclusif de 1763 à 1789,* vol. 1 (Paris: Presses universitaires de France, 1972), 198.

5. Ibid., 1:193.

6. Ibid., 1:196.

7. Ibid., 1:199, 202; Michèle Duchet, *Anthropologie et histoire au siècle des lumières: Buffon, Voltaire, Rousseau, Helvétius, Diderot,* 2nd ed. (Paris: Albin Michel, 1971), 132.

8. Ibid.

9. Gwynne Lewis, *France, 1715–1804: Power and the People* (New York: Pearson Longman, 2004), 185.

10. Pierre Pluchon, *Histoire de la colonisation française* ([Paris]: Fayard, 1991), 577. For Freeport legislation in the British Empire, see Allan Christelow, "Contraband Trade Between Jamaica and the Spanish Main, and the Freeport Act of 1766," *HAHR* 22 (1942): 309–43.

11. Robert Louis Stein, *The French Sugar Business in the Eighteenth Century* (Baton Rouge: Louisiana State University Press, 1988), 72–73.

12. "Recensement général, 1754"; "Recensement général de la colonie de St. Domingue pour l'année 1771," ANOM, G1/509, no. 30; Michel-Rolph Trouillot, "Motion in the System: Coffee, Color, and Slavery in Eighteenth-Century Saint-Domingue," *Review* 5 (1982): 337, 344.

13. Tarrade, *Le commerce colonial,* 2:739.

14. George Metcalf, *Royal Government and Political Conflict in Jamaica, 1729–1783* (London: Longman, 1965), 167.

15. Edward Long, *History of Jamaica . . . ,* 3 vols. (London: T. Lowndes, 1774), 2:263.

16. Ibid., 1:4, 42.

17. For a different view, see Sean X. Goudie, *Creole America: The West Indies and the*

Formation of Literature and Culture in the New Republic (Philadelphia: University of Pennsylvania Press, 2006), 67–70.

18. Long, *History of Jamaica*, 1:5–6.

19. Michael Craton, "Reluctant Creoles: The Planters' World in the British West Indies," in *Strangers Within the Realm: Cultural Margins of the First British Empire*, ed. Bernard Bailyn and Philip D. Morgan (Chapel Hill: University of North Carolina Press, 1991), 333–35; Long, *History of Jamaica*, 2:76–77.

20. Trevor Burnard, *Mastery, Tyranny, and Desire: Thomas Thistlewood and His Slaves in the Anglo-Jamaican World* (Chapel Hill: University of North Carolina Press, 2004), 59–63; Douglas Hall, "Botanical and Horticultural Enterprise in Eighteenth-Century Jamaica," in *West Indies Accounts: Essays on the History of the British Caribbean and the Atlantic Economy in Honour of Richard Sheridan*, ed. Roderick A. McDonald (Kingston: University of the West Indies Press, 1996), 101–25.

21. Julie Flavell, *When London Was Capital of America* (New Haven, Conn.: Yale University Press, 2010), 21–23, 249–50; Flavell, "Decadents Abroad; Reconstructing the Typical Colonial American in London in the Late Colonial Period," in *Old World, New World: America and Europe in the Age of Jefferson*, ed. Leonard J. Sadosky et al. (Charlottesville: University of Virginia Press, 2010), 32–60; Andrew Jackson O'Shaughnessy, "The West India Interest and the Crisis of American Independence," in McDonald, *West Indies Accounts*, 126.

22. Trevor Burnard, "'Passengers Only': The Extent and Significance of Absenteeism in Eighteenth-Century Jamaica," *Atlantic Studies* 1 (2004), 178–95.

23. Ibid.

24. For Kingston, see Emma Hart and Trevor Burnard, "Kingston, Jamaica and Charleston, South Carolina: A New Look at Comparative Urbanization in Plantation Colonial British America," *Journal of Urban History* 39 (2013): 214–34; and Burnard, "'The Grand Mart of the Island': Kingston, Jamaica in the Mid-Eighteenth Century and the Question of Urbanisation in Plantation Societies," in *A History of Jamaica, from Indigenous Settlement to the Present*, ed. Kathleen Monteith and Glen Richards (Kingston: University of the West Indies Press, 2002), 225–41.

25. For increasing land prices, see David Ryden and Ahmed Reid, "Sugar, Land Markets and the Williams Thesis: Evidence from Jamaica's Property Sales, 1750–1810," *Slavery and Abolition* 34 (2013): 401–24.

26. Burnard, *Mastery, Tyranny, and Desire*, 56.

27. Letter from Mumbee Goulbourn, 5 January 1790, Goulburn Papers, 304/J/1/1 (1), Surrey Record Office, Woking.

28. J. R. Ward, "The Profitability of Sugar Planting in the British West Indies, 1650–1834," *Economic History Review* 31 (1978): 197–213; Richard Sheridan, *Sugar and Slavery: An Economic History of the British West Indies, 1624–1775* (Bridgetown: Caribbean Universities Press, 1974); David Eltis, Frank W. Lewis, and David Richardson, "Slave Prices, the African Slave Trade and Productivity in the Caribbean," *Economic History Review* 58

(2005): 673–700; Noel Deerr, *The History of Sugar*, 2 vols. (London: Chapman and Hall, 1949), 1:193–204; David Beck Ryden, *West Indian Slavery and British Abolition, 1783–1807* (New York: Cambridge University Press, 2009), 217–27.

29. Accounts of the Spring estate, 1772–75, Ashton Court Collection, Woolnough Papers, AC/WO/16/89, 91, 93, 95, 97, Bristol Record Office, Bristol; Trevor Burnard, "From Periphery to Periphery: The Pennants' Jamaican Plantations, 1771–1812 and Industrialization in North Wales," in *Wales and Empire, 1607–1820*, ed. H. V. Bowen (Manchester: Manchester University Press, 2011), 124–26.

30. Some Americans also shared Taylor's opinion that Jamaica might be able to survive the cessation of American trade in a war. John Adams wrote in his diary on 17 August 1774 that although "none could pretend to foresee the effect of a Non Exportation to the West Indies, Jamaica was said to be the most independent part of the World. They had their own Plantane for bread. They had vast forests and could make their own Heading, Staves and Hoops. They could raise their own Provisions." Entry, 17 August 1774, *Diary of John Adams*, ed. L. H. Butterfield (Cambridge, Mass.: Harvard University Press, 1961), 2:101.

31. Lewis, *France, 1715–1804*, 192–94.

32. Tarrade, *Le commerce colonial*, 1972, 1:81n60.

33. [Hardy], "Histoire des derniers mouvements qui ont agité la colonie francaise établie dans l'isle StDomingue," ed. Gabriel Debien, 1770, p. 9, Archives départementales de la Gironde, Collection Chatillon 61J 15, piece 7.

34. Vergé-Franceschi, *Les officiers généraux*, 4:2357–67; on the land grants, see Vignols, "Early French Colonial Policy," 137.

35. [Hardy], "Histoire des derniers mouvements," 10–12; M. L. E. Moreau de Saint-Méry, *Loix et constitutions des colonies françoises de l'amérique sous le vent*, vol. 5 (Paris: Chez l'auteur, 1784), 226.

36. Frostin, *Les révoltes blanches*, 345.

37. [Charles-Bon Martin] des Paillières, "Représentations que présente au Roi l'assemblée des deux conseils supérieurs à Saint-Domingue,15 mars 1764," ANOM, C9A, microfilm 5345, reel 104.

38. Sécretariat d'Etat à la Marine, "Léger, substitut du procureur général au Conseil supérieur de Port-au-Prince à Saint-Domingue," 1764, ANOM, E 273; Moreau de Saint-Méry, *Loix et constitutions*, 1784, 4:874.

39. Moreau de Saint-Méry, *Loix et constitutions*, 5:200.

40. Frostin, *Les révoltes blanches*, 345, cites Robert, Count d'Argout, ANOM C9A, rec. 126, 26 January, 14 February 1765 letters to the Naval Minister; also rec. 134, 6 November 1768 letter to Bongars.

41. Moreau de Saint-Méry, *Loix et constitutions*, 5:214.

42. Ibid., 5:337.

43. Ibid., 5:214.

44. Ibid., 5:217.

45. [Hardy], "Histoire des derniers mouvements," 13.

46. Moreau de Saint-Méry, *Loix et constitutions*, 5:226.
47. Frostin, *Les révoltes blanches*, 346.
48. Moreau de Saint-Méry, *Loix et constitutions*, 5:337–40.
49. Richard Archer, *As If an Enemy's Country: The British Occupation of Boston and the Origins of Revolution* (New York: Oxford University Press, 2010); Neil Longley York, "Rival Truths, Political Accommodation, and the Boston 'Massacre,'" *Massachusetts Historical Review* 11 (2009): 57–95; and Benjamin L. Carp, *Defiance of the Patriots: The Boston Tea Party and the Making of America* (New Haven, Conn.: Yale University Press, 2010).
50. Moreau de Saint-Méry, *Description*, 912, 1299.
51. Differing interpretations can be found in T. R. Clayton, "Sophistry, Security, and Socio-Political Structures in the American Revolution; or, Why Jamaica Did Not Rebel," *Historical Journal* 29 (1986): 319–44; Andrew Jackson O'Shaughnessy, *An Empire Divided: The American Revolution and the British Caribbean* (Philadelphia: University of Pennsylvania Press, 2000), ch. 6; and Metcalf, *Royal Government and Political Conflict in Jamaica*, 186–87.
52. Woody Holton, *Forced Founders: Indians, Debtors, Slaves, and the Making of the American Revolution in Virginia* (Chapel Hill: University of North Carolina Press, 1999), 95–105; Allan Kulikoff, *Tobacco and Slaves: The Development of Southern Cultures in the Chesapeake, 1680–1800* (Chapel Hill: University of North Carolina Press, 1986), 118–61; T. H. Breen, *Tobacco Culture: The Mentality of the Great Tidewater Planters on the Eve of Revolution* (Princeton, N.J.: Princeton University Press, 1985), chs. 4 and 5; Jacob M. Price, *Capital and Credit in British Overseas Trade: The View from the Chesapeake, 1700–1776* (Cambridge, Mass.: Harvard University Press, 1980), 405–17; Bruce A. Ragsdale, *A Planter's Republic: The Search for Economic Independence in Revolutionary Virginia* (Madison: University of Wisconsin Press, 1996). For the 1772 credit crisis, see Richard B. Sheridan, "The British Credit Crisis of 1772 and the American Colonies," *Journal of Economic History* 20 (1960): 161–86.
53. Jack P. Greene, "Liberty, Slavery and the Transformation of British Identity in the Eighteenth-Century West Indies," *Slavery and Abolition* 21 (2000): 1–31; and Greene, *The Constitutional Origins of the American Revolution* (New York: Cambridge, 2011), 135, 170. See also Trevor Burnard, "'A Matron in Rank, a Prostitute in Manners . . .': The Manning Divorce of 1741 and Class, Race, Gender, and the Law in Eighteenth-Century Jamaica," in *Working Out Slavery, Pricing Freedom: Perspectives from the Caribbean, Africa and the African Diaspora*, ed. Verene Shepherd (London: St. Martin's, 2002), 133–52.
54. Nicholas Bourke, *The Privileges of the Island of Jamaica Vindicated* (Kingston: A. Aikman, 1765), 44, 48, 57, 66.
55. O'Shaughnessy, *Empire Divided*, 142.
56. Burnard, *Mastery, Tyranny, and Desire*, 94.
57. O'Shaughnessy, *Empire Divided*, 105, 128.
58. Thomas Iredell to James Iredell, [n.d. 1770]; Thomas Iredell to James Iredell, 8

January 1775, St. Dorothy, Jamaica, in *The Papers of James Iredell*, vol. 1, *1767–1777*, ed. Don Higginbotham (Raleigh: North Carolina Division of Archives and History, 1976), 54, 280.

59. "Report of Committee on the State of the Island," 18 December 1773, JAJ.

60. O'Shaughnessy, *Empire Divided*, 46.

61. Ibid., 44.

62. Ibid., 141–42.

63. Jonathan Dull, *The French Navy and American Independence: A Study of Arms and Diplomacy, 1774–1787* (Princeton, N.J.: Princeton University Press, 1975), 13; Bernard Bailyn, *To Begin the World Anew: The Genius and Ambiguities of the American Founders* (New York: Viking, 2003), 96.

64. Governor Basil Keith, "Answers to queries relative to the island of Jamaica," CO 137/70/88–90.

65. Vergé-Franceschi, *Les officiers généraux*, 4:2365–70.

66. Anne Pérotin-Dumon, *La ville aux îles, la ville dans l'île: Basse-Terre et Pointe-à-Pitre, Guadeloupe, 1650–1820* (Paris: Karthala, 2001), 345.

67. "Précis de ce qui s'est passé lors de l'installation du Gouverneur-Général au Conseil," *Affiches américaines*, 21 February 1770, 89.

68. Jean-André Tournerie, "Un projet d'école royale des colonies en Touraine au XVIIIe siècle," *Annales de Bretagne et des pays de l'Ouest* 99 (1992): 34–36, 41.

69. Guillaume-Pierre-François Delamardelle, *Réforme judiciaire en France* (Paris: Desenne, 1806) 5; Delamardelle, "Etat des services," ANOM, E 251; Jean-André Tournerie, "Un projet d'école royale des colonies," *Annales de Bretagne et des pays de l'Ouest*, 54.

70. "Précis de ce qui s'est passé lors de l'installation du Gouverneur-Général au Conseil," 87.

71. Yvan Debbasch, *Couleur et liberté. Le jeu de critère ethnique dans un ordre juridique esclavagiste* (Paris: Dalloz, 1967), 83–85.

72. John D. Garrigus, *Before Haiti: Race and Citizenship in Saint-Domingue* (New York: Palgrave Macmillan, 2006), 44–49.

73. Ibid., 86; Moreau de Saint-Méry, *Loix et constitutions*, 2:397.

74. Ibid., 4:825–31.

75. Julien Raimond, "Mémoire," 1786, 189, ANOM, F3 91, fols. 179–92.

76. Moreau de Saint-Méry, *Loix et constitutions*, 5:214.

77. Julien Raimond, *Véritable origine des troubles de S.-Domingue: Et des différentes causes qui les ont produits* (Paris: Chez Desenne ... Bailly ... , 1792), 4. It seems likely that this was an exaggeration, but a study of marriage contracts in the 1760s and 1780s shows that free colored brides and grooms were well represented among the wealthiest marriages in Saint-Domingue's southern peninsula. Garrigus, *Before Haiti*, 179–81.

78. Desdorides, "Remarques," 4; see also Garrigus, *Before Haiti*, 153–56.

79. Dominique Rogers and Stewart R. King, "Housekeepers, Merchants, Rentières: Free Women of Color in the Port Cities of Colonial Saint-Domingue, 1750–90," in

Women in Port: Gendering Communities, Economies, and Social Networks in Atlantic Port Cities, 1500–1800, ed. Douglas Catterall and Jodi Campbell (Leiden: Brill, 2012), 375–79; Garrigus, *Before Haiti*, 56–66, 72–73.

80. Jacques Houdaille, "Trois paroisses de Saint-Domingue au XVIIIe siècle," *Population* 18 (1963), 100.

81. Moreau de Saint-Méry, *Loix et constitutions*, 5:285.

82. Ibid., 5:290; Garrigus, *Before Haiti*, 83–85.

83. Moreau de Saint-Méry, *Loix et constitutions*, 5:498–50.

84. Ibid., 5:610.

85. Ibid., 5:855–56.

86. Ibid., 5:258.

87. Pluchon, *Histoire de la colonisation*, 392, 395.

88. Michel-René Hilliard d'Auberteuil, *Considérations sur l'état présent de la colonie française de Saint-Domingue: Ouvrage politique et législatif présenté au ministre de la marine*, vol. 1 (Paris: Grangé, 1776), 2:27.

89. Gabriel Debien, "L'Esprit de St. Domingue (un manuscrit de Lory)," *Revue de la société haïtienne d'histoire, de géographie, et de géologie*, October 1958, 21–22.

90. Jean-François Marmontel, *Contes moraux* (s.l. s.n., 1764), 331.

91. Hilliard d'Auberteuil, *Considérations*, 1776, 1:165.

92. Desdorides, "Remarques," 35.

93. Daniel Livesey, "The Decline of Jamaica's Interracial Households and the Fall of the Planter Class, 1733–1823," *Atlantic Studies* 9 (2012): 107–23.

94. For accounts of Raimond's early life, see Mercer Cook, "Julien Raimond, Free Negro of St. Domingo," *Journal of Negro History* 26 (1941): 139–70; Jacques de Cauna, "Julien Raimond un quarteron d'origine landaise à la tête de la révolution des gens de couleur à St.-Domingue," in *Actes du colloque, "Les Landes et la Révolution,"* ed. Bernadette Suau (Mont-de-Marson: Conseil général des Landes, 1992), Luc Nemours, "Julien Raimond, le chef des gens de couleur et sa famille," *Annales historique de la révolution française* 23 (1951): 257–62; and Garrigus, *Before Haiti*, 21–50.

95. Dominique Rogers, "Les libres de couleur dans les capitales de Saint-Domingue: Fortune, mentalités et intégration à la fin de l'Ancien Régime (1776–1789)" (Ph.D. diss., Université de Bordeaux III, 1999), 367; for Raimond's first descriptions of the discrimination faced by free coloreds, see Garrigus, *Before Haiti*, 218–21.

96. The term "mulatto safety hatch" is from Carl N. Degler, *Neither Black nor White; Slavery and Race Relations in Brazil and the United States* (New York: Macmillan, 1971), 95.

Chapter 8. The American Revolution in the Greater Antilles

1. Christopher Leslie Brown, *Moral Capital: Foundations of British Abolitionism* (Chapel Hill: University of North Carolina Press, 2006).

2. See Steven Wise, *Though the Heavens May Fall: The Landmark Case That Led to*

the End of Human Slavery (Cambridge, Mass.: Harvard University Press, 2005); David Waldstreicher, *Slavery's Constitution: From Revolution to Ratification* (New York: Hill and Wang, 2009), ch. 1; James Oldham, "New Light on Mansfield and Slavery," *Journal of British Studies* 27 (1988): 45–68; David Brion Davis, *The Problem of Slavery in the Age of Revolution, 1770–1823* (New York: Oxford University Press, 1975), 470–500; William R. Cotter, "The Somerset Case and the Abolition of Slavery," *History* 79 (1994): 31–56; Srividhya Swaminathan, "Developing the West Indian Proslavery Position After the Somerset Decision," *Slavery and Abolition* 24 (2003): 40–60; Swaminathan, *Debating the Slave Trade: Rhetoric of British National Identity, 1759–1815* (Burlington, Vt.: Ashgate, 2009); Brown, *Moral Capital*, 95–101; "Forum: Somerset's Case Revisited," *Law and History Review* 24 (2006): 601–72.

3. A Planter [Edward Long], *Candid Reflections Upon the Judgement lately awarded by The Court of King's Bench in Westminster-Hall, on what is commonly called The Negroe-Cause* (London: T. Lowndes, 1772); Samuel Estwick, *Considerations on the Negroe Cause Commonly So Called, Addressed to the Right Honourable Lord Mansfield . . .* , 2nd ed. (London: J. Dodsley, 1773); and Samuel Martin Sr., *A Short Treatise on the Slavery of Negroes in the British Colonies* (Antigua, 1775).

4. Long, *Candid Reflections*, 73.

5. Ibid., 40; Estwick, *Considerations on the Negroe Cause*, xvi.

6. Martin, *Short Treatise*, 11.

7. Brown, *Moral Capital*; and Wise, *Though the Heavens Might Fall*.

8. Estwick admitted that slavery was to some extent "inconsistent with the principles of the Constitution" of England while Long agreed that slavery ought to be, even if it was not so in fact, in his opinion, "repugnant to the spirit of the English laws." Long, *Candid Reflections*, 59; Estwick, *Considerations on the Negroe Cause*, 86; Brown, *Moral Capital*, 98.

9. Vincent Carretta, *Equiano, the African: Biography of a Self-Made Man* (New York: Penguin, 2005), 204–15. For a different view, see James Sidbury, *Becoming African in America: Race and Nation in the Early Black Atlantic* (New York: Oxford University Press, 2007), 39–66.

10. Sue Peabody, *There Are No Slaves in France: The Political Culture of Race and Slavery in the Ancien Régime* (New York: Oxford University Press, 1996), 92, 94–95, 106–19.

11. Madeleine Dobie, *Trading Places: Colonization and Slavery in Eighteenth-Century French Culture* (Ithaca: Cornell University Press, 2010), 256.

12. Jean-François de Saint-Lambert, "Ziméo," in *Les Saisons, poème* (Amsterdam, 1769), 226–59; Dobie, *Trading Places*, 256–60.

13. Michèle Duchet, *Anthropologie et histoire au siècle des lumières: Buffon, Voltaire, Rousseau, Helvétius, Diderot*, 2nd ed. (Paris: Albin Michel, 1971), 166.

14. Yves Charbit, "The Political Failure of an Economic Theory: Physiocracy," trans. Arundhati Virmani, *Population* (English ed.) 57 (2002): 855–83.

15. Edward Seeber, *Anti-Slavery Opinion in France During the Second Half of the Eighteenth Century* (New York: Greenwood, 1937), 103–5.

16. Dobie, *Trading Places*, 256–86.

17. Gene E. Ogle, "'The Eternal Power of Reason' and 'The Superiority of Whites': Hilliard d'Auberteuil's Colonial Enlightenment," *French Colonial History* 3 (2003): 37–39.

18. Michel-René Hilliard d'Auberteuil, *Considérations sur l'état présent de la colonie française de Saint-Domingue: ouvrage politique et législatif présenté au ministre de la marine*, vol. 1 (Paris: Grangé, 1776), 1776, 1:149.

19. Ibid., 1:235–36.

20. Ibid., 1:319–24.

21. Ibid., 1:130–35.

22. Ibid., 2:80–88.

23. For Hilliard's subsequent career, see Ogle, "The Eternal Power of Reason," 38–40; for the impact on free men of color, see Julien Raimond, "Memoire to Castries," 1786, ANOM, F3 91, fol. 192.

24. Jean-Barthélemy-Maximilien Nicolson, *Essai sur l'histoire naturelle de l'isle de Saint-Domingue avec des figures en taille-douce* (Paris: Gobreau, 1776), 126–27.

25. Hilliard d'Auberteuil, *Considérations*, 1:288.

26. Andrés Poëy y Aguirre, *Table chronologique de quatre cents cyclones qui ont sévi dans les Indes* (Paris: Imprimerie administrative de Paul Dupont, 1862), 13–14.

27. Jean-Barthélemy-Maximilien Nicolson, *Essai sur l'histoire naturelle de l'isle de Saint-Domingue avec des figures en taille-douce* (Paris: Gobreau, 1776), 128–29.

28. M. L. E. Moreau de Saint-Méry, *Description topographique, physique, civile, politique et historique de la partie française de l'isle Saint-Domingue* (Paris: Société de l'histoire des colonies françaises, 1958), 1321.

29. Charles Frostin, "Saint-Domingue et la révolution américaine," *Bulletin de la Société d'Histoire de la Guadeloupe* 22 (1974): 78.

30. Poëy y Aguirre, *Table chronologique*, 15; Nicolson, *Essai sur l'histoire naturelle*, 129.

31. Citations from Frostin, "Saint-Domingue et la révolution américaine," 80–81.

32. Ibid., 79.

33. Ibid., 82.

34. See Michael Craton, *Testing the Chains: Resistance to Slavery in the British West Indies* (Ithaca, N.Y.: Cornell University Press, 1982), 172–78; Richard B. Sheridan, "The Jamaican Slave Insurrection Scare of 1776 and the American Revolution," *Journal of Negro History* 3 (1975): 290–308.

35. Keith to Lord George Germain, 1 August 1776, CO 137/71/201.

36. Craton, *Testing the Chains*, 175, 177.

37. B. W. Higman, *Proslavery Priest: The Atlantic World of John Lindsay, 1729–1788* (Kingston: University of the West Indies Press, 2011).

38. Jack P. Greene, "Liberty, Slavery and the Transformation of British Identity in the Eighteenth-Century West Indies," *Slavery and Abolition* 21 (2000): 1–31.

39. Simon Taylor to Chaloner Arcedeckne, 26 June 1781, Jamaican Estate Papers, Vanneck MSS, bundle 2/9, Cambridge University Library.

40. For a good example of how this played out late in the period of abolition, see Emilia Viotta da Costa, *Crowns of Glory, Tears of Blood: The Demerara Slave Rebellion of 1823* (New York: Oxford University Press, 1994), 278–90.

41. *Danish American Gazette*, 20 November 1776, cited in Andrew Jackson O'Shaughnessy, *An Empire Divided: The American Revolution and the British Caribbean* (Philadelphia: University of Pennsylvania Press, 2000), 138; Thomas Johnson Jr. to Samuel Purviance Jr., 13 June 1775, Philadelphia, in *Letters of Delegates to Congress, 1774–1789*, ed. Paul H. Smith (Washington, D.C.: Library of Congress, 1976), 1:483.

42. The petition caused consternation in Britain. The Earl of Dartmouth, secretary of state for America, was furious at "so indecent, not to say criminal, Conduct of the Assembly." He made sure the petition was not debated in Parliament. O'Shaughnessy, *Empire Divided*, 139.

43. Alexander Hamilton, "A Full Vindication of the Measures of Congress etc December 15, 1774" and "The Farmer Refuted, 23 February 1775," in *The Revolutionary Writings of Alexander Hamilton*, ed. Richard B. Vernier (Indianapolis: Liberty Fund, 2008), 1–48.

44. "Address and Petition of the Assembly of Jamaica to the King," 31 December 1776, CO 140/56.

45. George III to Lord North, 11 June 1778, in *The Correspondence of King George the Third from 1760 to December 1783*, ed. Sir John William Fortescue (London: Macmillan, 1928), 4:350.

46. O'Shaughnessy, *Empire Divided*, 137–212; Piers Mackesy, *The War for America, 1775–1783* ([1964]; Lincoln: University of Nebraska Press, 1993), 225–36, 301–37, 446–59.

47. Robert Morris to John Paul Jones, 5 February 1777, in *Letters of Delegates to Congress, 1774–1789*, ed. Paul H. Smith (Washington, D.C.: Library of Congress, 1976), 6:221. See also William M. Fowler Jr. *Rebels Under Sail: The American Navy During the Revolution* (New York: Scribner, 1976), 99–101.

48. Keith to George Germain, 6 June 1776, CO 137/71.

49. George Metcalf, *Royal Government and Political Conflict in Jamaica, 1729–1783* (London: Longman, 1965), 198, 201–3.

50. Address to the King of the Gentlemen, Freeholders, Merchants and Inhabitants of the Town of Savannah la Mar and the Parish of Westmoreland in the County of Cornwall, 21 February 1777, CO 137/72/86.

51. O'Shaughnessy, *Empire Divided*, 207–8; Jack P. Greene, *Peripheries and Center: Constitutional Development in the Extended Polities of the British Empire and the United States 1607–1788* (Athens: University of Georgia Press, 1986), 209–10. For Ireland, see James Kelly, *Henry Flood: Patriots and Politics in Eighteenth-Century Ireland* (Dublin: Four Courts, 1998), 311–24.

52. J. H. Parry, "Eliphalet Fitch: A Yankee Trader in Jamaica During the War of Independence," *History* 40 (1955): 85–92; Selwyn H. H. Carrington, *The British West Indies During the American Revolution* (Dordrecht: Foris Publications, 1988), 73–74. For smuggling, see Wim Klooster, "Inter-Imperial Smuggling in the Americas, 1600–1800," in *Soundings in Atlantic History: Latent Structures and Intellectual Currents*, ed. Bernard Bailyn and Patricia L. Denault (Cambridge, Mass.: Harvard University Press, 2009), 141–80.

53. Carrington, *British West Indies During the American Revolution*. For the convoy system, see Richard Pares, *Yankees and Creoles: The Trade Between North America and the West Indies Before the American Revolution* (London, 1956).

54. Mackesy, *War for America*, 186–89, 219–22.

55. David Syrett, "The West India Merchants and the Conveyance of the King's Troops to the Caribbean, 1779–1782," *Journal of the Society for Army Historical Research* 14 (1967): 169–76.

56. [Samuel Estwick], *Considerations on the Present Decline of the Sugar Trade* (London, 1782), 53; Metcalf, *Royal Government and Political Conflict in Jamaica*, 208; O'Shaughnessy, *Empire Divided*, 197.

57. Stephen Conway, *The British Isles and the War for American Independence* (Oxford: Oxford University Press, 2000); Linda Colley, *Britons: Forging the Nation*, 3rd ed. (New Haven, Conn.: Yale University Press, 2009); David A. Bell, *The Cult of the Nation in France: Inventing Nationalism, 1680–1800* (Cambridge, Mass.: Harvard University Press, 2003).

58. Mackesy, *War for America*, 183–84; Brendan Simms, *Three Victories and a Defeat: The Rise and Fall of the First British Empire, 1714–1789* (London: Allen Lane, 2007); and Stephen Conway, "The British Army, 'Military Europe,' and the American War of Independence," *WMQ* 67 (2010): 69–100. The best modern treatment of George III is Andrew O'Shaughnessy, "'If Others Will Not Be Active, I Must Drive': George III and the American Revolution," *Early American Studies* 2 (2004): 1–47; and O'Shaughnessy, *The Men Who Lost America: British Leadership, the American Revolution, and the Fate of the Empire* (New Haven, Conn.: Yale University Press, 2013), 17–46.

59. Samuel Jones to Shelburne, 10 September 1779, Lord Shelburne Papers, 1663–1797, vol. 78, William L. Clements Library, University of Michigan, Ann Arbor, Michigan.

60. O'Shaughnessy, *Empire Divided*, 189.

61. Thomas Dancer, *A Brief History of the late Expedition against Fort San Juan . . .* (Kingston, 1791), 20, 22. See J.R. McNeill, *Mosquito Empires: Ecology and War in the Greater Caribbean, 1620–1914* (New York: Cambridge University Press, 2010), 189–91; and John Sugden, *Nelson: A Dream of Glory, 1758–1797* (New York: Henry Holt, 2004), 148–75.

62. O'Shaughnessy, *Empire Divided*, ch. 8; Metcalf, *Royal Government and Political Conflict in Jamaica*, 201–25.

63. For complaints about the cost of martial law, which Thistlewood estimated at

variously £1,800, £3,000, and £4,000 per day, see entries of 15 and 26 October 1778; 3 and 13 September 1779, TT.

64. "De la Nouvelle-Yorck," *Affiches américaines*, May 1766, Supplement edition, sec. Amérique, 177–78; "Angleterre," *Affiches américaines*, 20 November 1776, sec. Nouvelles d'Europe, 555–57, 591–92.

65. Roopnarine John Singh, *French Diplomacy in the Caribbean and the American Revolution* (Hicksville, N.Y.: Exposition, 1977), 35–36.

66. Jonathan R. Dull, *The Age of the Ship of the Line: The British and French Navies, 1650–1815* (Lincoln: University of Nebraska Press, 2009), 99, 101.

67. Frostin, "Saint-Domingue et la révolution américaine," 83–84.

68. Ibid., 90–92.

69. Ibid., 87–88.

70. Gabriel Debien, "À Saint-Domingue avec deux jeunes économes de plantation (1774–1788)," *Revue de la société d'histoire de géographie d'Haïti* 16 (1945): 65.

71. Debien, "À Saint-Domingue," 48, 50.

72. Joseph Hamon, *Le chevalier de Bonvouloir, premier émissaire secret de la France auprès du Congrès de Philadelphie avant l'indépendance américaine* (Paris: Jouve, 1953); Jonathan R. Dull, *Benjamin Franklin and the American Revolution* (Lincoln: University of Nebraska Press, 2010), 51.

73. Maurice Lever and Susan Emanuel, *Beaumarchais: A Biography* (London: Macmillan, 2009), 135–36.

74. Singh, *French Diplomacy in the Caribbean*, 154, 163, 169, 171; Brian N. Morton and Donald C. Spinelli, *Beaumarchais and the American Revolution* (Lanham, Md: Lexington Books, 2003), 109, 159; Louis Léonard de Loménie, *Beaumarchais et son temps: Études sur la société en France au 18e siècle d'après des documents inédits*, vol. 2 (Paris: C. Lévy, 1880), 151.

75. Jonathan R. Dull, *The French Navy and American Independence: A Study of Arms and Diplomacy, 1774–1787* (Princeton, N.J.: Princeton University Press, 1975), 101, 112, 119.

76. Frostin, "Saint-Domingue et la révolution américaine," 97–98.

77. Debien, "À Saint-Domingue," 33.

78. Frostin, "Saint-Domingue et la révolution américaine," 100.

79. *Affiches américaines*, 30 March 1779.

80. Debien, "À Saint-Domingue," 54.

81. John D. Garrigus, *Before Haiti: Race and Citizenship in Saint-Domingue* (New York: Palgrave Macmillan, 2006), 208–9.

82. Debien, "À Saint-Domingue," 54.

83. Garrigus, *Before Haiti*, 206–9.

84. "Du Cap," *Affiches américaines*, May 1780, sec. Nouvelles politiques, 133.

85. Garrigus, *Before Haiti*, 210.

86. Debien, "À Saint-Domingue," 54.

87. Frostin, "Saint-Domingue et la révolution américaine," 100.

88. Dull, *Age of the Ship of the Line*, 106.

89. *Affiches américaines*, 6 June 1780.

90. Mackesy, *War for America*, 229; O'Shaughnessy, *Empire Divided*, 230–37; McNeill, *Mosquito Empires*, 134, 167, 186, 248, 258.

91. Laurent François Le Noir, Marquis de Rouvray, "Mémoire pour la création d'un corps de gens de couleur levé en mars 1779 à St. Domingue," 1779, 1, ANOM, DFC 3, piece 103.

92. O'Shaughnessy, *Empire Divided*, 232; 3 May 1782, TT.

93. See, for example, Jane Kamensky and Edward Gray, eds., *The Oxford Handbook of the American Revolution* (New York: Oxford University Press, 2012).

94. Girod-Chantrans, *Voyage d'un Suisse*, 81.

95. Dull, *Age of the Ship of the Line*, 112–14.

96. Mackesy, *War for America*, 429–30, 454–59; O'Shaughnessy, *Empire Divided*, 230–37. Thistlewood gives detailed descriptions of French battleships in Bluefield Harbor, 22–24 July 1782, TT.

97. Girod-Chantrans, *Voyage d'un Suisse*, 245–46.

98. For battle casualties in North America, see John Shy, *A People Armed and Numerous: Reflections on the Military Struggle for American Independence*, 2nd ed. (Ann Arbor: University of Michigan Press, 1990), 249–50.

99. *Affiches américaines*, 8 May 1782, 167.

100. Debien, "À Saint-Domingue," 58.

101. Singh, *French Diplomacy in the Caribbean*, 57.

102. Jonathan R. Dull, *Benjamin Franklin and the American Revolution*, 116.

103. "Documents relating to the Ship Zong," REC/19, National Maritime Museum, Greenwich, London (commonly called the Sharp Transcript) and "Answers of William Gregson (January 1784) and James Kelsall (November 1783)," E112/1258/173, NA.

104. For the *Zong* as a case study in the workings of a new financial capitalism, see Ian Baucom, *Specters of the Atlantic: Finance Capital, Slavery, and the Philosophy of History* (Durham, N.C.: Duke University Press, 2005).

105. The landing at Black River is suspicious. Most western slave ships went to Montego Bay, and the *William*, the sister ship of the *Zong*, sold its slaves at Kingston. Our belief is that it was connected to the arrival in Westmoreland and St. Elizabeth of £25,426 relief money from the British government. Edward Woollery, in his attack on the commissioners giving out relief money, claimed that John Wedderburn deliberately delayed the day for giving out money for a fortnight so that favored people could make "agreements" in which they got discounts for providing sums in cash. Wedderburn denied such an accusation. 18 December 1783, *JAJ*.

106. James Walvin, *The Zong: A Massacre, the Law, and the End of Slavery* (New Haven, Conn.: Yale University Press, 2011).

107. Ottabah Cuguano, *Thoughts and Sentiments on the Evil and Wicked Traffic of the Commerce of the Human Species* (London, 1787), 111–12.

108. It is important to note, however, that the *Zong* case became notorious only after

the mobilization of abolitionism, in 1787-88. Seymour Drescher, "The Shocking Birth of British Abolitionism," *Slavery and Abolition* 33 (2012): 575-76.

109. In 1788, Clarkson wrote of the *Zong* that it was an event "unparalleled in the memory of man . . . and of so black and complicated a nature, that were it to be perpetuated to future generations . . . it could not possibly be believed." Thomas Clarkson, *Essay on the Slavery and Commerce of the Human Species* (London, 1788), 99. See also John Newton, *Thoughts upon the African Slave Trade* (London, 1788), 11.

110. Walvin, *The Zong*, 211.

Chapter 9. Recovery and Consolidation in the 1780s

1. Andrew Jackson O'Shaughnessy, *An Empire Divided: The American Revolution and the British Caribbean* (Philadelphia: University of Pennsylvania Press, 2000), 235-37; James Robertson, *Gone Is the Ancient Glory: Spanish Town, Jamaica, 1534-2000* (Kingston: Ian Randle, 2005), 125-29.

2. P. J. Marshall, *Remaking the British Atlantic: The United States and the British Empire After American Independence* (Oxford: Oxford University Press, 2012), 178; Seymour Drescher, "The Shocking Birth of British Abolitionism," *Slavery and Abolition* 33 (2012): 582-84.

3. Paul Cheney, *A Colonial Cul-de-Sac: Plantation Life in Wartime Saint-Domingue, 1775-1782* (Chicago: University of Chicago Press, forthcoming).

4. We are not the only scholars to use a mechanical rather than a biological metaphor to describe Saint-Domingue in this period. It suggests a dynamic interplay between men, institutions, and economic imperatives that reflects the particular ambience of the Enlightenment, when the legitimate domain of the machine permeated the whole of human discourse and knowledge. James E. McClellan III and François Regourd, *The Colonial Machine: French Science and Overseas Expansion in the Old Regime* (Brepols, Belgium: Turnhout, 2011), 19-21.

5. Lowell Joseph Ragatz, *The Fall of the Planter Class in the British Caribbean, 1763-1833* (New York: Century Co., 1928), vii-viii; Eric Williams, *Capitalism and Slavery* (Chapel Hill: University of North Carolina Press, 1944), 120.

6. Seymour Drescher, *Econocide: British Slavery in the Era of Abolition* 2nd ed. (Chapel Hill: University of North Carolina Press, 2010); John J. McCusker, "The Economy of the British West Indies, 1763-1790: Growth, Stagnation, or Decline?" in McCusker, *Essays on the Economic History of the Atlantic World* (London: Routledge, 1997), 330.

7. Nicholas Draper, *The Price of Emancipation: Slave-Ownership, Compensation and British Society at the End of Slavery* (Cambridge: Cambridge University Press, 2010), 100-101.

8. See Drescher, *Econocide*; David Ryden, "Does Decline Make Sense? The West Indian Economy and the Abolition of the British Slave Trade," *Journal of Interdisciplinary History* 31 (2001): 347-74; Christer Petley, ed., "Special Issue: Rethinking the Fall of the Planter Class," *Atlantic Studies* 9 (2012): 1-123; and J. R. Ward, *British West Indian*

Slavery, 1750–1834: The Process of Amelioration (New York: Oxford University Press, 1988). The most sustained defense of the strong version of the Williams thesis is Selwyn Carrington, "Econocide—Myth or Reality—The Question of West Indian Decline, 1783–1806," *Boletin de Estudias Latinoamerica y del Caribe* 36 (1984): 13–48.

9. J. R. Ward, "The Profitability of Sugar Planting in the British West Indies, 1650–1834," *Economic History Historical Review*, 2nd ser., 31 (1978): 207–9; Richard B. Sheridan, "The Crisis of Slave Subsistence in the British West Indies During and After the American Revolution," *WMQ*, 3rd ser., 33 (1976): 615–41.

10. Lowbridge Bright, Bristol, to William Savage, Jamaica, 1 August 1779, in *The Bright-Meyler Papers: A Bristol-West India Connection 1732–1837*, ed. Kenneth Morgan (Oxford: Oxford University Press, 2007), 511.

11. Trevor Burnard, "Et in Arcadia Ego: West Indian Planters in Glory, 1674–1784," *Atlantic Studies* 9 (2012): 19–40.

12. Jean Tarrade, *Le commerce colonial de la France à la fin de l'ancien régime: l'évolution du régime de l'exclusif de 1763 à 1789* (Paris: Presses universitaires de France, 1972), 2:776.

13. Thésée, *Négociants bordelais et colons de Saint-Domingue: Liaisons d'habitations: La maison Henry Romberg, Bapst et Cie, 1783–1793* (Paris: Geuthner, 1972), 240.

14. Tarrade, *Le commerce colonial*, 2:774.

15. See Éric Saugéra, *Bordeaux, port négrier: Chronologie, économie, idéologie, XVIIe–XIXe siècles* (Paris: Karthala, 1995), 233.

16. Thésée, *Négociants bordelais et colons*, 85.

17. Gabriel Debien, "L'Esprit de St. Domingue (un manuscrit de Lory)," *Revue de la société haïtienne d'histoire, de géographie, et de géologie*, October 1958, 14.

18. See, for example, Guillaume Raynal, *Histoire philosophique et politique des établissemens & du commerce des Européens dans les deux Indes, nouvelle édition*, 2nd ed., vol. 5 (La Haye: Gosse fils, 1773), 57–64.

19. Thésée, *Négociants bordelais et colons*, 91, 242.

20. Tarrade, *Le commerce colonial*, 2:756.

21. Thésée, *Négociants bordelais et colons*, 53, 66, 96–97, 102, 104, 241.

22. On the financial impact in France of the French Revolution, see the difficulties of the Dolle and Raby families of Grenoble, who borrowed money from Romberg et Bapst. Pierre Léon, *Marchands et spéculateurs dauphinois dans le monde antillais du XVIIIe siècle: Les Dolle et les Raby* (Paris: Les Belles Lettres, 1963), 131–36.

23. Albane Forestier, "A 'Considerable' Credit in the Late Eighteenth Century French West Indies Trade: The Chaurands of Nantes, *French History* 25 (2011): 48–68. For colonial agents, see R. Charles, "Enquêter autrement avec les milieux négociants dominguois sous l'Ancien Régime," *Cahiers des Anneaux de la Mémoire* 6 (2004): 155–82.

24. Jacques de Cauna, "Les comptes de la sucrerie Fleuriau: Analyse de la rentabilité d'un plantation de Saint-Dominigue au XVIIIe siècle," in *Commerce et plantation dans la Caraïbe au XVIIIe et XIXe siècles*, ed. Paul Butel (Bordeaux: Maison des Pays Ibériques, 1992), 143–67.

25. Forestier, "A 'Considerable' Credit," 59.

26. Ibid., 48–68.

27. Pierre Ulric Dubuisson and Lenoir, Marquis de Rouvray, *Lettres critiques et politiques sur les colonies & le commerce des villes maritimes de France* (Genève: s.n., 1785), 122–25; and M. L. E. Moreau de Saint-Méry, *Loix et constitutions des colonies françoises de l'Amérique sous le Vent* (Paris: Quillau, 1784), 1784, 5:701, 729.

28. Jacques de Cauna, *Au temps des isles à sucre: Histoire d'une plantation de Saint-Domingue au XVIIIe siècle* (Paris: Karthala, 2003), 259.

29. La Lorie, "[Unaddressed letter of 15 August 1780]," 8, AN, T548-2, Causans papers.

30. La Lorie, "Rapport pour mes cohéritiers et Mr. D'hillou de notre compte courant depuis 12 May 1781," AN, T548-2 Causans papers.

31. For a particularly compelling account of the relentless capitalist logic of the plantation system in nineteenth-century Louisiana, see Richard Follett, *The Sugar Masters: Planters and Slaves in Louisiana's Cane World, 1820–1860* (Baton Rouge: Louisiana State University Press, 2006).

32. Drescher, *Econocide*, 7. The literature on this question is enormous. The best Marxist account of the rise and fall of slavery is Robin Blackburn, *The American Crucible: Slavery, Emancipation and Human Rights* (London: Verso, 2011). See also Dale Tomich, *Through the Prism of Slavery: Labor, Capital, and World Economy* (Lanham, Md.: Rowman and Littlefield, 2004). The best non-Marxist approach is David Brion Davis, *Inhuman Bondage: The Rise and Fall of Slavery in the New World* (New York: Oxford University Press, 2006). See also Seymour Drescher, *The Mighty Experiment: Free Labor vs. Slavery in British Emancipation* (New York: Oxford University Press, 2002).

33. B. W. Higman, *Writing West Indian Histories* (Basingstoke: Macmillan, 1999); Roger Anstey, *The Atlantic Slave Trade and British Abolition, 1760–1810* (London: Macmillan, 1975); and Christopher Leslie Brown, *Moral Capital: The Foundations of British Abolitionism* (Chapel Hill: University of North Carolina Press, 2006).

34. Walter Johnson critiques scholars' tendency to use the concept of "agency" in works on slavery, arguing that it is based on anachronistic assumptions about slaves' material and social conditions. Walter Johnson, "On Agency," *Journal of Social History* 37 (2003): 113–24; and Johnson, "Agency: A Ghost Story," in *Slavery's Ghost: The Problem of Freedom in the Age of Emancipation*, ed. Richard Follett (Baltimore: Johns Hopkins University Press, 2011), 8–30.

35. Clarke to Sydney, 10 September 1785, 5 November 1786, CO 137/85/188, CO 137/86/142–43.

36. 4 and 8 October, 31 December 1780, TT.

37. William Beckford, *A Descriptive Account of the Island of Jamaica* (London: T. and J. Egerton, 1788), 1:129–30.

38. Matthew Mulcahy, *Hurricanes and Society in the British Greater Caribbean, 1624–1783* (Baltimore: Johns Hopkins University Press, 2006), 180–86.

39. Richard Sheridan, "The Crisis of Slave Subsistence in the British West Indies During and After the American Revolution," *WMQ*, 3rd ser., 33 (1976): 615–41.

40. Speech in the House of Lords, 21 February 1780, *St. James Chronicle*, 19–22 February 1780.

41. John Kelly to Chaloner Arcedeckne, 9 April 1782, Vanneck MSS, bundle 2/10, Cambridge University Library; John Vanheelen to Joseph Foster Barham, 9 and 27 April 1782, Barham Papers, C.357/1, Bodleian Library, University of Oxford.

42. O'Shaughnessy, *Empire Divided*, 173.

43. David Geggus, "The Enigma of Jamaica in the 1790s: New Light on the Causes of Slave Rebellion," *WMQ*, 3rd ser., 44 (1987): 274–99.

44. Simon Taylor to Chaloner Arcedeckne, 5 July 1789, Vanneck MSS, bundle 2/15, Cambridge University Library; Hector McNeill, *Observations on the Treatment of the Negroes, in the Island of Jamaica* (London: G. G. J. and J. Robinson, and J. Gore, 1788), 39; Alured Clarke to Lord Sydney, 5 November 1786, CO 137/86; Copy of Minutes of Council, 7 August 1784, CO 137/84; "Extract of Two letters from Jamaica to a Gentleman now in London [1787]," BT6/76, NA; "Second Report: Presented the 12th Day of November," in *Two Reports from the Committee of the Honourable House of Assembly of Jamaica, on the Subject of the Slave Trade* (London: Stephen Fuller, Agent for Jamaica, 1789), 13–15.

45. [Anon], *An Essay Concerning Slavery and the Danger Jamaica is ex-to from the too great Number of Slaves* (London: Charles Corbett, 1746), 38.

46. Simon Taylor to Chaloner Arcedeckne, 17 April 1796, Vanneck Mss., 2/21, Cambridge University Library; Thomas Barritt to Nathaniel Phillips, 28 April 1796, Slebech Papers, Mss. 11,572, National Library of Wales, Aberystwyth.

47. Justin Roberts, *Slavery and the Enlightenment in the British Atlantic, 1750–1807* (New York: Cambridge University Press, 2013).

48. Gilbert Mathison, *Notices Respecting Jamaica in 1808, 1809, 1810* (London, 1811), 30–31.

49. Justin Roberts, "Uncertain Business: A Case Study of Barbadian Plantation Management, 1770–93," *Slavery and Abolition* 32 (2011): 253–55; and Heather Cateau, "The New 'Negro' Business: Slave Hiring in the British West Indies," in *In the Shadow of the Plantation: Caribbean History and Legacy*, ed. Alvin O. Thompson (Kingston: University of the West Indies Press, 2002), 100–120.

50. Robert William Fogel, *Without Consent or Coercion: The Rise and Fall of American Slavery* (New York: W. W. Norton, 1989), 124.

51. David Patrick Geggus, "Sugar and Coffee Production and the Shaping of Slavery in Saint Domingue," in *Cultivation and Culture: Labor and the Shaping of Slave Life in the Americas*, ed. Ira Berlin and Philip D. Morgan (Charlottesville: University of Virginia Press, 1993), 75.

52. James Thomson, *A Treatise on the Diseases of Negroes, As they Occur in the Island of Jamaica: With Observations on the Country Remedies* (Kingston: Alex. Aikman, 1820). See Kenneth Morgan, "Slave Women and Reproduction in Jamaica, c. 1776–1834,"

History 91 (2006): 251; B. W. Higman, *Slave Populations of the British Caribbean 1807-1834* (Baltimore: Johns Hopkins University Press, 1984); and Michael Tadman, "The Demographic Cost of Sugar: Debates on Slave Societies and Natural Increase in the Americas," *AHR* 105 (2000): 1534–75.

53. B. W. Higman, *Plantation Jamaica 1750–1850: Capital and Control in a Colonial Economy* (Kingston: University of the West Indies Press, 2005), 166–226; Alexander X. Byrd, *Captives and Voyagers: Black Migrants Across the Eighteenth-Century British Atlantic World* (Baton Rouge: Louisiana State University Press, 2008), chs. 3 and 4. The quotations come from *The Letters of Simon Taylor of Jamaica to Chaloner Arcedeckne, 1765–1775*, ed. Betty Wood (London: Royal Historical Society, Camden Misc., 2002), 14, 53, 71, 87. See also Betty Wood and T. R. Clayton, "Slave Birth, Death and Disease on Golden Grove Plantation, Jamaica, 1765–1810," *Slavery and Abolition* 6 (1985): 99–121.

54. Simon Taylor to Chaloner Arcedeckne, 30 January, 11 June, 29 October 1782, Vanneck MSS, bundle 2/10, Cambridge University Library.

55. Keith P. Herzog, "Naval Operations in West Africa and the Disruption of the Slave Trade During the American Revolution," *American Neptune* 55 (1995): 42–48.

56. Ragatz, *The Fall of the Planter Class*, 165–66.

57. *TSTDB*.

58. *TSTDB*; O'Shaughnessy, *Empire Divided*, 166. Slave prices dropped severely in Jamaica. In 1774, 2,026 slaves in seven shipments were sold in Jamaica for an average price of £51.09. Between 1778 and 1783, 1,598 slaves in five shipments were sold in Jamaica for an average price of £40.81. The reduction in value was 20.1 percent. For Gold Coast slaves only, 1,401 slaves were sold in five shipments in 1774–75 for an average price of £55.86. Between 1776 and 1783, 1,012 slaves from three ships were sold for £39.71 each. The reduction was 29 percent. *TSTDB*.

59. Jacob M. Price, *Capital and Credit in British Overseas Trade: The View from the Chesapeake, 1770–1776* (Cambridge, Mass.: Harvard University Press, 1980); John J. McCusker, "Growth, Stagnation, or Decline? The Economy of the British West Indies, 1763–1790," in *The Economy of Early America: The Revolutionary Period, 1763–1790*, ed. Ronald Hoffman (Charlottesville: University Press of Virginia, 1988), 299.

60. Figures derived from Edward Long Papers, Add. MSS, 12,431, fol. 219, British Library, as adjusted in Ryden, *West Indian Slavery and British Abolition*, 300–301. Slave trade figures come from *TSTDB*. See also Herbert S. Klein, "The English Slave Trade to Jamaica, 1782–1808," *Economic History Review* 31 (1978): 25–45.

61. "Second Report"; Simon Taylor to Chaloner Arcedeckne, Vanneck MSS, 3A/1788/3, Cambridge University Library.

62. Simon Taylor to Chaloner Arcedeckne, 1 May 1788, Vanneck MSS, bundle 2/14, Cambridge University Library.

63. Testimony of Capt. Thomas Lloyd, 25 February 1791 in HCSP, 82:147.

64. Vincent Brown, *The Reaper's Garden: Death and Power in the World of Atlantic Slavery* (Cambridge, Mass.: Harvard University Press, 2008).

65. Dubuisson and Lenoir, Marquis de Rouvray, *Lettres critiques*, 122.

66. Moreau de Saint-Méry, *Loix et constitutions*, 6:701.

67. "Mémoire sur la maladie épizootique pestilentielle de l'île Saint-Domingue, par M. Worlock, médecin-inoculateur associé du Cercle," in *Recherches, mémoires et observations sur les maladies épizootiques de Saint-Domingue*, ed. Charles Arthaud (Cap-François: Imprimerie royale, 1788), 162–79; David M. Morens, "Characterizing a 'New' Disease: Epizootic and Epidemic Anthrax, 1769–1780," *American Journal of Public Health* 93 (2003): 888–89; Justin Girod-Chantrans, *Voyage d'un Suisse dans différentes colonies d'Amérique pendant la dernière guerre* (Neuchatel: Société typographique, 1785), 259.

68. Lenoir, Marquis de Rouvray, "Reflexions."

69. Moreau de Saint-Méry, *Loix et constitutions*, 5:805.

70. Ibid., 5:802, 6:3.

71. Ibid., 6:257–58.

72. Ibid., 6:429.

73. James H. Sweet, *Domingos Álvares, African Healing, and the Intellectual History of the Atlantic World* (Chapel Hill: University of North Carolina Press, 2011).

74. Moreau de Saint-Méry, *Loix et constitutions*, 4:353.

75. Herbert S. Klein and Ben Vinson, *African Slavery in Latin America and the Caribbean*, 2nd ed. (New York: Oxford University Press, 2007), 162–64.

76. Sue Peabody, "'A Dangerous Zeal': Catholic Missions to Slaves in the French Antilles, 1635–1800," *French Historical Studies* 25 (2002), 79.

77. [Anon.], *Rélation d'une conspiration*; Moreau de Saint-Méry, *Loix et constitutions*, 1784, 4:326.

78. Girod-Chantrans, *Voyage d'un Suisse*, 199.

79. Peabody, "'A Dangerous Zeal,'" 89–90.

80. Since the eighteenth century the term "vaudoux" or "vodou," the current accepted orthography in Haiti, has been used more by outsiders than by those in Haiti who "serve the mysteries." Kate Ramsey, *The Spirits and the Law: Vodou and Power in Haiti* (Chicago: University of Chicago Press, 2011), 40.

81. Moreau de Saint-Méry, *Description*, 64.

82. Ibid., 64–69.

83. Moreau de Saint-Méry, *Description*, 64–68; David Patrick Geggus, "The Bois Caïman Ceremony," in Geggus, *Haitian Revolutionary Studies* (Bloomington: Indiana University Press, 2002), 81–92.

84. Moreau de Saint-Méry, *Description*, 51–52, 69, 99; Ramsey, *Spirits and the Law*, 39–40.

85. Pierre Pluchon, *Vaudou, sorciers, empoisonneurs: de Saint-Domingue à Haïti* (Paris: Karthala, 1987), 66–67.

86. Neufchâteau is cited at length in Gabriel Debien, "Assemblées nocturnes d'esclaves: La Marmelade, 1786," *Annales historiques de la Révolution française* 44 (1972): 273–84.

87. M. Thomas J. Desch-Obi, *Fighting for Honor: The History of African Martial Art*

Traditions in the Atlantic World (Columbia: University of South Carolina Press, 2008), 145.

88. Debien, "Assemblées nocturnes," 276.

89. Desch-Obi, *Fighting for Honor*, 145–46.

90. Ramsey, *Spirits and the Law*, 39, 274.

91. McClellan, *Colonialism and Science*, 177–78.

92. Charles Arthaud, *Discours prononcé à l'ouverture de la première séance publique du Cercle des Philadelphes, tenue au Cap-François le 11 mai 1785* (Paris, 1785).

93. Debien, "Assemblées nocturnes," 281.

94. Pluchon, *Vaudou, sorciers*, 67.

95. Moreau de Saint-Méry, *Description*, 275.

96. Michel Taillefer, *La franc-maçonnerie toulousaine sous l'Ancien Régime et la Révolution, 1741–1799* (Paris: ENSB; CTHS, 1984), 92–92; Susan Buck-Morss, "Hegel and Haiti," *Critical Inquiry* 26 (2000): 860.

97. David Geggus, "Haitian Voodoo in the Eighteenth Century: Language, Culture, Resistance," *Jahrbuch für Geschichte von Staat, Wirtschaft und Gesellschaft* 28 (1991): 21–49; Ramsey, *Spirits and the Law*, 40–43.

98. *Morning Chronicle*, 11 and 28 May 1787; *London Chronicle*, 20–22 September 1787, cited in Drescher, "Shocking Birth," 583.

99. Eugene Genovese, *From Rebellion to Revolution: Afro-American Slave Revolts in the Making of the New World* (Baton Rouge: Louisiana State University Press, 1979). See also Michael Craton, *Testing the Chains: Resistance to Slavery in the British West Indies* (Ithaca, N.Y.: Cornell University Press, 1982); Mary Turner, *Slaves and Missionaries: The Disintegration of Jamaican Slave Society, 1787–1834* (Urbana: University of Illinois, 1982); David Barry Gaspar, *Bondmen and Rebels: A Study of Master-Slave Relations in Antigua with Implications for Colonial British America* (Baltimore: Johns Hopkins University Press, 1985); Michael Mullin, *Africa in America: Slave Acculturation and Resistance in the American South and the British Caribbean, 1736–1831* (Urbana: University of Illinois Press, 1992); Emilia Viotti da Costa, *Crowns of Glory, Tears of Blood: The Demerara Slave Rebellion of 1823* (New York: Oxford University Press, 1994); Claudius Fergus, *Revolutionary Emancipation: Slavery and Abolitionism in the British West Indies* (Baton Rouge: Louisiana State University Press, 2013); Gelien Matthews, *Caribbean Slave Revolts and the British Abolitionist Movement* (Baton Rouge: Louisiana State University Press, 2006). The most influential, theoretical contribution to studies of slave resistance in the British West Indies and one that accords well with Genovese's argument is Hilary McD. Beckles, "Caribbean Anti-Slavery: The Self-Liberation Ethos of Enslaved Blacks," *Journal of Caribbean History* 22 (1988): 1–19. Michael Craton sees slave resistance as more opportunistic in nature and, in the early nineteenth century, as often counterproductive. Craton, "Proto-Peasant Revolts? The Late Slave Rebellions in the British West Indies, 1816–1832," *Past and Present* 85 (1979): 99–125; and Craton, "Emancipation from Below? The Role of the British West Indian Slaves in the Emancipation Movement, 1816–1834," in *Out of Slavery: Abolition and After*, ed. Jack Hayward (London: Frank Cass, 1985), 119.

100. Kit Candlin, *The Last Caribbean Frontier, 1795–1815* (London: Palgrave Macmillan, 2012), ch. 1; Edward L. Cox, "Fedon's Rebellion 1795–96: Causes and Consequences," *Journal of Negro History* 67 (1982): 7–19.

Chapter 10. The Ancien Régime in the Greater Antilles

1. [Marc-Antoine] Avalle, *Tableau comparatif des productions des colonies françaises aux Antilles, avec celles des colonies anglaises, espagnoles et hollandaises de l'année 1787 à 1788* (Paris: Goujon fils, [1799]), fig. Tables VIII and IX.

2. Ward Barrett, *The Efficient Plantation and the Inefficient Hacienda* (Minneapolis: Associates of the James Ford Bell Library, 1979), 22. David Geggus says the mean area planted in cane in Saint-Domingue's northern plain in the late 1780s was one hundred hectares; see "Sugar and Coffee Production and the Shaping of Slavery in Saint Domingue," in *Cultivation and Culture: Labor and the Shaping of Slave Life in the Americas*, ed. Ira Berlin and Philip D. Morgan (Charlottesville: University of Virginia Press, 1993), 75.

3. This is converted from the range of eight thousand to ten thousand French pounds per carreaux given in François Girod, *La vie quotidienne de la sociéte créole: Saint-Domingue au XVIIIe siècle* ([Paris]: Hachette, 1972), 46. These figures are borne out by the survey of plantations in Natacha Bonnet, "L'organisation du travail servile sur la sucrerie domingoise au XVIIIe siècle," in *L'esclave et les plantations: De l'établissement de la servitude à son abolition*, ed. Philippe Hroděj (Rennes: Presses universitaires de Rennes, 2008), 144. See also David P. Geggus, "Slave Society in the Sugar Plantation Zones of Saint Domingue and the Revolution of 1791–93," *Slavery and Abolition* 20 (1999): 31–46.

4. Barrett, *Efficient Plantation*, 19.

5. Ibid., 22.

6. Geggus, "Cultivation and Culture," 75. These ratios are created by taking the total size of the slave force and dividing it by the land area planted in sugarcane.

7. Geggus, "Slave Society in the Sugar Plantation Zones of Saint Domingue and the Revolution of 1791–93," 35, table 3.

8. Barrett, *Efficient Plantation*, 4. See Samuel Martin, *An Essay upon Plantership: Humbly Inscrib'd to All the Planters of the British Sugar-Colonies in America* (Antigua: Printed by T. Smith, 1750); equivalents for the French Antilles began to appear only at the end of the century. See Alexandre Casaux, *Essai sur l'art de cultiver la canne et d'en extraire le sucre* (Paris: Clousier, 1781); Jacques-François Dutrône de La Couture, *Précis sur la canne et sur les moyens d'en extraire le sel essentiel* (Paris: Chez Duplain, 1790). For an example of the deep private discussions of planting and refining techniques, see Jacques de Cauna, *Au temps des isles à sucre: Histoire d'une plantation de Saint-Domingue au XVIIIe siècle* (Paris: Karthala, 2003), 178–88.

9. For technology in Jamaica, see Veront Satchell, *Sugar, Slavery and Technological Chamge, Jamaica 1760–1830* (Saarbrucken: VDM Verlag, 2010).

10. Michel-René Hilliard d'Auberteuil, *Considérations sur l'état présent de la colonie française de Saint-Domingue: ouvrage politique et législatif présenté au ministre de la marine*, (Paris: Grangé, 1776), 1:23.

11. James E. McClellan and François Regourd, "The Colonial Machine: French Science and Colonization in the Ancien Regime," *Osiris*, 2nd ser., 15 (2000): 31–50.

12. James E. McClellan III, *Colonialism and Science: Saint-Domingue in the Old Regime* (Baltimore: Johns Hopkins University Press, 1992), 74; and [François Auguste de] Genton, "Essai de minérologie de l'isle de Saint-Domingue dans la partie françoise," *Observations et mémoires sur la physique, sur l'histoire naturelle et sur les arts* 31 (July 1787): 173–77.

13. McClellan, *Colonialism and Science*, 71–75; M. L. E. Moreau de Saint-Méry, *Description topographique, physique, civile, politique et historique de la partie française de l'isle Saint Domingue* (Paris: Société de l'histoire des colonies françaises, 1958), 819–41.

14. McClellan, *Colonialism and Science*, 74.

15. Gabriel Debien, *Les esclaves aux Antilles françaises, XVIIe–XVIIIe siècles* (Basse-Terre: Société d'histoire de la Guadeloupe, 1974), 163–69.

16. Belin de Villeneuve and M. L. E Moreau de Saint-Méry, *Mémoire sur un nouvel équipage de chaudières à sucre: Pour les colonies, avec le plan dudit equipage* (Paris: Chez Hardouin et Gattey, 1786), 1–9.

17. On coffee processing, see Debien, *Les esclaves*, 168–69; Pierre Joseph Laborie, *The Coffee Planter of Saint Domingo* (London: T. Cadell and W. Davies, 1798), 92.

18. However, see Veront Satchell, "Innovations in Sugar-Cane Mill Technology in Jamaica, 1760–1830," in *Working Slavery, Pricing Freedom: Perspectives from the Caribbean, Africa and the African Diaspora*, ed. Verene Shepherd (Kingston, Jamaica: Ian Randle, 2002), 93–111.

19. Jean Tarrade, *Le commerce colonial de la France à la fin de l'ancien régime: l'évolution du régime de l'exclusif de 1763 à 1789*, vol. 2 (Paris: Presses universitaires de France, 1972), 2:510–15. On Dubuc's goal in Martinique, see Dutrône de La Couture, *Précis sur la canne et sur les moyens d'en extraire le sel essentiel*, xiii, note 1.

20. Lafon was a French Huguenot born to a banking family in the Netherlands who returned to France at midcentury. After selling wine and provisions to the colonies, he got involved in the slave trade in the 1770s and bought a colonial plantation during the American Revolutionary War, sending his younger son there in 1778 to manage it.

21. Dutrône de La Couture, *Précis sur la canne*, 214, 222, 224.

22. See the financially disastrous agricultural experiments of the Marquis d'Hanache in Françoise Thésée, *Négociants bordelais et colons de Saint-Domingue: Liaisons d'habitations: La maison Henry Romberg, Bapst et Cie, 1783–1793* (Paris: Geuthner, 1972), 102.

23. Cauna, *Au temps des isles à sucre*, 179–80.

24. Ibid., 181.

25. See David Geggus's excellent overview of Caradeux's multiple reputations in Geggus, "The Caradeux and Colonial Memory," in *The Impact of the Haitian Revolution in the Atlantic World*, ed. Geggus (Columbia: University of South Carolina, 2001), 231–46.

26. C. Malenfant, *Des colonies, et particulièrement de celle de Saint-Domingue* (Paris: Audibert, 1814), 157–58, 196.

27. J. R. Ward, "The Profitability of Sugar Planting in the British West Indies, 1650–1734," *Economic History Review*, 2nd ser., 31 (1978): 207.

28. Trevor Burnard, "From Periphery to Periphery: The Pennants' Jamaican Plantations, 1771–1812 and Industrialization in North Wales," in *Wales and Empire, 1607–1820*, ed. H. V. Bowen (Manchester: Manchester University Press, 2011), 122–26.

29. Christer Petley, *Slaveholders in Jamaica: Colonial Society and Culture during the Era of Abolition* (London: Pickering and Chatto, 2009), 28.

30. C.O. 137/87.

31. "Statistics of Jamaica, 1739–1775," Long Papers, Add. MSS 12,435, fol. 41, British Library.

32. "Statistics of Jamaica, 1739–1775," Add. MSS 12,345, fo1. 41, Long MSS, British Library; "Slaves on York Estate, 1 January 1778," Gale-Morant Papers, 3/c, University of Exeter Library, Devon, England; Inventory of John McLeod, Inventories 1B/11/56/72–76, JA.

33. Bryan Edwards, *The History, Civil and Commercial, of the British Colonies in the West Indies*, 2d ed., 3 vols. (London; J. Stockdale, 1793), 2: 213.

34. Ibid., 2: 261, 285–89, 307.

35. Cited in P. Étienne Herbin de Halle, *Statistique générale et particulière de la France et de ses colonies, avec une nouvelle description topographique physique, agricole, politique, industrielle et commerciale de cet état*, 7 vols. (Paris: Buisson, 1803), 7: 46.

36. *TSTDB*.

37. David P. Geggus, "The French Slave Trade: An Overview," *WMQ* 3d. ser., 58 (2001): 131; Robert Louis Stein, *The French Slave Trade in the Eighteenth Century: An Old Regime Business* (Madison: University of Wisconsin Press, 1979), 33.

38. *TSTDB*.

39. In 1776, the French had about sixty ships of the line, and in 1787, despite the kingdom's looming financial crisis, the French Navy still had sixty-two ships of the line. Dull, *The French Navy and American Independence*, 19, 66, 336–37.

40. Bernardo de Sá Nogueira de Figueiredo Sá da Bandeira, *Facts and Statements Concerning the Right of the Crown of Portugal to the Territories of Molembo, Cabinda, Ambriz, and Other Places on the West Coast of Africa, Situated Between the Fifth Degree Twelve Minutes, and the Eighth Degree of South Latitude* (London: Fitch, 1877), 8–9; on Drouin, see Laure Pineau-Defois, "Un modèle d'expansion économique à Nantes de 1763 à 1792: Louis Drouin, négociant et armateur," *Histoire, économie et société* 23 (2004): 327.

41. Tarrade, *Le commerce colonial*, 2:623–28.

42. Pierre Ulric Dubuisson and Lenoir, Marquis de Rouvray, *Lettres critiques et politiques sur les colonies & le commerce des villes maritimes de France* (Genève: s.n., 1785), 242, 248, 250.

43. Ibid., 83.

44. Bonnet, "L'organisation du travail servile," 143.

45. Dubuisson and Lenoir, Marquis de Rouvray, *Lettres critiques*, 242, 248, 250.
46. Thésée, *Négociants bordelais et colons*, 85.
47. *TSTDB*.
48. Albane Forestier, "Principal-Agent Problems in the French Slave Trade: The Case of Rochelais Armateurs and Their Agents, 1763–1792" (MSc Global History, London School of Economics, 2005), 16–17.
49. Geggus, "Saint-Domingue on the Eve of the Haitian Revolution," 7.
50. *TSTDB*.
51. M.L.E. Moreau de Saint-Méry, *Loix et constitutions des colonies françoises de l'Amérique sous le Vent* (Paris: Quillau, 1784), 6:655–27.
52. B. W. Higman, *Plantation Jamaica 1750–1850: Capital and Control in a Colonial Economy* (Kingston: University of the West Indies Press, 2005).
53. Maurice Begouën Demeaux, *Stanislas Foäche, 1737–1806: Négociant de Saint-Domingue* (Paris: Société française d'histoire d'outremer1982), 109.
54. Debien, *Les esclaves*, 486.
55. Moreau de Saint-Méry, *Loix et constitutions*, 6:918–28.
56. Moreau de Saint-Méry, *Loix et constitutions*, 6:667.
57. Moreau de Saint-Méry, *Description*, 1002.
58. Jean Tarrade, "L'administration coloniale en France à la fin de l'Ancien Régime: Projets de réforme," *Revue Historique* 229 (1963): 121; Tournerie, "Un projet d'école royale des colonies," *Annales de Bretagne et des pays de l'Ouest* 99 (1992): 50.
59. Guillaume-Pierre-François Delamardelle, *Eloge funèbre du comte d'Ennery et réforme judiciaire à Saint-Domingue* (Port-au-Prince: Mozard, 1789), 61–83.
60. Ibid., 60.
61. Malick W. Ghachem, *The Old Regime and the Haitian Revolution* (New York: Cambridge University Press, 2012), 173, 196–201.
62. Ibid., 131–33.
63. Malenfant, *Des colonies, et particulièrement de celle de Saint-Domingue*, 185, tells an elaborate story of Caradeux killing his enslaved sugar refiner by having him buried up to his neck and throwing a heavy rock at his head.
64. In Brazil in the early 1730s the Vodou healer Domingo Álvarez was sold away from one plantation in Recife because he was suspected of poisoning slaves and livestock. Then in 1738 he was used to find a poisoner on another plantation. The owner of that estate subsequently purchased him, at a 25 percent premium, to use him as a healer. James H. Sweet, *Domingos Álvares, African Healing, and the Intellectual History of the Atlantic World* (Chapel Hill: University of North Carolina Press, 2011), 69–70, 100–101.
65. . Nicolas Louis François de Neufchâteau, "Undated, unsigned manuscript," n.d., AN 27 AP 12 papiers de François de Neufchâteau; Lemay, "Letter, dated Plaisance 17 September 1785" (Plaisance, September 1785), AN 27 AP 12, papiers de François de Neufchâteau.
66. Lejeune, "Un Griffe Créole De d'Ouanaminthe," *Affiches américaines* (Cap Français, 8 May 1781), no. 19, p. 176, from *Le Marronnage à Saint-Domingue (Haïti):*

Histoire, mémoire, technologie, http://www.marronnage.info/fr/lire.php?type=annonce&id=5958.

67. Cited in Evelyne Camara, Isabelle Dion, and Jacques Dion, *Esclaves, Regards de blancs, 1672–1913, Collection Archives Nationales d'outre-mer* (Marseille: Images en Manoeuvres Editions, 2008), 134.

68. Neufchâteau, undated, unsigned manuscript.

69. Pierre Pluchon, *Vaudou, sorciers, empoisonneurs: de Saint-Domingue à Haïti* (Paris: Karthala, 1987), 199–200.

70. François Barbé-Marbois, "Letter to Versailles, signed 29 août 1788," 8 August 1788, ANOM, F3 150 fol. 88.

71. Camara, Dion, and Dion, *Esclaves*, 134.

72. Ghachem, *Old Regime*, 200–203; 211–55, 265.

73. David Patrick Geggus, *Haitian Revolutionary Studies* (Bloomington: Indiana University Press, 2002), 91; on the tensions between native-born creole ex-slaves and Africans in the rebel army, see the 12 December 1791 letter of Jean-François and Biassou translated in Laurent Dubois and John D. Garrigus, *Slave Revolution in the Caribbean, 1789–1804: A Brief History with Documents* (New York: Bedford / St. Martin's, 2006), 100–101.

74. This statement compresses a large—and growing—literature on the advent of the Haitian Revolution. See Jeremy D. Popkin, "The French Revolution's Royal Governor: General Blanchelande and Saint Domingue, 1790–92," *WMQ*, 3rd ser., 71 (2014): 203–28; and Popkin, "Saint-Domingue, Slavery, and the Origins of the French Revolution," in *From Deficit to Deluge: The Origins of the French Revolution*, ed. Thomas E. Kaiser and Dale K. Van Key (Stanford, Calif.: Stanford University Press, 2011), 220–48. An extensive account of these years is found in Oliver Gliech, *Saint-Dominguer und die Französische Revolution: Das Ende der wessen Herrschaft in einer karibischen Plantagenwirtschaft* (Cologne: Böhlau, 2011). See also John D. Garrigus, "Vincent Ogé *jeune* (1757–1791): Social Class and Free Colored Mobilization on the Eve of the Haitian Revolution," *Americas* 68 (2011): 33–62.

75. For the sudden rise of abolitionism in Britain in 1787–88, see Seymour Drescher, "The Shocking Birth of British Abolitionism," *Slavery and Abolition* 33 (2012): and J. R. Oldfield, *Transatlantic Abolitionism in the Age of Revolution: An International History of Anti-Slavery, c. 1787–1820* (Cambridge: Cambridge University Press, 2013), chs. 1–3.

76. Linda Colley, *Britons: Forging the Nation*, 3rd ed. (New Haven, Conn.: Yale University Press, 2009), 350–60; C. A. Bayly, *Imperial Meridian: The British Empire and the World, 1780–1830* (London: Longman, 1989); Stephen Conway, *The British Isles and the War for American Independence* (Oxford: Oxford University Press, 2000); Dror Wahrman, *The Making of the Modern Self: Identity and Culture in Eighteenth-Century England* (New Haven, Conn.: Yale University Press, 2004).

77. Nicholas B. Dirks, *The Scandal of Empire: India and the Creation of Imperial Britain* (Cambridge, Mass.: Harvard University Press, 2009), 32–34; Tillman W. Nechtman, *Nabobs: Empire and Identity in Eighteenth-Century Britain* (Cambridge: Cambridge University Press, 2010).

78. Maya Jasanoff, "Revolutionary Exiles: The American Loyalist and French Émigré Diasporas" in *The Age of Revolutions in Global Context, c. 1760–1840*, ed. David Armitage and Sanjay Subrahmanyam (London: Palgrave Macmillan, 2010), 53.

79. Troy Bickham, *Making Headlines: The American Revolution as Seen Through the British Press* (DeKalb: Northern Illinois University Press, 2009), 253; Wahrman, *Making of the Modern Self*, 223, 246, 264; Dirks, *Scandal of Empire*; Conway, *British Isles and the War for American Independence*, 354.

80. Mimi Sheller, *Consuming the Caribbean: From Arawaks to Zombies* (London: Routledge, 2003), 114.

81. Trevor Burnard and Richard Follett, "Caribbean Slavery, British Antislavery and the Cultural Politics of Venereal Disease," *Historical Journal* 55 (2012): 1–25.

82. Madeleine Dobie, *Trading Places: Colonization and Slavery in Eighteenth-Century French Culture* (Ithaca, N.Y.: Cornell University Press, 2010), 202, 241–42.

83. Ian Baucom, *Specters of the Atlantic: Finance Capital, Slavery and the Philosophy of History* (Durham N.C.: Duke University Press, 2005), 66.

INDEX

Abercromby, James, 59
abolitionism, 191, 192, 226, 234, 263–264, 267; Saint-Domingue and, 265–266; *Somerset* case and, 193–194; *Zong* slave ship incident and, 321n108. *See also* moral concerns about slavery; treatment of slaves
absentees/transatlantic brokers, 129, 170, 222, 225, 253–255
Abstract of the Evidence (planters' testimony), 234
Adams, John, 202, 312n30
Affair of the Poisons (1676–82, France), 108, 112
Affiches américaines (broadside), 64, 207, 210, 214–215
age demographics, 9, 60, 249
Aguy (slave), 127
Alico (slave in Tacky's revolt), 125, 301n105
Álvarez, Domingo, 237
American Revolutionary War, 19, 192–218; Battle of the Saintes, 213–215, 219; blockades and convoys during, 204, 210, 219, 222, 232; Britain's reactions after, 264; British navy in, 20, 203–205, 207–209, 212–214, 219; British strategy during, 203–206; death of soldiers, disease and, 206, 212, 214, 215; France and, 20, 203, 205, 206–215, 226; France and Spain's entry into conflict, 203, 205–207, 226; French navy in, 207–208, 214; French support for North America in, 209; investments following, 221–225; Jamaica and North American relationship preceding, 173, 177–181, 201–205; natural disasters in Saint-Domingue during, 198–200; plans to attack Jamaica during, 212–215, 217; smuggling during, 204–205, 207–210; *Zong* incident and, 215–218, 321n105, 321n108, 322n109. *See also* North America
Les Amis des Noirs (French anti-slavery society), 265, 266

anthrax, 235–236
Antigua, 87
Arcedeckne, Chaloner, 172
Archard de Bonvouloir, Julien Alexandre, 209
d'Argout, Robert, Count, 211, 212
Arlequin mulâtresse sauvé par Macandal (pantomime), 80
Arnaudeau, Jean-Baptiste, 247–248
Arthaud, Charles, 241
Artibonite River, 246
Assam (confessed poisoner), 118
Atkins, Jonathan, 12
Augier, Jane, 154
Augier, Mary, 154
Augier, William, 154
Austria, 85
Avalle, Marc-Antoine, 244

Bapst, Christophe, 222–223
Barbados, 4, 9, 12, 62, 87
Barbé-Marbois, François, 250, 259, 260
Battle of the Saintes, 213–215, 219
Battle of Ticonderoga, 128
Bayly, Zachary, 125
Beauharnais, François de, 89
Beaumarchais, Pierre Caron de, 209
Beaumont, Regnaud de, 209–211
Beauvernet, Louis de, 43–48
Beckford, Ballard, 125–126
Beckford, William, 36–37, 71, 142, 144, 150; hurricanes and, 226–227; Pitt and, 87
Beckford, William II, 21
Behn, Aphra, 145
Belin de Villeneuve, Paul, 246–247
Bellanton, Zabeau, 63
Belzunce, Vicomte de, 90, 157
Benin, 110
Besselère, overseer, 209
Bickham, Troy, 264

Bicknell, John, 141
Bilby, Kenneth, 134
blacks: in Beauvernet's vignettes, 45–48; flour peddlers, 118–119; marriage and, 68–69; population statistics of, 32–33, 38, 249, 282n37. *See also* free people of color; racial classification; racial legislation; slaves; whites
black women: free, 63, 67, 68, 76–77, 80, 185; as mistresses, 68, 69, 70, 72, 76, 77, 151, 185; sex and, 68–70; in theater, 77–79
blockades and convoys: during American Revolution, 204, 210, 219, 222, 232; drought during, 200; private firms allowed through, 222; during Seven Years' War, 93, 106, 112–117, 219; during War of Austrian Succession, 84. *See also* smuggling; trade monopoly, French
Blue Mountains (Jamaica), 12
La Borde, Jean-Joseph, 165, 167
Bory, Gabriel de, 90, 157
Boucherie brothers, 247–248
Bourbon Reforms, 96
Bourke, Nicholas, 17, 144, 148, 160, 175, 178
Boydell, John, 37
Braddock, Edward, 85
Braithwaite, Kamau, 49
Braudel, Fernand, 50
Brazil, 4, 6
bread, 114–115
Brett, Curtis, 58
Bright, Henry, 58, 86
Bright, Lowbridge, 221
Brissot, Jacque-Pierre, 265, 266
Britain: colonial investment by, 98–99; conflicts with France before Seven Years' War, 82–84; Jamaican planters' loyalty to, 68, 99, 218, 263–264; Jamaica's importance to, 20, 21, 24, 37, 59, 99; moral concerns about slavery in, 24, 99, 100, 138–141, 192–194, 216–218, 233, 263–265; negotiations after Seven Years' War, 18–19, 82, 92–96, 292n67; news of slave revolts, 140, 146, 228; North American rebellion and, 178–179; race relations in, 21–22, 138–139, 153; social class in, 150; strategy during American Revolution, 203–206; strategy during Seven Years' War, 86–92, 156–157, 165, 292n60, 292n67; views of Jamaican colonists' patriotism in, 99, 138–139, 263–264, 267; witchcraft in, 134.

See also France; Jamaica; Jamaica and Saint-Domingue, compared; Saint-Domingue
British navy, 200; during American Revolution, 20, 203–205, 207–209, 212–214, 219; protection of Jamaica by, 20, 179–181, 303n122; during Seven Years' War, 85–86, 92–93, 112–114, 116, 117; during War of Austrian Succession, 83–84. *See also* French navy; military/militia
Brodie, Hugh, 227
Brown, Vincent, 49, 127, 141
brutality of slave masters. *See* treatment of slaves
Buffon, Count de, 153, 166
Bulman, Isabella, 227
Burke, Edmund, 95
Burke, William, 95
Byrd, Alexander, 230

cacao production, 34
Caillon (planter), 258
Campbell, Archibald, 206
Canada, 82, 89, 92, 94, 167, 181
Cap Français, Saint-Domingue, 13, 51–55, 57, 59–60, 63, 64, 114
Cap Français Council, 13, 17, 159, 161, 258, 261; court merging, 256–257, 262–263; Desmé-Dubuisson on, 162; military to civilian authority, 158; militia revolt, 175; racial legislation, 186; 1784 estate laws, 256, 263; slave religion and, 236, 237, 239–241. *See also* government, Saint-Domingue; imperial reforms, Saint-Domingue
Carabasse, agent, 209
Caradeux, Nicolas, 247–248, 253, 258, 332n63
Carenan, Paul, 185
Carroll, Charles, 59
cartography/maps: of cities, 51; Craskell/Simpson Map of Jamaica, 12–14, 27–32; of Kingston, 56–57; of Saint-Domingue, 32, 51, 52
Cassarouy (planter), 258
casualties. *See* death
Ceded Islands, 98
Césaire, Aimé, 49
Chaillau (militia commander), 258
Chamber of Agriculture (Jamaica), 61
Charlevoix, Pierre-François-Xavier de, 42, 112, 187, 297n39
Chasseurs volontaires/royaux (free colored

army unit, Saint-Domingue), 157, 158, 210–212
Chaurand merchant house, 22, 223–224
Chevalerie, Bacon de la, 67
Choiseul, Duc de, 146, 156, 158–159; colonial loyalty and, 96–98; Delamardelle and, 182; Dubuc and, 165–166; imperial reforms and, 157, 161–162, 174, 207; during Seven Years' War, 89, 91, 94
Christianity: Jesuits, 54, 109, 122, 135, 237, 297n39; macandal talismans and, 109, 110–111; martyrdom and, 141; missionaries, 42–43, 110, 122, 135, 237, 297n39. *See also* slave religion
churches, 53–54
Clarke, Alured, 226, 228–229
Clarkson, Thomas, 265, 266, 322n109
Clavière, Etienne, 265
Clay, Lauren, 77, 80
climate theory, 100
Clinton, Henry, 212
clothing of slaves, 45, 60–61, 146, 186, 254
Clugny de Nuits, Jean-Etienne Bernard de, 120–121, 146, 156–158
Code Noir (1685, Saint-Domingue), 17, 53–54, 183, 254
Coercive Acts, 178, 200
coffee production, 14, 35, 167, 247
Collingwood, Luke, 215–217
Collins, David, 41
colonial loyalty, Jamaica, 99, 138–139, 164, 179, 192, 263–264, 267
colonial loyalty, Saint-Domingue, 138, 154–155, 167, 173, 179; free colored people and, 187, 190; imperial reforms and, 82, 89–90, 96–97, 162, 182–183, 186, 265; Kourou colony and, 97–98, 146–147; *vs.* trade interests during Seven Years' War, 84, 87–89; trade liberalization and, 96–98, 261. *See also* trade monopoly, French
Compagnie des Indes, 34
comparative history, 8–9, 11, 271n20
Condamine, Charles Marie de La, 51
Congo, 109–111
Conseils Supérieurs (Saint-Domingue), 17; meeting location of, 54
Considérations sur l'état présent de . . . Saint-Domingue (Hilliard), 196–198
Consolidated Slave Act (1809), 133
Conway, Henry Seymour, 144

Conway, Stephen, 99
Cope, John, Sr., 128, 170
Coronio, Stephen, 208
Costen, Mary, 227
Cotes, Thomas, 119, 129
Coupées, Colas Jambes, 108
Courtin, Sébastien Jacques, 108–110
Craskell, Thomas, 12, 27
Craskell/Simpson Map of Jamaica (1763), 12–14, 27–30; cartouche scenes in, 31–32
Craton, Michael, 102, 122
creole population (island-born whites) of Saint-Domingue, 10, 100, 197
Crèvecoeur, J. Hector St. John de, 25–26
crime/punishment of masters, 17, 254–258
crime/punishment of slaves, 130–132, 197–198; Hanover slave plot, 200–201; public displays, 130–131; sugar planters' authority in, 101–102, 120–121, 135, 154, 256, 257. *See also* executions
Croll, Mary, 227
Cuba, 91–92, 94, 96
Cubbah (slave), 127
Cuguano, Ottobah, 217
cultivation process and difficulties, 6–7, 28, 171, 270n13; irrigation, 35, 45, 234, 244–246

Dalling, Sir John, 206
dance, 78; slave religion and, 121, 135, 238–240
Dancer, Thomas, 206
Danish American Gazette, 201–202
Davies (slave), 131
Day, Thomas, 141
death, 9, 25; body disposal, 57; of colonists, 67–68, 97–98; in slave revolts, 127–128. *See also* executions; food-borne illness; poisonings
death of slaves: high rates of, 171; from overwork/famine, 41–42, 228–230, 232–233; poisonings and, 105–106; slave trade to replace, 230, 248–249, 252, 266
death of soldiers, disease and: during American Revolution, 206, 212, 214, 215; in Saint-Domingue, 55; during Seven Years' War, 88, 92, 128; during War of Austrian Succession, 83
deaths blamed on poisoning, 109, 235–236; food-borne illness and, 111, 115–119; Médor's confession and, 104–105; of slaves and livestock, 105–106. *See also* poisonings
Deficiency Act, 129

Delamardelle, Guillaume, 182–183, 185, 190, 257–258
Delavaud (planter), 106, 107, 296n23
Delisle, Guillaume, 13
DePauw, Cornelius, 153
Desch-Obi, Thomas, 240–241
Desdorides, Lieutenant Colonel, 26, 48, 54, 71, 185, 188, 275n5
Desmé-Dubuisson, Claude, 155–156
Desmé-Dubuisson, Jean-Pierre, 155–156, 159–162, 174
Dessalines, Jean-Jacques, 20
Dessources (planter), 121, 258
Dickie, Henry, 227
Diderot, Denis, 166
disease. *See* illness/disease
Dobie, Madeleine, 265
Dominica, 89, 98
Dom Pèdre, 239, 240
Dorrill, William, 39
Doyle, William, 24
Drax, Henry, 40
Drescher, Seymour, 221, 225
drought, 199–200, 210, 224–225, 229
Drouin, Louis, 251
Dubuc, Jean-Baptiste, 165–166, 251–253
Duke of Bedford, 93–94
Dull, Jonathan, 208
Dunn, Richard, 48
Dupont de Nemours, Pierre Samuel, 195–196, 225
Dutrône La Coutûre, Jacques-François de, 247
Dying Negro (Day and Bicknell), 141

earthquakes, 52–53, 198–199
economic decline thesis, 219–222, 224–225, 274n62
education, 152, 182
Edwards, Bryan, 69, 125, 131, 142–144, 147, 301n105; on Jamaican population statistics, 250
d'Ennery, Victor-Thérèse Charpentier, Count, 208
Ephémérides (journal), 195
ergotism, 116–117
escaped slave populations. *See* maroon populations
escaped white sailors/indentured servants, 14–15

Essai sur les colonies françaises (Saintard), 64–65
Essay Concerning Slavery (anonymous), 148–150, 155
Essay upon Plantership (Martin), 5
d'Estaing, Charles-Henri Théodat, Count, 13, 160–162, 167, 173, 174; during American Revolution, 210–212; free colored militia and, 184; Léger and, 175
estate sales, 224
Estwick, Samuel, 193, 316n8
executions: after Tacky's revolt, 122, 123, 125, 139, 141, 146, 301n105; cruelty of, 42; legal right to perform, 138; of Macandal, 103–106, 118, 119, 121; moral concerns about, 139, 141; of poisoners, 108, 236, 257; of rebel slaves, 122, 123, 125, 127, 130–131, 146, 200. *See also* crime/punishment; death
export statistics: from France's colonies, 35–36, 43, 95, 167, 244, 246; from Jamaica, 35, 37–38, 59, 84, 171, 221, 250, 289n11; sugar prices and, 95, 221, 242, 246, 249

famine, 112–113, 228–229, 243
Févret de Saint-Mémin sugar estate, 43–49
Fitch, Eliphalet, 204–205
Fleuriau plantation, 224, 225, 247
flood control, 246
Flour Act/Provisions Act, 114, 119
flour shipments, 114–118. *See also* food-borne illness from spoiled flour
Foäche, Martin, 64
Foäche firm, 225
food-borne illness, 256; from spoiled flour, 103, 106, 111, 115–119, 235
food supply, 64, 199–200, 235, 298n56, 298n57; provisions prices, 119, 210, 221, 227–228; slaves' gardens, 199, 229, 232, 254
food supply during Seven Years' War, 111–122; black market, 115; blockades, 113–116; in Cap Français, 114; flour, 114–118; local food, 114–115; spoiled flour, 103, 106, 111, 115–119, 235
Forestier, Albane, 223–224
Forrest, Arthur, 124, 126, 128, 130, 301n100
Fortune (slave), 131
Fouchard, Jean, 102
Fournier, Daniel, 31
France: after American Revolution, 219; during American Revolution, 20, 203, 205–215, 226; claim to Saint-Domingue by, 14, 32, 33;

conflicts with Britain before Seven Years' War, 82–84; emigration from, 97, 187; flour from, 116; Freemasonry in, 65–67, 284n81; imports to, 34, 167; irrigation in, 35; moral concerns about slavery in, 24, 100, 145–147, 194–198, 254, 264–265; negotiations after Seven Years' War, 18–19, 88–89, 92–95, 292n60, 292n67; North American support by, before American Revolution, 180–181; parlement difficulties in, 174, 175; poisoning fears in, 108; race relations in, 22, 265; religion in, 54, 66; Saint-Domingue culture and, 52, 77, 80; Saint-Domingue's importance to, 20, 22–24, 81, 93, 164–165, 192, 215; settlement schemes from, 70–71; during Seven Years' War, 82, 85–87, 89–91, 113, 128–130, 135–136; slave trade from, 34. *See also* Britain; French navy; Jamaica and Saint-Domingue, compared; Saint-Domingue; trade monopoly, French

Franco-American alliance (1778), 181

Franklin, Benjamin, 52, 92, 99, 181, 195

Fraser, Hannah, 227

Freemasons, 65–67, 237, 284n73, 284n81

free people of color, Jamaica, 148–149; demographics of, 62; invisibility of, 32; maroon populations, 14, 38, 146; marriage and, 68–69, 76–77; population of, 58, 62, 97, 249–250; racial legislation regarding, 62, 68, 137, 150–154, 183, 186–187, 189, 190; social class and, 101, 190–191; urban life of, 50, 62; wealth of, 18, 137, 151–152, 187, 189–191

free people of color, Saint-Domingue, 262; *Chasseurs volontaires* (army unit of), 157, 158, 210–212; demographics of, 61–63; flour and, 118; Hilliard on, 197–198; in *maréchaussée*, 255; maroon populations, 14; marriage and, 68–69, 76–77, 314n77; in militia, 157, 158, 184, 187–190, 210–212, 219; population of, 55, 58, 62, 63, 97, 183; racial legislation regarding, 62, 121, 137, 183–187, 189–190; sexualized, 76–77; social class and, 101, 189–191, 265; theater and, 78; urban life of, 50, 61–63, 78; wealth of, 18, 62–63, 137, 187, 189–191, 197

free-soil tradition, 193, 194

free women of color, 63, 67, 80, 185; occupations of, 76–77; passing of, 68; prostitution and, 76

French Guiana, 97–98, 147

French navy, 15, 251, 331n39; during American Revolution, 207–208, 214; attempts to increase power of, 24, 96; during Seven Years' War, 85–87, 92–93, 115; during War of Austrian Succession, 83–84. *See also* British navy; France; military/militia

French Revolution, 19, 262–263

Frostin, Charles, 35–36

Fryer, Peter, 153

Fuller, Rose, 170–171

Fuller, Stephen, 151, 170–171

Galbaud du Fort, Philippe, 159–160

gang system of slave labor, 4–5, 7, 40, 230, 234

Geggus, David, 61, 110, 228, 329n2; on slave numbers, 43, 60, 245, 253

gender, 63, 67, 71, 78, 284n73; planters seen as effeminate, 264–265; of slaves, 41, 230, 249–250. *See also* men; women

Gentleman's Magazine, 140, 146

George III, King of England, 82, 203, 205

Ghachem, Malick, 256

Gibraltar, 181, 212

Girod-Chantrans, Justin, 70–72, 213–214, 237

Golden Grove Estate (Jamaica), 172–173, 228, 230–231

Gordon, Adam, 57

Goulbourn, Mumbee, 171

government, Jamaica, 167–168; centralization of, 13; influence on Britain, 21, 168, 170–171; military spending in, 205; North American rebellion and, 177–181. *See also* imperial reforms, Jamaica; Jamaican Assembly; racial legislation, Jamaica

government, Saint-Domingue, 17, 32, 156–162, 164–165, 173–177; court merging, 256–257, 263; decentralized, 13–14, 52, 54–55; military to civilian, 156–159; militia revolt, 161–162, 167, 174–177, 181–184, 187. *See also* Cap Français Council; imperial reforms, Saint-Domingue; racial legislation, Saint-Domingue

Graham, Jacob, 249

Grasse, François Joseph de, Admiral, 206–207, 212–213, 217

Greater Antilles colonies. See Jamaica; Jamaica and Saint-Domingue, compared; Saint-Domingue

Greene, Jack, 144

Grenada, 98, 210

Grenville, William, Lord, 242

Groves, John, 123, 142
Guadeloupe, 82, 95–96, 104, 154; inheritance laws in, 183; missionaries in, 237; slave trade in, 88–89, 95, 96, 128

Haiti, 12, 13, 280n67, 327n80; establishment of, 19, 20
Haitian Revolution, 20, 23, 191, 236, 260, 262; slave religion and, 235
Hamilton, Alexander, 202
Hamilton, Elizabeth, 227
Hamilton, John, 102
Hancock, David, 98
Handler, Jerome, 134
Hanley, Richard, 215–216
Hanover slave plot (1776), 200–201, 204
Haran, Joseph-Charles, 62
Harrison, Thomas, 205
Haughton, Richard, 170
Havana, 91–92, 94, 96
Hay, Michael, 56–57
Helyar, Cary, 28–29
Hibbert, Thomas, 57
Highmore, Joseph, 73
Higman, Barry, 172, 226, 230
Hilliard d'Auberteuil, Michel-René, 70, 76, 188, 195–198, 245, 254
Hispaniola, 15, 33
Histoire des deux Indes (Raynal et al.), 94, 98, 100, 195, 267–268
History of Jamaica (Leslie), 145
History of Jamaica (Long), 153
Hood, Samuel, Viscount, 213
Hotten, Jean-Baptiste, 22–23
housekeeping, 76–77
Howe, William, 212
humanitarianism, 138, 146, 196–197, 243, 264–265. *See also* moral concerns about slavery
hunger/starvation, 112, 210, 228–229, 231, 243
hurricanes, 4, 199, 226–228, 242, 318n42

illness/disease: food-borne, 103, 106, 111, 116–117, 235, 256; Kourou colony disaster, 97–98, 147; of slaves, from overwork, 41, 42, 230; from unsanitary body disposal, 57; venereal disease, 69; yellow fever, 88, 92. *See also* death; Macandal poisoning episode; poisonings
imperial reforms, 94, 96
imperial reforms, Jamaica, 40, 99, 164, 261; free people of color and, 154, 186, 190. *See also* government, Jamaica; racial legislation, Jamaica
imperial reforms, Saint-Domingue, 158–159, 164, 207, 256–260; antimilitia movement, 161–162, 174, 175, 186; colonial loyalty and, 82, 89–90, 96–97, 162, 182–183, 186, 265; court merging, 256–257, 262–263; free people of color and, 190; military to civilian government, 156–159; 1784 estate law, 254–256, 262, 263; trade liberalization, 166–167, 173, 251. *See also* Cap Français Council; government, Saint-Domingue; racial legislation, Saint-Domingue
indentured servants, 31, 149
indigo production, 34, 84, 167
industrialization, 7–8
inheritances, 183, 261; racial legislation and, 62, 151–152, 185–187, 189, 308n57
integrated sugar plantations, 3–4, 266; adoption of, in Jamaica, 31; adoption of, in Saint-Domingue, 34–35; definition of, 4; development and spread of, 11–12; profitability of, 5, 37; slaves' productivity on, 40. *See also* sugar plantations
interracial marriage, 67, 68, 185, 187, 197
interracial sex, 68–70, 115–116, 151, 185
investors, 22–23, 164–165, 245; after American Revolution, 221–225; after Seven Years' War, 98–99, 167
Iredell, James, 179
Iredell, Thomas, 179
Ireland, 204
irrigation, 35, 45, 234, 244–246

Jamaica, 27–32; British military protection of, 20, 129–130, 132, 179–181, 261–262, 303n122; Craskell/Simpson map of, 12–14, 27–32; export statistics from, 35, 37–38, 59, 84, 171, 221, 250, 289n11; Freemasons in, 65; importance of, to Britain, 20, 21, 24, 37, 59, 99; Kingston, 13, 55–58, 250; natural disasters in, 220, 226–228; plans to invade, 212–215, 217; prosperity in, 167–173, 242, 248–249; total wealth/value of, 21, 37, 59, 99, 221; urban life in, 50, 55–59. *See also* Britain; government, Jamaica; imperial reforms, Jamaica; Jamaica and Saint-Domingue, compared; Kingston, Jamaica; racial legislation, Jamaica; Saint-Domingue; slave population statistics, Jamaica; Tacky's slave revolt (1760, Jamaica); whites, Jamaica

Jamaica (slave), 126, 127; decapitation of, 130
Jamaica and Saint-Domingue, compared, 8–19; administration location, 13–14; British and French claims to, 14–15; culture and creativity, 48–49, 64, 74, 267–268; free people of color, 62, 97, 183, 184, 186–191; gubernatorial authority, 16–18; interracial marriage, 68; maroon populations of, 14; materialism, 24; mercantile policy, 15–16, 63–64; moral reputation of, 25–26, 68–69, 76, 99, 267; physical characteristics of islands, 9, 11–14, 28, 32, 244; population statistics, 9–11, 14, 32–33, 34, 35, 52, 58; racial legislation, 18, 19, 62, 68, 137, 154, 190, 261; rural or urban living, 52, 58, 62; Seven Years' War and, 18–19; slave population statistics, 11, 34, 35, 100, 253; slave religion, 101, 134–135; slave trade to and through, 15, 251, 253; treatment of slaves, 8–11, 17, 42–43, 145, 150, 226, 260–261; wealth disparity, 9; wealth/value of, 3, 21–24, 37, 98, 192, 244; women, 70–72. *See also* Jamaica; Saint-Domingue
Jamaican Assembly, 16, 177–178, 303n127; call for British troops by, 129–130, 132; combativeness of, 143–145; North American rebellion and, 202, 203, 206; racial legislation by, 68, 147–148, 150–151; relief funds and, 227; on slave population numbers, 233; Tacky's revolt and, 132
Jean (arrested maroon), 240–242
Jean (arrested slave "poisoner"), 236
Jean-Baptiste (slave), 258–259
Jérôme Poteau (arrested slave), 240–242
Jesuits, 54, 109, 122, 135, 237, 297n39
Johnson, Samuel, 92, 140, 169
Johnson, Thomas, Jr., 202
Jones, John Paul, 203
judges, 159, 174, 178, 182

Keith, Basil, 181, 200, 204, 205
Kelly, John, 228, 231, 233–234
Keppel, George, 92
Kerversau, François-Marie-Périchou, 104
Kingston (slave), 131
Kingston, Jamaica, 13, 55–58, 250
Kingué, Marie (slave), 258–259
Knight, Franklin, 14
Knight, James, 127
Knowles, Charles, 55–56, 74, 84, 148
Kourou colony, 97–98, 146–147, 293n75

"L." (militia reestablishment opposer), 176
Labat, Jean-Baptiste, 34, 42–43, 145, 146
Laborde, Jean-Joseph de, 22
labor force size, 38, 245
Laffon de la Débat, André-Daniel, 247, 330n20
Lambert (free black), 118–119
Lambert, Claude-Ange, interim intendant, 113, 115, 120
land value, 171
Laporte-Lalanne, Jean-Baptiste de, intendant, 120
law/legislation. *See* government, Jamaica; government, Saint-Domingue; imperial reforms, Jamaica; imperial reforms, Saint-Domingue; Jamaican Assembly; racial legislation, Jamaica; racial legislation, Saint-Domingue; slave religion, criminalization of
Ledoux, Claude-Nicholas, 43
Léger, Jean-Baptiste-Pierre, 175–177, 182, 183
Lejeune, Nicholas, 10, 257–259
Lejeune, Nicholas (son), 259–260
Lerond, G., 53
Leslie, Charles, 40, 141, 145
Lesser Antilles colonies, physical characteristics of, 9, 11–12
Lesuire, Robert Martin, 139
Letters from an American Farmer (Crèvecoeur), 25–26
Lindsay, John, 201
Livesey, Daniel, 152
Lloyd, Richard, 91
Lloyd, Thomas, 234
London Chronicle, 139–140
London Evening Post, 139
Long, Edward, 305n3; on Guadeloupe slaves, 128; on Kingston architecture, 57; on legality of slavery, 316n8; on Lyttleton, 168; on Mansfield, 193; racism of, 69, 153; on sugar planters, 10, 40, 168–169; on Tacky's revolt, 123, 124, 127, 131, 132; on William Trelawny, 168–169; on women, 69–70, 72, 75, 76, 80
La Lorie heirs, 225
Lory (colonist), 188, 222
Louise, Elizabeth Alexandrine (Minette), 79–80
Louis XV, King of France, 158–160, 162, 165, 174
Louis XVI, King of France, 181, 207, 247
Louverture, Toussaint, 66, 104
La Luzerne, César Henri Guillaume, de, Count, 246

Lyttleton, William, 130, 143–144, 160, 168, 174, 178

Macandal, François (escaped slave), 11, 100, 101, 234, 296n14; capture of, 105, 118; charms distribution by, 110–111; execution of, 103–106, 118, 119, 121; theatrical adaptations of, 80
Macandal poisoning episode, 101–122, 132, 145, 235; death count of, 105. *See also* food supply during Seven Years' War; poisonings
macandals (talismans), 104, 109–111, 133, 258; criminalization of, 121
Mackesy, Piers, 203
magnetism/mesmerism, 234, 241
Magon, René, 161, 175
Malenfant, Charles, 248
Manchioneal slave revolt, 126–127
Le Maniel maroons, 14
Mansfield, William Murray, Earl of, 192–194, 216
manumission, 103, 107, 149, 183, 186, 197
maréchaussée constabulary for hunting escaped slaves, 154–155, 255
Maré plantation, 252
Marmontel, Jean-François, 188
maroon populations, 14, 38, 146. *See also* free people of color
Maroon War (1733–39), 102
marriage, 70, 75–76; free people of color and, 68–69, 76–77, 314n77; interracial, 67, 68, 185, 187, 197; racial legislation and, 151
martial law, 228, 319n63
Martin, Samuel, 5, 41, 193
Martinès de Pasqually, Jacques de Livron, 67
Martinique, 4, 9, 154, 209, 247; development of, 12; Dubuc and, 165–166; inheritance laws in, 183; poisoning in, 103, 108; during Seven Years' War, 82, 87, 88, 89–90
Marx, Karl, 225, 324n32
materialism, 24–26, 53–54, 81, 267
Mathison, Gilbert, 41, 229
McClellan, James, 65, 246
McNeill, Hector, 228
McNeill, John, 92
Médor (slave), 110, 296n14, 298n43; confession of, 103–107, 117. *See also* Macandal, François (escaped slave); Macandal poisoning episode; poisonings
men, 41, 249; Freemasons and, 67; marriage and, 69–70. *See also* gender; women

Mercure de France, 105, 118, 120
Mesmer, Anton, 241
metropolitan opinions about Jamaica and Saint-Domingue, 243; after Tacky's revolt, 139–141, 201; during American Revolution, 208; British increased disdain, 99, 100, 139–141, 263; colonial loyalty and, 96–97; French increased disdain, 99–100. *See also* moral concerns about slavery
Mexico, 94
Meyler, Jeremiah, 85
military/militia, 92–93, 302n107; in France, 210; free people of color and, 184, 187, 189–190, 210–212, 219; in Jamaica, laws enforcing, 132; martial law and, 228; provisions for, 114. *See also* British navy; French navy
military presence: British protection of Jamaica, 20, 129–130, 132, 179–181, 261–262, 303n122; in Saint-Domingue, 55, 158, 262; slave quietude and, 228
militia in Saint-Domingue, 156–157; revolt in, 161–162, 167, 174–177, 181–182, 183, 184, 187
Minette (Elizabeth Alexandrine Louise), 79–80, 388n144
Minorca, 87
missionaries, 42–43, 110, 122, 135, 237, 297n39. *See also* Jesuits
mistresses, 68, 69, 70, 72, 76, 77, 151, 185
mixed-race people: Hilliard's proposal for, 197; marriage and, 69–70; racial legislation and, 121, 138, 185; Raimond, 184, 189–190, 262, 265, 266. *See also* passing
molasses, 6, 166–167
money lending, 59, 165
Montesquieu, Charles-Louis de Secondat, 145
Moore, Henry, 28, 74, 151, 302n107; Tacky's revolt and, 126, 127, 129–130
moral concerns about slavery, 226, 243, 316n8; development of, in Britain, 24, 99, 100, 138–141, 192–194, 216–218, 233, 263–265; development of, in France, 24, 100, 145–147, 194–198, 254, 264–265; Dubuc and, 252; in Saint-Domingue, 253–254; *Somerset* case, 192–194; Trelawny and, 150; *Ziméo,* 146, 194–195; *Zong* incident, 216–217, 218, 263. *See also* abolitionism
moral reputation: of Jamaica and Saint-Domingue, 25–26, 68–69, 76, 99, 267; materialism, 25–26, 53–54, 81, 267; of Phillips, 72–73, 75; of women, 70–72

Morange (plantation attorney), 255
Moreau de Saint-Méry, M. L. E., 10, 54, 55; on court merging, 256; hurricane in Saint-Domingue and, 199; on judges in Saint-Domingue, 159; on Macandal, 105, 122; on marriage, 68; slave religion and, 109, 238–239, 241–242; spoiled flour and, 118; on urban life, 53, 61, 77; on women, 72, 75, 76, 78–80
Morgan, Philip, 50
Morin, Etienne, 66
Morris, Robert, 203
Moses Bom Saam (Maroon chief), 146
Motmans family, 45, 48
mountains, 12–13
mulattos. *See* mixed-race people
Mure, Robert, 227
muscovado, 16, 35, 171
mycotoxins, 116–117

Naipul, V. S., 49
natural disasters: in Jamaica, 220, 226–228; in Saint-Domingue, 198–200, 224–225
Natural History of Jamaica (Sloane), 145
Navigation Acts (1651, England), 16
navy. *See* British navy; French navy
Necker, Mme. de (Suzanne de Curchod), 166
Needham, Henry, 74
Neufchâteau, François de, 240, 241
Nicolson, Jean-Barthélmy-Maximilien, 114–115, 198–199
nkangi kiditu (bound crucifixes), 109–110
Nolivos, Count de, 182, 183, 190
North, Frederick, Lord, 99
North America: commercial importance *vs.* West Indies, 95, 99; concerns about West Indies in, 201–202; flour from, 119; Jamaican trade with, 173, 220; Saint-Domingue trade with, 119, 166–167, 200, 207–210; trade restrictions on, 114, 119. *See also* American Revolutionary War
North American rebellion: Jamaica and, 173, 177–181, 202–204, 204–205, 206
Nouveau voyage aux isles de l'Amérique (Labat), 34

obeah, 127, 132–135, 238, 303nn127, 128. *See also* slave religion
Observations on the Increase of Mankind (Franklin), 195

"Ode to the Sable Venus" (Teale), 69
"Of Public Spirit in Regard to Public Works" (Savage), 140
Ogé *jeune,* Vincent, 62, 262, 266
O'Gorman, Victoire-Arnauld-Martin, 248
Olivier, Vincent, 211
Olyphant, John, 143–144
Oronoko (Behn), 145
O'Shaughnessy, Andrew, 99, 178, 203, 303n122

Palo Mayombe religion, 240–241
passing, racial legislation and: in Jamaica, 147–148, 150–151; in Jamaica *vs.* Saint-Domingue, 18, 68, 137, 154, 190; in Saint-Domingue, 184. *See also* racial legislation
Paton, Diana, 108, 134
Le Patriotisme américain (Petit), 155
Payne, Charles, 227
Penet, Pierre, 209
Pennant, Richard, 172
Penrhyn Estates (Jamaica), 172, 249
Perrier engine, 246
Le Pers, Jean-Baptiste, 10, 297n39
Petit, Emilien, 53, 69, 71, 155, 156, 190, 286n97
Phélipeau, René, 51
Phillips, Teresia Constantia, 72–77, 79, 80
Pic de la Selle (Saint-Domingue), 12–13
Pitt, William, 85, 95, 205, 291n52; Martinique invasion, 88; peace negotiations and, 292n67; Saint-Domingue and, 90, 290n32; Tacky's revolt and, 126; Treaty of Paris and, 93–94, 292n60; war strategy of, 86–89, 92
Pittis, William, 70
Pliarne, Emmanuel de, 209
Plombard, merchant at Cap Français, 208
Pluchon, Pierre, 122
poison, slaves' meanings of, 106, 107, 297
poisonings, 101, 332n64; accusations of, 117–118, 235, 259–260; confessions to, 106, 260; detection of, by slaves, 258; executions and, 108, 236, 257; illness from spoiled flour, as attributed to, 103, 106, 111, 116–117, 235; legislation about, 108, 112, 120, 120–121, 135, 235, 256, 257–260; Médor's confession to, 103, 104–105, 106–107, 117; regional distribution of, 113, 115–116, 117–118; slaves' fear of, 122. *See also* deaths blamed on poisoning

poisonings, whites' fear of, 11, 104–106, 110, 111, 234–236, 258–259; accusations, 145; in France, 108; regional arrests, 117–118
Pompey (slave), 127
Port-au-Prince, Saint-Domingue, 13, 51–53, 55, 60
Port Royal, Jamaica, 30–31
Portugal, 251
Potofsky, Allan, 22–23
Le Pour et le Contre (Prévost), 146
poverty, 9, 188, 227
Praslin, Duc de, 165
pregnancy, labor efficiency and, 41, 230
Prévost, Antoine-François, 146
Price, Charles, 28, 169
Price, Charles, Jr., 144
Price, Charles, Sr., 144
Price, Francis, 28
price of slaves, 35, 171, 196, 232, 242, 326n58
price of sugar, 95, 221, 242, 246, 249
privateering/piracy, 30–31, 33, 87, 91, 200, 224; during American Revolution, 208–209
productivity of slaves. *See* sugar plantations, profit maximization on
productivity of sugar plantations, 37, 171–172; decline thesis and, 224–225; technology in Saint-Domingue and, 244–248. *See also* sugar plantations
profitability analysis, 98, 242; integrated sugar plantations, 37, 40; in Jamaica, 171–172, 220–221, 248–249; in Jamaica *vs.* Saint-Domingue, 244–245
property ownership, 64; absentees/transatlantic brokers, 129, 170, 222, 225, 253–255; by colonial administrators in Saint-Domingue, 165; by free people of color, 62, 63
prostitution, 70, 76
provisions prices, 119, 210, 221, 227–228. *See also* food supply

Québec, 89, 93

race relations: in Britain, 21–22, 138–139, 153; in France, 22, 265
racial classificatory systems, 100, 137, 265; biological racism and, 18, 69–70, 153, 186; caste consciousness and, 149–150, 154, 164, 197–198, 261; marriage and, 68. *See also* social class; whiteness
racial legislation, Jamaica, 3, 153–154, 162–163; enforcement of, 152–153; inheritances and, 62, 151–152, 186–187, 189, 308n57; passing and, 18, 68, 137, 147–148, 150–151, 154, 190. *See also* free people of color, Jamaica; whites, Jamaica
racial legislation, Saint-Domingue, 3, 154, 162–163, 183–186, 190–191, 261–263; Code Noir, 17, 53–54, 183, 254; inheritances and, 185–187, 189; passing and, 18, 68, 137, 184, 190. *See also* free people of color, Saint-Domingue; whites, Saint-Domingue
racism, biological, 18, 69–70, 153, 186
Ragatz, Lowell, 220
Raimond, Julien, 184, 189–190, 262, 265, 266
Ramsay, James, 153
rattooning, 6, 245
Raynal, Abbé, 2, 20, 166
rebel slaves, 134; abolitionism and, 141; executions of, 122, 123, 125, 127, 130–131, 146, 200; in Tacky's revolt, 102, 123–132, 139, 142. *See also* Macandal, François (escaped slave); Médor (slave)
Recherches philosophiques sur les Américains (DePauw), 153
Regourd, François, 246
religion, 53–54, 66, 67. *See also* Christianity; slave religion
rent, 64
reproduction rates of slaves, 230
Richardson, James, 171
Rivière, Pierre-Paul Mercier de la, 89
Robertson, George, 36–37
Rockingham, Charles Wentworth-Watson, Marquess of, 99, 228
Rockingham administration, 144
Rodney, George, 207, 213, 219
Rogers, Dominique, 63, 190
Rohan-Montbazon, duc de, 165, 174–177, 181–182, 184
Romberg, Bapst et Cie, 221–224
Romberg, Henry, 222–223
rootlessness, 64–65
Ross, Robert, 227
Rouvray, Laurent-François Le Noir, Marquis de, 211, 212
Royal African Company, 31
rum production, 6, 167
Ryden, David, 221, 232–233

Saintard, Pierre-Louis de, 65
Saint-Domingue, 32–38, 322n4; Cap Français,

13, 51–55, 57, 59–60, 63, 114; France's claims to, 14, 32, 33; Freemasons in, 65–66; importance of, to France, 20, 22–23, 24, 81, 93, 164–165, 192, 215; maps of, 32, 51, 52; natural disasters in, 198–200, 224–225; prosperity in, 164–167, 173–177, 219; role during American Revolution, 207–209; total wealth/value of, 24, 164–165, 196, 244; urban life in, 50–55, 60, 64. *See also* Cap Français Council; colonial loyalty, Saint-Domingue; France; free people of color, Saint-Domingue; government, Saint-Domingue; imperial reforms, Saint-Domingue; Jamaica; Jamaica and Saint-Domingue, compared; racial legislation, Saint-Domingue; slave population statistics, Saint-Domingue; whites, Saint-Domingue

St. Jago de la Vega, Jamaica, 55

Saint-Lambert, Jean-François de, 146, 194–195, 254

Saint-Marc Assembly, 262

Saint-Mémin, Charles, 44

Saint-Mémin, Févret de, 44–45

Saint-Ouen, Bertrand de, 246

St. Vincent, 98

Salem witch trials, 116–117

Santo Domingo, 14

Sartine, Antoine de, 207–208

Saunier, Eric, 67

Savage, Richard, 140

Scott, Sarah, 141

Scottish Rite Freemasonry, 66, 67

segregation, in theater, 78

settlement schemes, 70–71, 97–98, 148

Seven Years' War, 81, 82–100; Battle of Ticonderoga in, 128; beginning of, 84–85; blockades and convoys during, 93, 106, 112–117, 219; British/French conflicts preceding, 82–84; British investments following, 98–99, 167; British navy in, 85–86, 92–93, 112–114, 116, 117; British strategy during, 86–92, 156–157, 165, 292n60, 292n67; death of soldiers, disease and, 88, 92, 128; France in, 82, 85, 86–87, 89–91, 113, 128, 129–130, 135–136; free colored militia during, 184; French actions following, 96–98; French navy in, 85–87, 92–93, 115; Guadeloupe in, 88–89; Havana in, 91–92, 94, 96; Jamaica in, 90–91; Macandal poisoning episode and, 112–119; Martinique in, 82, 87, 88, 89–90;

negotiations following, 18–19, 82, 88–89, 92–96, 292n60, 292n67; Pitt and, 86–88, 89, 90, 92, 93–94; Saint-Domingue in, 90–91, 112–114, 219, 290n32; Tacky's revolt in context of, 127–129; territorial control changes, 92

Sharp, Granville, 192, 194, 216–217

Sheridan, Richard, 58–59, 171, 227–228, 232

Shickle, John, 172

shipping, 45

Simon (slave), 127

Simpson, James, 12, 27

skilled trades, 149

slave labor, as expendable capital investment, 7–8, 40–41, 230–231, 266

slave management, 172–173, 220, 225, 266–267

slave mastery, 143

slave population statistics, Jamaica, 11, 100; age and gender in, 249–250; increase in, 31, 34, 35, 38, 95, 232–233, 249–250, 277n35; in Kingston, 57

slave population statistics, Saint-Domingue, 11, 100, 245, 329n6; increase in, 34, 35, 52, 110, 145, 250–253; poisonings and, 105; in Port-au-Prince, 55, 60

slave prices, 35, 171, 196, 232, 242, 326n58

slave rebellion. *See* slave revolts

slave religion, 236–238, 240–243; community-building through, 234–235, 237; Freemasons and, 67; obeah, 127, 132–135, 238, 303nn127, 128; poisonings and, 236; religious objects in, 104, 109–111, 121, 133, 240–241, 258; Tacky's revolt and, 132–135; views of, in Jamaica *vs.* Saint-Domingue, 101–102, 261

slave religion, criminalization of, 101, 121–122, 240–241, 303n127; Code Noir, 54; dance and, 121, 135, 238–239, 240; obeah, 133–135, 238

slave religion, Macandal episode and, 103; Courtin's memorandum and, 108–109; Médor's confession, 107; talismans and, 110–111

slave revolts, 11, 18, 226, 242–243, 295n6, 328n99; Britons' opinions of, 139–141; as consequence of treatment of slaves, 131–132, 141, 146; fear of, 138, 145, 201, 260–261; Hanover slave plot, 200–201, 204; as lower priority than colonial loyalty, 146–147, 154–155; news of, in Britain, 140, 146, 228; quietude of during times of hardship, 227–228;

slave revolts (*cont.*)
  seen as inevitable, 103; of 1791, 260–263; slave religion and, 239; slaves in defense of plantations during, 123–124; before Tacky's revolt/Macandal poisoning, 102, 122; in urban areas, 61. *See also* poisonings; Tacky's slave revolt (1760, Jamaica)
slave ships, 208, 232, 321n105
slave trade: in Guadeloupe, 88–89, 95, 96, 121; number of slaves on shipments, 84, 232, 250; profitability of, 58, 170–172, 252–253; smuggling in, 208; during War of Austrian Succession, 83–84; *Zong* incident and, 215–217, 218
slave trade, dependence on, 1, 42, 266–267; British and French metropolitan disdain for, 100; Kourou colony as attempt to subvert, 97–98, 147; to replace death of slaves, 230, 248–249, 252, 266
slave trade, Jamaica, 58–59, 251, 253; decline in, during American Revolution, 231–233; by free people of color, 148–149; to Hispaniola, 15; move from public to private, 31; number of slaves on ships, 84, 232; wealth from, 58, 170–172, 220
slave trade, Saint-Domingue, 15, 196, 242; in Cap Français, 59–60; by free people of color, 63; increase in, 250–251; move from public to private, 34; number of slaves on ships, 84, 250; Romberg, Bapst et Cie, 222–223; wealth from, 219–220, 252–253
Sloane, Hans, 145
Smith, Adam, 28, 225
smuggling, 15, 16, 251; during American Revolution, 204–205, 207–210; of indigo, 34; during Seven Years' War, 113–115, 117–119; of slaves, 59, 60
social class, 101, 265; caste society, 149–150, 154, 164, 197–198, 261; merchants, 58–59, 64; by race rather than by wealth, 190–191, 267; racial legislation and, 137, 189, 261; sugar planters as elite, 4, 31
soil exhaustion, 245
Somerset, James, 192–194
Somerset v. Steuart (1772), 153, 192–194
Spain, 33, 90, 180–181, 203, 205, 290n32
Spanish American trade, 171
Spanish colonists, 18, 83, 152
Spirit of the Leaves (Montesquieu), 145
spoilage danger, 116–117; black market sales, 115–116; of sugar, 6. *See also* food-borne illness
Spragg, William, 132
Spring Estate (Jamaica), 172
Stamp Act, 98–99, 144, 177, 178
Stanhope, Lovell, 151–152
Stanton, Robert, 86
steam engines, 246
Steuart, Charles, 192
stick fighting, 240
Stubbs, Robert, 215
sugar plantations, 275n12; art depicting, 31–32, 36–37, 43–48; beginning of cultivation in Saint-Domingue, 34–35; capital investment needed for starting, 34; increase in number of, 37–38, 43, 96, 171; natural disasters and, 224–225; productivity of, 37, 171–172, 224–225, 244–248; risks in starting operation of, 28–29; technological improvements in, 245–247. *See also* integrated sugar plantations
sugar plantations, profit maximization on, 4–7, 220, 234, 245, 266–267; arable land clearing for, 251–252; gang labor and, 4–5, 7; investments for, 224; slaves' productivity and, 40–41, 172, 225, 229–232; technology for, 245–247
sugar planters: hospitality of, 187–188; imperial influence of, in Jamaica, 98–99; Jamaican loyalty to Britain, 68, 99, 218, 263–264; Long on, 10, 168–169; power of, to interrogate and punish slaves, 101–102, 120–121, 135, 154, 256, 257; self-assessments of *vs.* Britons' views of, 264–265; as social elite, 4, 31. *See also* treatment of slaves
sugar prices, 95, 221, 242, 246, 249
sugar production, 34, 245–246; after American Revolution, 221; description of process, 5–6, 329n3; molasses, 6, 166–167; muscovado, 16, 35, 171; white sugar, 35
Sweet, James, 236–237
syndic law, 158

Tacky (slave), 127, 302n115; decapitation of, 126, 130
Tacky's slave revolt (1760, Jamaica), 11, 62, 100, 101, 122–133, 136, 302n116; aims of, 123, 125; British newspapers' accounts of, 139–140, 146; Britons' opinions of white colonists after, 139–141; in context of Seven Years' War, 127–129; executions after, 122, 123, 125, 139,

141, 146, 301n105; explanations for, 127, 132, 135; organization of, 125–126, 127; racial legislation after, 147; rebels in, 102, 123–132, 139, 142; response to, 102–103, 130, 132, 139, 201; secrecy in, 102, 123, 125, 301n100; severity of, 122–123, 201; slave "loyalty" during, 123–125; slave religion and, 132–135; as unexpected/surprising, 125–127. *See also* slave revolts
tafia, 167
Tannenbaum, Frank, 8–9
Targe, Jean-Baptiste, 146
Tarrade, Jean, 221, 224
tassou (dried beef), 235, 236, 256
taxes, 16, 178; on manumission, 183; in Saint-Domingue, 158, 160
Taylor, John, 30–31, 39, 70, 312n30
Taylor, Simon, 172–173, 178–179, 201, 246; treatment of slaves by, 228, 230–231, 233–234
Teale, Isaac, 69, 70
technology, 7, 225, 245–248
Télémaque (arrested slave), 240, 241
theater, 51–52, 74, 75, 77–80
Thésée, Françoise, 221, 223, 224
Thistlewood, Thomas, 171, 204, 319n63; American Revolution and, 206–207, 213; hurricanes and, 226–227; North American rebellion and, 178, 203; on obeah, 132–133; during Seven Years' War, 90–91, 93; on Tacky's revolt, 123–125, 127–132; treatment of slaves by, 10, 39–40, 141–142; wealth of, 169–170
Thomas, Dalby, 31
Thomson, James, 230
Three Finger'd Jack (rebel slave), 228
torture, 107, 121, 131
Touche, Louis-Charles Levassor de la, 89
Townshend Acts (1767-70), 99, 178
trade. *See* blockades and convoys; slave trade; trade monopolies
trade monopoly, British, 15–16, 96
trade monopoly, French, 15–16, 64, 85, 192; during American Revolution, 207–210, 222; colonial loyalty and, 96–98, 261; liberalization of, 96–98, 166–167, 173, 251, 261; molasses and, 166–167; natural disasters and, 199; as obstacle to growth, 16, 197; slaves in Guadeloupe/Martinique, 88–89
Transatlantic Slave Trade Data Base, 95
transportation, 13, 33, 45

treatment of slaves: death from, 41–42, 228–229, 230, 232–233; by English *vs.* French, 8–11, 17, 42–43, 145, 150, 260–261; gender and, 41, 72; humanitarianism and, 138, 146, 196–197, 243, 264–265; for increased productivity, 40–42, 172, 266; legal protection from, 258; malnutrition/neglect of, 112, 115, 199, 226, 229, 231, 233, 254; revolts as consequence of, 131–132, 141, 146; urban *vs.* rural, 60–61; by women, 72; *Zong* slave ship incident, 215–217, 218, 263, 321n105, 321n108, 322n109. *See also* abolitionism
treatment of slaves, Jamaica, 38–43, 147; Britons' view of, 138–140, 140–142, 150; at Golden Grove, 230–231; by Grove, 142; legislation on, 17; slaves' workload, 41–42; by Taylor, 228, 230–231, 233–234; by Thistlewood, 10, 39–40, 141–142
treatment of slaves, Saint-Domingue: Beauvernet's vignette's depictions of, 45–48, 49; by Caradeux, 248, 253, 258, 332n63; court merging and, 257–258; crime/punishment and, 120–121; legislation on, 17, 146, 253–258; by Lejeune father and son, 10, 259–260
Treaty of Paris (1763), 92–95, 292n60
Treaty of Ryswick (1697), 33
Treaty of Versailles (1783), 215
Treil, Nadau de, 88
Trelawny, Edward, 84, 150, 168
Trelawny, William, 168–169, 170, 174, 182

unemployment, 188–189
urban life, 50–81; in Cap Français, 60, 64; city planning, 52–53; Freemasons in, 65–67; for free people of color, 50, 61–63, 78; in Jamaica, 50, 55–59; racial demographics of, 57–58, 64; rent, 64; rootless feeling in, 64–65; in Saint-Domingue, 50–55, 60, 64; for slaves, 60–61; theater, 51–52, 74, 75, 77–80; for women, 50, 67–80

Vaivre, Guillamin de, 183
Vanhee, Hein, 109
Vanhellen, John, 228
Vassall, Florentius, 178
Vaudoux, 135, 238–239, 327n80
Vaudreuil, Joseph Hyacinthe François-de-Paule de Rigaud, Count de, 248
Vergennes, Charles Gravier, Count de, 207–208, 210

Vernon, Edward, 84
"A View in the Island of Jamaica, of Fort William Estate, with Part of Roaring River, belonging to William Beckford, Esqr. Near Savannah La Marr, Westmoreland, 1778" (Robertson), 36
violence. *See* treatment of slaves
Voltaire (François-Marie Arouet), 153, 195

Wager (Apongo) (rebel slave leader), 127, 128–129, 130–131, 301n100
Ward, J. R., 172, 221, 248–249
War of Austrian Succession (1744-48), 83–84, 113–114
Washington, George, 85
Watt, James, 246
wealth, 3, 9; of free people of color, 18, 62–63, 137, 151–152, 187, 189–191, 197; hierarchy among whites and, 142, 150; Pitt's attacks on French sources of, 86–87
wealth, Jamaica, 3, 38, 167–173, 221; Beckford and, 36; importance of, to Britain, 21, 24, 37, 59, 99; merchant class and, 58–59; planter class and, 4, 31; racial legislation and, 18, 151–152
wealth, Saint-Domingue, 3, 37; importance of, to France, 21, 22–23, 24, 196; merchant class and, 59; racial legislation and, 18; as threat to Britain, 85
Weech, Harry, 142, 170
West Africa, 87, 92
West African captives (Congos), 103, 109–110, 111
Westmoreland slave revolt, 124, 126, 127, 128, 141
Westmoreland sugar estates, 36–37
Whatley, Robert, 85
whiteness, 164, 186, 261; Hilliard's proposal for, 197–198; racial legislation and, 3, 18, 19, 143, 147, 153–154, 162, 190
whites, Jamaica: age and gender demographics, 249–250; hierarchy of, 142, 150; hospitality to other whites, 142–143; male opinions of female, 70–71, 72; marriage and, 68–70; population statistics, 31, 32, 38, 57–58, 63, 142, 249–250; as predominantly slave owners, 143
whites, Saint-Domingue: in Beauvernet's vignettes, 45–48; island-born (creoles), 10, 100, 197; male opinions of female, 70, 71–72; population statistics, 33, 43, 55, 63, 97, 283n65; racial legislation and, 186; theater and, 78; unemployment of, 188–189
white sugar production, 35
white women, 70–72, 80
Williams, Eric, 220
Wilson, Kathleen, 22, 69, 74, 75
Wimpffen, Alexandre-Stanislas de, 26, 71–72
Witter, Norwood, 132
women, 67–80, 249; attributed to protection from poisoning, 115–116; Craskell/Simpson map, 32; free women of color, 63, 67, 68, 76–77, 80, 185; Long on, 69–70, 72, 75, 76, 80; marriage and, 68–69; Minette, 79–80, 388n144; Moreau on, 72, 75, 76, 78–79, 80; Phillips, 72–77, 79, 80; pregnant slaves and, 41, 230; property ownership by, 63; reputation/opinions of, 70–72; in theater, 77–80; treatment of, 41, 71–72; urban life for, 50, 67–80; white, 70–72, 80. *See also* men
women of color: free, 63, 67, 68, 76–77, 80, 185; as mistresses, 68, 69, 70, 72, 76, 77, 151, 185; sex and, 68–70; in theater, 77–79
Woodbine, Catherine and George, 227
Worthy Park plantation, 28, 245

yellow fever, 88, 92

Zabeau (prosecuted slave "poisoner"), 236
"Ziméo" (Saint-Lambert), 146, 194–195, 254
Zong slave ship incident, 215–217, 218, 263, 321n105, 321n108, 322n109

ACKNOWLEDGMENTS

We first thought about this book a decade ago, when Trevor traveled to Texas to speak at the Webb Lecture series, hosted annually by the Department of History at the University of Texas at Arlington. Each of us had published a book (Trevor on Jamaica; John on Saint-Domingue) exploring changes in racial thinking in the Greater Antilles. We had each witnessed the beginnings of those earlier projects while we were friends in graduate school at Johns Hopkins University, but in the ensuing twenty years we had followed each other's research from afar. The writing partnership we forged in 2006 has made each of us a part of the other's daily and weekly work life for ten years now.

By design, this is very much a jointly authored book, composed and edited in different hemispheres. Each of us wrote from his own sources, frequently responding to the other's findings and questions. Each of us then poured over the resulting text, editing and organizing the text one after the other and then discussing the results from our different time zones. It has been very enjoyable process, though sometimes a bit cumbersome, especially as one of us has been in Britain and then in Australia and the other has been in Florida and then in Texas.

Working separately, each of us could have written a comparative history that superficially resembled *The Plantation Machine* in about half the time. But we believe that this book, hammered out in partnership and friendly negotiation by two specialists with many years of experience in different archives, will be far more valuable to readers.

As in all scholarship, we rely on the work done by others, much of which we have acknowledged in our notes. We are grateful for the help we have had along the way in helping us understand Jamaica and Saint-Domingue better.

Trevor started this project in earnest when a fellow at the National Humanities Center in Durham, North Carolina. He is grateful for the opportunity to have spent a very pleasant year at that oasis of learning. Material support has been given by his previous employer, the University of Warwick, and by his present employer, the University of Melbourne. John benefited from the financial support of the Department of History and the College of Liberal Arts at the University of Texas at Arlington. The UTA History Department brown bag and the Dallas Area Social History Group provided valuable feedback for the chapter on Macandal. Similarly, Jeremy Popkin and Jim Sidbury helped by reading sections of the manuscript. We have also profited from the comments of anonymous readers along the way, and from the insights of Simon Newman.

We are delighted that this work comes out in the Early Modern Americas series at the University of Pennsylvania Press, where Peter Mancall has been a great series editor. Bob Lockhart has been very helpful in getting this book into press.

Both of us thank our respective families for their support and encouragement—Deborah, Nicholas, and Eleanor for Trevor; and Susan and Paco for John.

www.ingramcontent.com/pod-product-compliance
Lightning Source LLC
Chambersburg PA
CBHW020635230426
43665CB00008B/189